Exuberant Life

Exuberant Life

An Evolutionary Approach to Conservation in Galápagos

WILLIAM H. DURHAM

OXFORD
UNIVERSITY PRESS

OXFORD
UNIVERSITY PRESS

Oxford University Press is a department of the University of Oxford. It furthers
the University's objective of excellence in research, scholarship, and education
by publishing worldwide. Oxford is a registered trade mark of Oxford University
Press in the UK and certain other countries.

Published in the United States of America by Oxford University Press
198 Madison Avenue, New York, NY 10016, United States of America.

Library of Congress Control Number: 2020952645
ISBN 978–0–19–753151–8

DOI: 10.1093/oso/9780197531518.001.0001

1 3 5 7 9 8 6 4 2

Printed by Sheridan Books, Inc., United States of America

To Ryan
celebrating your place in the tree of life

Contents

List of Figures

Chapter 1

Chapter 2

Chapter 3

Chapter 4

Chapter 5

Chapter 6

Chapter 7

Chapter 8

Chapter 9

Chapter 10

Preface

The natural history of this archipelago is very remarkable: it seems to be a little world within itself; the greater number of its inhabitants, both vegetable and animal, being found nowhere else.
Charles Darwin (*Narrative of the Surveying Voyages of His Majesty's Ships Adventure and Beagle*)

The book before you is a contemporary synthesis of what we know about the evolution of the curiously wonderful organisms of Galápagos, of how they are faring in the tumultuous world of human-induced change, and how evolution can guide our efforts today for their conservation. The book stems from the conviction that the more people know and appreciate the story Galápagos tells, with key subplots told by a diverse set of the archipelago's species, the better our chances of successfully managing the challenges facing the archipelago and ensuring its sustainable future.

The book's impetus runs back to events earlier in my life. In 1975, while a graduate student, I wrote a proposal to the Smithsonian Tropical Research Institute to join their expedition to Galápagos. I proposed to study a potential mutual relationship between giant tortoises and their key food, the prickly pear cacti of the genus *Opuntia*. I wanted to ask whether tortoises, who readily consume the large, fleshy fruits of the cactus, thereby enhance its seed dispersal, building on the suggestion of Dawson (1966, 212). Accepted, I excitedly found funding, equipment, and was ready to go when, alas, the expedition was postponed because a health issue befell its organizer. In the meantime, I finished my degree, took a job, started a family, and kept busy with other work on evolution (Durham 1991). The expedition was never reorganized, but still I dreamed of getting to Galápagos.

Meanwhile, educational tourism in Galápagos was flourishing, and the chance to visit finally came when Stanford University's Travel-Study Program invited me to lead a tour. Arriving at last, I found Galápagos even more fascinating than I had imagined. Happily, evidence today confirms my earlier curiosity: tortoises do enhance *Opuntia* seed dispersal and, indeed, they do much more, as described in Chapter 3. A second discovery of that visit was even more thrilling. I grew up in northern Ohio, and one of my childhood delights was raising sunflowers in summer and trying various experiments to get them to grow taller. One year, on a tip from a neighbor, I planted seeds amid crushed chicken eggshells and raised a sunflower to a whopping 6.5 ft (2 m)—it took first place at the county fair. What a surprise, years later, to find in Galápagos a closely related species—an endemic daisy—growing 10 times higher (to 65 ft or 20 m) in a genuine *forest* of daisies! With that I was convinced that something very special about Galápagos had shaped its flora and fauna in ways that seemed to break all records.

That trip taught me another key lesson about Galápagos. Because organisms adapted to this "little world within itself," per Darwin above, free of terrestrial predators or competitors, they live their lives fully in the open with their many wonderful details on display. Today's visitors can have a field day using principles of evolution and ecology to make sense of it all. For example, the wonderful adaptations of plants and animals to the islands consistently illustrate the principle of reproductive advantage. No matter how curious the adaptations— flightlessness, seaweed-eating, giant proportions, and more—*all of them* have some net survival and reproduction advantage to individual bearers in their Galápagos environment. For the fascinated observer, the challenge is figuring out that advantage. People find themselves caught in an inspiring feedback loop: the more sense one makes of life in Galápagos, the more one wants to learn. Galápagos becomes "a perennial source of new things," as Darwin wrote in a letter to his friend, botanist Joe Hooker, in 1846.[1]

A product of my own fascination with Galápagos, this book is designed to offer readers special insight into the archipelago by interrelating the themes of evolution and conservation. Each chapter focuses on one, or a few, iconic organisms. The chapters provide enough depth to give readers a "feeling for the organism" (Keller 1984), for how much we know and don't know about its evolutionary past, and for how that past relates to its present predicament. A sense of the whole, the big picture of Galápagos today, emerges from the sequence of cases. I hope it appeals to a broad audience who enjoy reading about science, including those who have visited, or plan to visit, the islands to see for themselves. Like a pair of binoculars, the book will help readers gain a closer, sharper focus on the flora and fauna of Galápagos. My goal is to help readers understand and enjoy what is there, how it came about, why it is imperiled today, and what can be done to help. Ideally, these pages will satisfy some of the curiosity of those who find themselves in that inspiring Galápagos feedback loop, eager to learn more.

I hope the cases here also prove useful to readers interested in ecology, evolution, conservation, and sustainability. Galápagos exhibits the same principles and processes found in many other places, but in especially dramatic, memorable ways. I hope the book will also reach readers who are genuinely concerned about the future of the planet's precious and precarious biodiversity. Galápagos is

[1] Charles Darwin to J. D. Hooker, 8 or 15 July 1846, Darwin Correspondence Project, Letter 986, https://www.darwinproject.ac.uk/letter/DCP-LETT-986.xml

a special case, in part due to the relative ease with which change can be seen and appreciated there. The same kinds of change are happening elsewhere, but they are often harder to see in larger populations, more complex ecosystems, more furtive fauna, or all three. In Galápagos, life is out in the open, showing with special clarity the impact of human activity upon flora and fauna.

What will you learn from this volume that you are unlikely to learn from another? Here are five sample puzzles that are among those pieced together in these pages, although not in exactly this order. I know of no other book that includes even three of these puzzles:

1. Why are the majority of Galápagos organisms so wonderfully *unusual*, ranging from delightful miniature penguins to forests of giant daisies? Not to mention the world's only cormorant with tatty vestigial wings and the only ocean-foraging lizard? Why are organisms so very different in this little group of islands, and how do their extraordinary features complicate conservation efforts?

2. Why are some native species of Galápagos so dark and drab (like marine iguanas in the northern islands) and others so *brightly colored* (like marine iguanas in the south, or the three species of boobies)? With so few colorful land plants—most native plants of Galápagos have only tiny white or yellow flowers—where do these colors come from? What are the implications of color for conservation?

3. Galápagos is home to three fascinating species of *flightless*, or near-flightless, birds: the aforementioned penguin and cormorant, and a diminutive land bird, the Galápagos rail. Did all three lose their former capacity for flight for similar reasons? Why aren't more Galápagos birds flightless or near-flightless? How do such birds fare amid the changes underway today? What extra measures does flightlessness entail for conservation?

4. Many Galápagos land organisms show *exuberant nonchalance*. A favorite example is the "parade" of male blue-footed boobies strutting around their mating territory, lifting bright turquoise feet repeatedly, even if no one is watching. If a potential mate does join in, males add sky-pointing to the parade: head back, beak straight up, and wings fully extended. Females often follow suit. All this and more they do with complete abandon, as if nothing else mattered. Red-footed boobies do their sky-pointing up in trees, with the same abandon: a real balancing act. Why such exaggerated, carefree displays? What do they imply about evolution? What are their conservation implications?

5. Galápagos often promotes the *diversification* of species, but not always. Ancestors of the giant tortoises floated out about 3.2 million years ago to the Galápagos Islands of that time. About the same time, ancestors of

the Galapagos rail, a small ground bird, flew out to the same place. Ages passed, islands came and went, and the tortoises colonized ten islands and diversified into 15 species with different shell shapes—a genuine "adaptive radiation." The rails, meanwhile, colonized seven islands and evolved to be near-flightless, but all in one species without even any subspecies. What's behind such divergent outcomes in the same place at the same time?

In addition to exploring these puzzles and more of the kind, a major take-home lesson of this book is that Galápagos is no longer the "little world within itself" that Darwin experienced. Now it is fully integrated into the globalizing planet, with over 35,000 inhabitants, nearly 300,000 annual visitors (in the years before the coronavirus pandemic), jet airports, luxury hotels, and more. It has become one big "social-ecological system" in which the human social aspects of the islands continually interact with its ecological aspects. How can we tell? As this book shows, the survival of each Galápagos organism is now a joint product of natural conditions *and* human influence. The archipelago's peculiarly beautiful flora and fauna are no less intriguing than in Darwin's time, and there are additional examples we marvel at today that Darwin did not observe (described in Chapter 8). But we see them today in a new light, challenged as never before by recent human impact in the islands—which only adds to our admiration. It also adds to our concern for their future, for surely their years of complete isolation are over. In a very real sense, we hold their future in our hands: *humanity has today a major new responsibility in Galápagos.*

Finally, while this book is about Galápagos, its relevance and insight extend well beyond these remote Pacific islands. Not only does biodiversity face similar challenges on other oceanic islands, but also there are counterpart lessons from evolution for conservation in other lands. I hope this book will help prompt parallel discoveries elsewhere.

Acknowledgments

Many things evolve in this world and a book manuscript is one of them. Through the many phases of "descent with modification" of the text and illustrations presented here, I benefited enormously from the efforts of many friends and family. I especially thank my dedicated research assistants Claire Menke, Samantha Selby, Autumn Albers, Julia Goolsby, and Marika Jaeger. Claire was the initial sounding board—the first to offer suggestions on many of the wild ideas that went on to become hypotheses here. In most instances, Samantha was the first to assemble data from the literature—typically a massive job—for a first test of the propositions. Autumn was the quality control and rigor specialist, editing tables and figures for accuracy, clarity, and scientific merit. She assembled the separate chapter drafts into one long document and helped design summary flowcharts for several of the chapters. Julia patiently proofed the whole manuscript, looking for everything from spelling mistakes and missing bibliographic entries to errors of logic. Marika was the artist behind most of the graphics, the editor who kept the text concise and flowcharts tidy, and the magician who brought it all close to *Chicago Manual of Style* conventions. I was fortunate to have the help of a young scholar so very talented at both science and art.

My gratitude is huge to each of these assistants, and to the colleagues, friends, and students who read and commented on various pieces of the manuscript: Ari Cartun, Jaime Chaves, Howard Chiou, Rodolfo Dirzo, Sibyl Diver, Rich Ellson, Caroline Ferguson, Mario Fernández-Mazuecos, Don and Margaret Ann Fidler, Harvey Fineberg, Lynn Gale, David Gilbert, Alex Hearn, Arturo Izurieta, Roger Lang, David Lansdale, Diane Matar, Joyce Moser, Darren Sears, Mitzi Spesny, Alana Springer, Shannon Swanson, Jose Urteaga, and Ciara Wirth. Bob Siegel and Susan Charnley read *all* the chapters and gave me helpful comments; Bob also provided the photograph used in Figure 8.1. Lila Thulin not only read all the chapters, but also proposed early drafts for many footnotes. Carter Hunt offered great suggestions on chapter names. Martin Wikelski provided corrections to the earlier published version of Figure 5.9. Frank Sulloway collaborated on a couple of research trips to the islands: one (described briefly in Chapter 4) with a team of first-rate researchers, and one in conjunction with software producer Doug Weihnacht—I am grateful to each companion on these expeditions. Lynn Gale provided indispensable help with statistics and graphics, especially in Chapter 8, where she painstakingly helped analyze population estimates for penguins and cormorants in relation to the swings of the El Niño-Southern Oscillation

(ENSO). I deeply appreciate the helpful feedback and advice of Jeremy Lewis and Bronwyn Geyer, editors at Oxford University Press, in preparing the final version of the manuscript.

Special thanks are due the members of my family for their patience and support of my Galápagos endeavors over several decades, including making many trips to the islands and building an over-sized personal library for this project. I am indebted to Robert Hess and Lysbeth Anderson for loaning me lovely quiet places for thinking and writing. I also thank the Charles Darwin Foundation, especially its staff and former Director, Dr. Arturo Izurieta, for kindnesses extended during the writing of this book. The manuscript was completed during the challenging days of the COVID-19 epidemic. Out of dedication to Galápagos, some authors wrote from quarantine to convey their permission to re-use artwork and figures here.

I will always be grateful to a MacArthur Prize Fellowship for encouraging pursuit of my goals, even if the rewards were destined to be long-term. I have also appreciated my faculty position at Stanford, half-time in Anthropology Department with responsibility for ecological and evolutionary anthropology, and half-time in the Program in Human Biology, where I taught ecology, evolution, and conservation for many years. I am grateful for what I learned during my years as Co-founder and Co-Director, with Martha Honey, of the Center for Responsible Travel (CREST). In addition, I benefited greatly from my affiliation with the Woods Institute for the Environment at Stanford, and from my work as Co-Director with Rodolfo Dirzo and Larry Crowder, and collaborators, in the Initiative for Osa and Golfito (INOGO), Costa Rica. I have been fortunate to work all of my career at the interface of the social and natural sciences.

Earlier versions of all chapters of this book were presented to various classes at Stanford and to conferences and meetings from 2014 to 2020, including talks at the Darwin Station and in Puerto Ayora in 2018. I appreciate the feedback of the respective audiences, who also inspired important modifications, but I alone bear responsibility for these contents.

I have learned so very much from Galápagos; this book represents a modest effort to give something back.

About the Companion Website

www.oup.com/us/exuberantlife

Oxford has created a website to accompany *Exuberant Life*. Readers will find further material to help you engage with the Galápagos.

1

Out of the Ordinary

Reviewing the facts here given, one is astonished at the amount of creative force, if such an expression may be used, displayed on these small, barren, and rocky islands; and still more so, at its diverse yet analogous action on points [of land] so near each other.

Charles Darwin (*Voyage of H.M.S. Beagle*)

Few would deny that the Galápagos Islands are a special place, like nowhere else on Earth. They've earned that reputation from flora and fauna that are downright peculiar—almost otherworldly—compared to related species elsewhere. Consider, for example, the marine iguana, "that strangest anomaly of outlandish nature" (Melville 1854, 17):[1] the only lizard that swims in the ocean to forage for food. Did you know that the marine iguana's skeleton also shrinks under duress and later regrows? A skeleton with the ability to shrink and regrow was unknown until researchers confirmed it in this Galápagos species. Or consider the islands' giant tortoises: each big volcano in the archipelago originally had its own tortoise species, although one extra-large volcano had two tortoise species. Among the longest-lived animals on Earth, they are also surprisingly diverse, each species sporting its own characteristic shell shape on volcanoes so close together you can often see several at once. Last, consider the beautifully odd daisies of Galápagos. The 15 species of the genus *Scalesia* are my personal favorites, all unique to this place. The plants are members of the same family as all daisies, Asteraceae, each sporting characteristic daisy-type flowerheads (themselves composites of many small flowers). However, four of the species are giants, forming full-blown *forests* in Galápagos, with canopies towering up to 20 m (65 ft) high! As a Galápagos visitor, you can walk in a forest of daisies. Your friends at home will think you are telling a tall tale.

True, other oceanic islands are volcanic and are, or were, isolated like Galápagos. Many have unusual flora and fauna as well. Other islands have species that have taken on giant proportions, akin to tortoises and giant daisies. Some have unusual birds and reptiles. So, what's the big deal about Galápagos? How are these islands different from other oceanic archipelagoes, like Hawai'i, Cape Verde, or the Canary Islands?

---

[1] From Herman Melville's novel *Las Encantadas*, 1854, included in Melville and Michelsohn 2011.

Exuberant Life. William H. Durham, Oxford University Press (2021). © Oxford University Press.
DOI: 10.1093/oso/9780197531518.003.0001

Three main aspects together make Galápagos stand out. First, human impact to date is strikingly less in Galápagos than in almost any other archipelago, owing in turn to two key factors. (1) There is no evidence for sustained human settlement in Galápagos before 1832. Among oceanic islands, that's unusual. Human activity has certainly had its impact in Galápagos since 1832, an impact growing alarmingly today. But less than 200 years of human settlement is just about the record for major islands and archipelagoes.[2] It means that some of the most vulnerable of Galápagos organisms are still there, struggling perhaps, but persisting. In places with longer human settlement, similarly fragile species have been gone for years.[3] (2) From its creation in 1959, the Galápagos National Park has protected 97% of the terrestrial surface of the islands, with human activity confined to 3%. Even with tourism booming, bringing more than 275,000 visitors annually before the pandemic (the 2018 tally), the park has clear and appropriate regulations, and generally effective enforcement. Together, the two factors mean that Galápagos shows *relatively little* human impact. Most native species are still there, happily, and most of the land animals retain little or no fear of humanity, allowing us to walk among them as just another species.

Second, Galápagos stands out for its vast collection of unusual organisms, a natural menagerie of extreme life forms: the world's only flightless cormorant; three colorful species of boobies, each with its own fascinating features; the world's only nocturnal gull; an owl that is mainly active by day; and a hawk that feeds like a scavenger. Not to mention the Galápagos penguin, a bird of Antarctic ancestry roosting on the equator, sometimes within a few hundred meters of greater flamingoes from warm tropical regions. Galápagos is almost too much to believe. Author Annie Dillard (1982, 112) put it well: Galápagos organisms are a "Hieronymus Bosch assortment of windblown, stowaway, castaway, flotsam, and shipwrecked creatures."[4] Charles Darwin visited Galápagos in 1835 at the youthful age of 26, years before authoring the works that made him famous. He said he was "astonished at the amount of creative force" displayed on the rocky little islands (in the quote that opens this chapter, 1845, 398).

The third reason Galápagos stands out among the world's biodiversity treasures is a special urgency. Because of its many unusual species, and because change is coming fast to a place with less than 200 years of human residence, there

---

[2] Ecuador annexed the islands on February 12, 1832 (coincidentally, Darwin's 23rd birthday), when Ecuadorian General José María Villamil began the first settlement on Floreana Island (Watkins and Oxford 2009). Lord Howe Island, near Australia, settled two years later, is believed to be the last land on Earth to be continuously occupied.

[3] For examples of Pacific islands with substantial human impact, see Steadman (2006) and Rauzon (2016). The latter describes a "wide array of ecological disasters" brought to islands by introduced organisms.

[4] Widely used in casual conversation, "creature" connotes a product of "creation," easily misinterpreted in an evolutionary context. "Organism" and "living being" are better terms.

are legitimate concerns that many unique species will soon be lost. Other authors have pointed to this urgency, often so effectively that there is little to be gained by my repeating the arguments (see especially Boersma et al. 2005; Watkins and Cruz 2007; Bassett 2009; Watkins and Oxford 2009). Galápagos is certainly not alone in having pressing conservation issues, for there are many other islands in the world—and, for that matter, many other habitats of diverse kinds—in danger of losing unique species. But for both reasons cited earlier—its vast collection of unusual life forms and relatively low historical human impact—there is a special urgency to conservation in Galápagos now that the islands are changing rapidly. Some have called it a "conservation crisis," emphasizing the pressing plight of Galápagos flora and fauna. But special attention is also warranted to social and economic issues in the archipelago. For instance, environmental education is surprisingly weak in a place that has taught the world so much about ecology and evolution. School-age children in Galápagos commonly know only the island where they live, but would love to see more of the archipelago. Many residents have little background with which to appreciate the flora and fauna around them and to understand why they warrant conservation efforts, and plenty of residents would love more training and tools to make sense of it all. Great strides are being made today (see Creamer and Cabot 2019), but there is still a long way to go. Thus, I prefer to speak not of a conservation crisis but of "sustainability urgency" in the islands, thereby giving more balanced valence to social and well as ecological components, and to the call to action. We think of humans as such a highly talented species, able to work wonders with our capacities and technologies. One real test of our talents is surely this: can we find ways to coexist for years to come with the uniquely wonderful organisms of Galápagos? Can we make Galápagos a model for the integrated sustainability of ecological and social systems? And can we use those lessons to help shape strategies for sustainable coexistence with the rest of life on Earth?

These three aspects of Galápagos explain my attention both to the diverse and unusual products of Darwin's "creative force," and to the efforts of people today to conserve those products and improve local livelihoods.[5]

[5] A fourth feature to set Galápagos apart—its unrivaled role in the history of science—is partially a product of the other three. It receives less emphasis here because that history has been thoroughly reviewed elsewhere. Beyond its key role in Darwin's discovery of evolution (see Sulloway 1982a, 1982b; Durham 2012), Galápagos research has yielded volumes about the pattern and process of evolution (Weiner 1995; Larson 2001; Valle and Parker 2012), conservation (see Walsh and Mena 2013; Quiroga and Sevilla 2017; Kelley et al. 2019), marine biology (Grove and Lavenberg 1997; Denkinger and Vinueza 2014), volcanology and plate tectonics (Harpp et al. 2014), El Niño-Southern Oscillation (Zhang et al. 2014; Thompson et al. 2017), and more. In 1957, Bowman argued that "No area on Earth of comparable size has inspired more fundamental changes in Man's perspective of himself and his environment" (R. Bowman, quoted in Nicholls 2014, 144).

The Foci of This Book

To my knowledge, this is the first book about Galápagos to attempt a systematic synthesis of *evolution and conservation*—that is, to explore in every chapter one or more organism's evolved adaptations to Galápagos and their implications for contemporary conservation efforts.[6] The text looks in some detail at each organism's ancestry, how and when it came to Galápagos, and how it has changed since its arrival. The discussion also considers each organism's closest relatives living elsewhere and asks how the Galápagos species differs, and why. Those differences will help illuminate both the environmental challenges organisms faced in colonizing Galápagos and the novel environmental challenges they now face as the archipelago changes at unprecedented rates. No doubt, all conservation efforts are carried out with at least a casual eye to the evolutionary history of the organisms concerned, and some efforts have even focused on evolutionary conservation per se.[7] This book aims to do the latter for Galápagos, offering a series of test cases of the argument that Galápagos conservation will fare better if we deliberately and explicitly integrate evolutionary thinking and analysis. Understanding an organism's evolution, this book argues, is a fundamental tool for its conservation.

But how, specifically, will this book help? First, it shows that the beautifully peculiar features of Galápagos organisms formed over time once their ancestors colonized Galápagos and evolved adaptations to its uncommon environmental conditions. For example, cormorants flew to Galápagos ages ago and found no terrestrial predators and a bountiful fish supply just a few wobbly steps away. Chapter 8 shows how evolution changed the cormorants over time to fit handsomely into these very special conditions, converting them to specialized divers without the means to fly. Second, the beneficial adaptations that emerged in this way, among cormorants and other colonists, have more recently become liabilities as Galápagos integrates into the globalizing world system. Splendidly adapted to their isolated Pacific outpost, the peculiar organisms of Galápagos are now vulnerable to the many changes humans have brought to the islands, including mammalian predators and competitors. Imagine a hungry feral dog coming upon a flightless cormorant sitting on its eggs, fully exposed on the seacoast, with only its beak for protection.

Third, an evolutionary perspective not only exposes endemic organisms' vulnerabilities, but also helps us appreciate their impressive resilience. To some extent, resilience was already a property of organisms that arrived in Galápagos,

---

[6] Steps in the same direction can be found in Wolff and Gardener (2012), De Roy (2016), Quiroga and Sevilla (2017), and Walsh and Mena (2013).

[7] Some classics include Purvis, Gittleman, and Brooks (2005), Carroll and Fox (2008), and Höglund (2009).

because without it they likely would not have survived the ocean crossing to colonize in the first place. But Galápagos has also promoted the evolution of added resilience arising from natural perturbations, such as repeated volcanic eruptions, the birth and death of whole islands, extensive lava flows, the diverse changes of El Niño events, La Niña droughts, recurrent floods, fires, the arrival of new colonizing species, and the natural extinction of established island species. Each of these natural perturbations created selection pressures among resident organisms (causing differential reproduction according to their features)[8] that favored diverse forms of resilience that this book explores. These evolved resiliences offer sources of hope for the survival and continuation of Galápagos species.

But we must always remember that the resilience of endemic species evolved in response to *natural* perturbations. Human activities have added greatly to the number, kind, magnitude, and duration of the perturbations experienced by local organisms. Signs abound that the evolved resilience of Galápagos organisms is challenged today as never before and is, in some cases, on the verge of being overwhelmed. The urgency of the situation in Galápagos stems from the shifting balance of native resilience and the challenges of our added, anthropogenic (human-induced) perturbations in the islands. There is evidence suggesting that Galápagos is near the "tipping point" for many species: a point where additional human-generated perturbations, even small ones, will overcome the capacity of endemic organisms to recover. But all is not lost: there is still time if we take appropriate action. The aim of this book is to make a contribution to that end.

Let me emphasize from the start that an evolutionary approach to conservation does not imply an evolutionary timescale for action. Far from it. I draw upon evolution to enlighten and inform us about the long-term history of Galápagos organisms and to help us understand how and why they came to be different from related species living elsewhere. The evolutionary approach offers us a way to appreciate the difference between the conditions that evolution has prepared the organisms for, and the conditions we create for them to live in today. It does not tell us what to do about that gap, nor about how quickly we must act (although there may be some hints). What we do, and how quickly we do it, is up to us. It helps us see what needs to be done, but we will have to do it, not "leave it to natural processes," or else much will be lost. The changes of today are too many and too fast.

---

[8] "Selection pressure" refers to a specific cause of natural selection, like predation or competition. Natural selection, further described in Appendix 1, refers to the preservation of a feature by its survival and reproductive benefit(s) in a given environment.

So What?

At this point you might find yourself asking, why does Galápagos matter? Why make a big deal of a tiny, isolated archipelago of extreme species? Is it worth all the trouble one must go through to make sense of it, let alone invest the time and effort for its conservation?

I like to think it matters for several reasons. First, Galápagos is loaded with lessons of broad significance for the study of life on Earth. It matters because it teaches us about important, general processes of life and of human impact on life, and it does so in a uniquely dramatic way. Because of its unusual ecological conditions before humanity arrived, including freedom from mammalian predators and competitors, Galápagos is a unique classroom that reveals things often hidden elsewhere. If you have the chance to visit, you'll see dramatic births and deaths, and extravagant mating rituals that are just another everyday occurrence. You'll see territorial behavior and fights over mating opportunities right before you, beside the trail or even right in the middle of it. You'll see sibling rivalry right on the beach and efforts of adult female sea lions to keep the peace. You'll see the immodest behavior of "great" and "magnificent" frigatebirds as they practice kleptoparasitism overhead (i.e., parasitism by theft), stealing precious food and nest-building materials from other species, or even from one another. You'll see and experience these things and many, many others. So much of daily life is out in the open in Galápagos, bare and exposed with an exuberance all its own. A place with all of this on display, so many unique species doing their unique things, surely has much to teach us about life on Earth if we will only pay attention. Darwin was just the first of many to appreciate that fact. His experience in Galápagos had a profound effect on his thinking and, later, it completely changed his view of life. If we listen closely to what the archipelago has to say, it may also change ours.

Second, Galápagos matters for its message about the kinship of life. The simple fact that organisms in Galápagos display little fear of humans even now, allowing us to walk beside them on the beaches, look deep into their nests by the trails, or swim with them underwater (as they often playfully blow bubbles while we try to keep up), reminds us that we are not "apart from nature." We're right in the thick of it, and you see it, hear it, and feel it in Galápagos. Reconnecting with life up close and personal can be quite enthralling for modern visitors. I have seen Galápagos inspire people to a whole new appreciation of our place in nature. It helps us to understand "how it is that all forms of life, ancient and recent, make together one grand system . . . connected by generation" (Darwin 1859, 344). It helps us to see and feel our place in the history of the whole. It helps us to understand that we humans are a tip on a branch of the "tree of life."

We, too, are survivors of an uninterrupted history going all the way back to the first organisms on Earth. Inevitably, a special feeling of connection, of oneness, emerges from a visit to Galápagos. Learning about Galápagos organisms gives us the opportunity to appreciate and marvel at the all-encompassing kinship of life, and to gain new perspectives on our place within it. In the end, an experience with the special living organisms of Galápagos gives us a better sense of our own humanity . . . of who *we* are, not just who *they* are.

Third, Galápagos matters because it is a "test bed." It is a test bed for seeing if we humans can find ways to coexist sustainably with the biodiversity endowment of the planet. On the one hand, Galápagos is small: its land surface area is slightly larger than that of Delaware, the second smallest state in the United States. One would think we might learn the intricacies of so small a test bed, learn to live with them sustainably, and then maybe spread the lessons to other regions. On the other hand, life in Galápagos is also especially fragile, for having existed in an impressively isolated state for so long. Will we be wise and caring enough to find ways to coexist with its fragility? Will we find ways to support a resident human population with a healthy and hardy lifestyle and similarly support fragile island biodiversity with the opportunity to evolve for years to come? The islands give us that chance: Galápagos is a proving ground for our future.

To make the test still more demanding, there are already a lot of humans there. The resident human population currently numbers over 35,000 on five inhabited islands (see Figure 1.1), not to mention up to eight times that many annual visitors. True, the archipelago retains some aspects of the "little world within itself," as Darwin so aptly put it, but it is now a little world with much less isolation than in the past, and where human and biophysical processes incessantly affect one another. Sometimes, the effects of human presence are indirect, mediated by plants and animals that humans have introduced to Galápagos, or by climate change or pollution. Sometimes they are direct, through efforts at clearing native vegetation or harvesting local marine resources, for example. Our human impact today is far-reaching and relentless. Galápagos offers us a microcosm in which to try to understand our impact on biodiversity, and in which to search for solutions that might also help in other contexts.

My hope is that the book will motivate all of us to work toward long-term solutions for human–wildlife coexistence, both in Galápagos and elsewhere. I often ask myself: if we cannot live sustainably with the wildlife of a tiny isolated archipelago way out in the Pacific Ocean, how will we ever achieve sustainability overall?

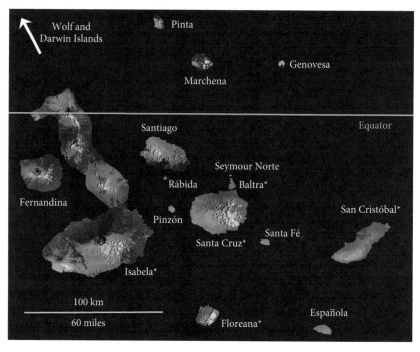

Figure 1.1. Galápagos from space. Landsat 8 took this mosaic of photographs in March 2015, during the warm and wet season. Clearly visible are humid zones (*green*), arid areas (*brown*), barren lava (*black*), and clouds (*white*) near the high elevations (low islands have neither clouds nor a humid zone). Not shown are Darwin and Wolf Islands, which lie 160 km (100 miles) and 125 km (73 miles), respectively, northwest from the top of Isabela. Each island is a single volcano except San Cristóbal, which is a fusion of two, and Isabela, which is a fusion of six, although the sixth—called "Ecuador"—is half-eroded (giving the island's seahorse shape its "mouth"). In clockwise order from Ecuador, the other five volcanoes are Wolf, Darwin, Alcedo, Sierra Negra, and Cerro Azul. Note that most western volcanos have a caldera (i.e., a collapsed magma chamber looking like large central "crater"). Asterisks indicate the five human-inhabited islands, although little or no human impact is visible from this altitude (705 km, 438 miles). Also not visible in this image is a small humid zone on Pinzón Island. *Source*: Courtesy of U.S. Geological Survey.

How This Book Is Different

Besides its explicit evolutionary approach to conservation, this book differs from others on Galápagos because: it is not a guidebook, though it works well in conjunction with one;[9] and it is not a coffee table book, though it does include a good number of pictures. This book is about the scientific study of terrestrial flora and fauna of Galápagos, with special attention to the connections between their past evolution and current conservation. My intention has been to place it well within the grasp of general adult readers who enjoy science. Where the book draws upon principles of evolution and conservation, I provide explanations in the text, supplemented with additional material in printed and online appendices (for example, where needed, the latter contain details on methods). I use a fair number of tables and graphs, but I explain each one in the text and accompanying legends. Like a trip to Galápagos, I'm sure this book will take some effort. But also like a trip to Galápagos, I hope it will give readers a new appreciation of life.

This book is also different from others on Galápagos because I prefer to use evolutionary explanation deliberately and consistently. While we may not be accustomed to evolutionary explanations in our daily lives, the principles and logic are not arduous. The more we use them, the more comfortable and compelling they become. Among other things, evolutionary explanations help us to understand and prioritize the various conservation threats and challenges organisms face. In a word, *we'll be better at conservation if we are better at evolution.* So, it will help us to practice even in reading.[10]

Finally, one of the most important features of this book is an explicit effort in each chapter to integrate social and ecological components of conservation, and thus to keep people in the picture at all times. My reasoning is simple: conservation always involves human effort of some kind, so people must logically be included, and evolution makes it clear that we humans are part of the "natural world." Now that the Anthropocene is upon us—the era in which humans exert a dominant influence over environment and climate—it is more important than ever that our analysis of conservation issues include human activity, both as contributing to problems and as integral to lasting solutions. For these reasons, I make repeated use of the framework of social–ecological systems, or SESs, introduced by Berkes and Folke (1998) and nicely elaborated by Ostrom

[9] Worthy guidebooks include Jackson (1993), Kricher (2006), Stewart (2006), Nicholls (2014), Moore and Cotner (2014), Litt (2018), and Lynch (2018). This book also works well in conjunction with the edited volume by De Roy (2016), which offers short field updates from many researchers cited herein.

[10] For a brief review of evolutionary principles, see Appendix 1, "A Primer of Evolution," which also provides concise definitions for terms and other tools for evolutionary thinking and analysis.

(2009) and Ostrom et al. (2012).[11] Here is a brief example to explain what the SES framework entails and how I find it helpful. The example serves also to introduce a wonderful Galápagos organism that receives a full look in Chapter 2.

How the SES Framework Helps

The largest bird of Galápagos is the waved albatross (Figure 1.2A). Like other albatrosses, this species spends a good portion of every year in long-distance foraging by flying over the sea, where it feeds on a varied diet of fish and squid. Also like other albatrosses, mature waved albatrosses come to land at some point each year to mate, incubate eggs, and care for chicks. However, in this case, the vast majority of all adults come ashore every year to one small area of Española, the southernmost island of Galápagos (Figure 1.2B). They come in March for mating, stay in April and May for egg laying, spend June and July in brooding (guarding and feeding new hatchlings), and devote the months from August to November to feeding the growing chicks until they fledge and fly off. During brooding, both parents are involved in protecting the egg and in feeding the chick if and when it hatches. Chicks are large, are sedentary at first, and have a big appetite, which they advertise by energetically begging from their parents (Figure 1.2A). So parents take turns: one goes out to sea to feed and to bring back rations for junior, while the other posts guard over the growing chick. The guarding parent uses its long beak to keep at bay the predatory birds of Española, especially hawks and frigatebirds, while the foraging parent uses its long beak to bring back fish and squid from areas close to the island (see Figure 1.2B).

The fish supply for waved albatrosses during brooding is influenced by a whole host of local ecological variables, including ocean currents' strength and direction, ocean depth, water temperature, photosynthesis rates in marine plankton, and natural dynamics among all species in the local food web, including other marine birds who compete for the same fish. The availability of albatross prey is also influenced by the density of marine predators like the endemic sailfin groupers (locally known as *bacalao,* a favorite human food) and by various shark species, including some who prey on albatrosses, too, when they land on the water to feed. But, importantly, the fish supply is also influenced today by a range of human activities in the archipelago, especially fishing. Fishing in the area has been carried out in various ways over the years, from small-scale subsistence fishing to large commercial enterprises, each with their own manner of boats and

---

[11] A similar interdisciplinary framework for interactive social and ecological systems is Coupled Human and Natural Systems (CHANS; see Liu et al. 2007). CHANS strives for explicit, formal models of social–ecological linkages, which are beyond the scope here.

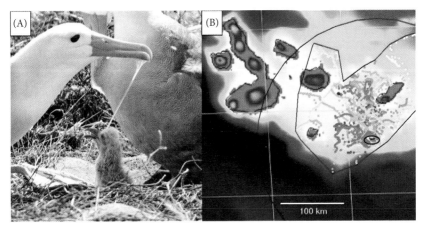

Figure 1.2. *A,* Waved albatross parents and their new hatchling during a break in feeding. With their small beaks, young chicks do better with regurgitated fish oil (visible here) and small pieces than with whole fish. Defenseless at this age, chicks require constant parental protection. *B,* During the chick's early weeks, adult albatrosses take turns foraging to feeding areas very close to Española Island (*black oval*). Flight paths are shown for 26 recorded foraging trips, each a different color, during one brooding season. All foraging took place within the black polygon, very near the colony (*black arc* is part of a circle of radius 165 km/100 miles, emphasizing the small size of the foraging area). Shades of blue represent depth of the ocean. *Source*: *B,* Reprinted by permission of author and publisher from Awkerman et al. (2005a, 292), © Inter-Research 2005.

technology, regulated to various degrees by a governance system with rules that specify how much can be taken, when, and by whom.

Thus, the food supply of waved albatrosses during the crucial brooding period is influenced not just by natural ecological processes and variables—what we can call the ecological system—as it was years ago. Today, it is also affected by human social processes and variables, such as fishing, market demand, and governance—what we can call the social system. And because food supply is crucial for waved albatross chicks, we can say with no exaggeration that the survival and reproduction of waved albatrosses is today a joint product of social and ecological processes and their interactions. The example lends itself well to the kind of systems diagram and analysis that was pioneered by Ostrom (see Figure 1.3). The diagram depicts key components of the interrelated SESs, implying that interactions within and between social and ecological systems are vital determinants of food supply for adults and chicks of the focal species. Because such diagrams serve us well in the chapters ahead, let's take a closer look what's entailed.

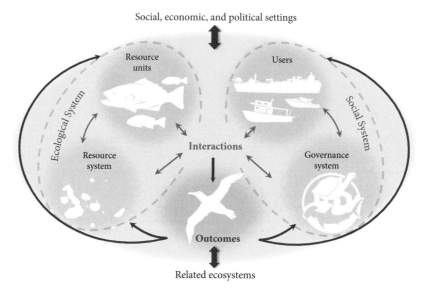

Social, economic, and political settings

Resource units

Users

Ecological System

Social System

Resource system

Interactions

Governance system

Outcomes

Related ecosystems

Figure 1.3. Model of a Galápagos Social–Ecological System (SES). Following Ostrom (2009, 419), this schema integrates the "relatively separable" ecological and social systems (*left* and *right*, respectively) related to a given resource (here, finfish of the archipelago). The social system includes heterogeneous *users,* represented by a diverse set of fishing boats, and a ***governance system,*** represented by the logo of the Galápagos National Park Directorate. The ecological system includes diverse ***resource units***, exemplified by the large economically valuable sailfin grouper amid populations of other fishes, and the larger ***resource system***, here the finfish fishery. ***Interactions*** of components are represented by *green arrows*, as they impact the focal species of our analysis (which varies by chapter)—here the waved albatross—generating ***outcomes*** shown at *bottom*. The survival and reproduction of the focal species is a key outcome for our purposes, linking social–ecological interactions with its ongoing evolution. Decreased waved albatross fertility is one such outcome, feeding back (*red arrows*) upon other components, where it may, or may not, trigger compensatory processes. All SESs are embedded within social, political, and economic settings (*top*) and include ties to other, related ecosystems (*bottom*), such as ecosystems of the open ocean. In the social system, diverse user icons represent inequality among users by class, ethnicity, race, gender, and intersections, complete with power differentials (implied by icon size). In the ecological system, diverse fish icons represent various competitors, predators, and symbionts as may exist for the resource, and the multiple interactions that come with them. *Source*: Adapted from Ostrom (2009).

Interactions of Social and Ecological Systems in Galápagos

Human activity hasn't always been such a prominent part of life in Galápagos. Until five centuries ago, the isolated archipelago was pristine, little affected by humans. Back then, one could have described it comprehensively and faithfully by speaking of only the ecological system: diverse habitats, resident organisms, and natural processes. Those days have long since ended. It's now a place where human influences interact continuously with habitats, organisms, and natural processes, impacting just about all flora and fauna directly or indirectly. It has been a major change in just 500 years, contributing to the sense of urgency today for sustainability.

Fortunately, the SES framework can help us understand these changes; it is a model ideally suited for these purposes because it gives center stage to the continual interactions of social and ecological processes. Ostrom's (2009) description of the framework seems especially germane for Galápagos. She says an SES consists of a "resource system (e.g., a coastal fishery), resource units ([fish]), users (fishers), and governance systems (organizations and rules that govern fishing on that coast). . . . [They] are relatively separable but interact to produce outcomes at the [larger system] level, which in turn feedback to affect these subsystems and their components, as well other larger or smaller SESs" (Ostrom 2009, 419).

A second intrinsic feature of the framework is that it requires the ecological *and* social sciences to interact to make sense of the overall dynamics, despite the fact that these sciences "developed independently and do not combine easily" (Ostrom 2009, 419). Indeed, the symmetry of components sets up the expectation of strong analytical tools and procedures from both sides to make it genuinely social–ecological, and that, in turn, encourages interdisciplinary teams, training, or both to do the analysis. Moreover, the framework reminds us that the ecological and social systems have interacting histories, sometimes going back centuries. This continual interaction of the social and ecological is the hallmark of the SES approach (Colding and Barthel 2019)—a good match, as we shall see, for Galápagos. The additional step we take here is integrating a "focal organism"—like the waved albatross—into the SES for each chapter, an organism affected by case-specific interactions among the other components, and whose survival and reproduction are key outcomes of the SES for our purposes.

The SES perspective is hardly news to the hard-working conservationists in Galápagos, who have been among the thought leaders of this approach (see, for example, Watkins and Cruz 2007; Tapia et al. 2008; González et al. 2008; and Castrejón et al. 2014). They have woven SES into the current management plan for protected areas of the archipelago (Dirección del Parque Nacional Galápagos 2014) and into a 10-year vision for local sustainable development (Consejo de Gobierno del Régimen Especial de Galápagos 2020). This book draws on their insights often, and it also takes advantage of recent efforts to more fully integrate social theory and social structure into the SES framework (see, for example, Brown 2014; Fabinyi et al. 2014) as befits a context with as much social, cultural, and economic diversity as Galápagos. But adding focal organisms to the SES and showing just how they are impacted by the social–ecological interactions have the further benefit of keeping us from "missing ecology" in our analysis, as Epstein et al. (2013) put it of some attempts at SES analysis. In the approach taken here, ecology and social science are both integral.

We should also note that an SES diagram is a deliberately simplified summary of a complex system of interactions at one point in time, while the analysis behind it is typically far more complex. Some critics have been quick to label the SES approach as advocating social science "light," or even as natural science passing as social science, because the integration looks so simple in diagrams.[12] For example, such models often represent social and ecological structure via diminutive icons, although they have great empirical importance and represent complex social and ecological feedback effects by deceptively simple arrows. One should remember these sketches are just heuristic devices to represent complex interactions that change over time in unique and context-specific ways.[13] The hope is that SES thinking will help us to take seriously "both sides of the coin," as Watkins and Oxford (2009) put it, and will help us guard against perspectives and policies set up to "govern humans and natural environments as separate," as has so often been done in the past (Lu et al. 2013, 84).

My hope is that "SES thinking" will also help us to draw together evolution and conservation in pragmatically useful ways. A fundamental argument here is that the very survival and reproduction of Galápagos organisms—their "evolutionary fitness"[14]—is now a product of both natural conditions and human activity. It is thus a key SES "outcome" measured in the focal species, like the waved albatross of Figure 1.3. The focal organism's fitness may be affected by introduced alien species, by habitat modifications humans have wrought, or by climate change's effects in the region, among other causes. No Galápagos organism escapes these effects, or certainly very few. So a focus on local organisms' fitness within an SES framework allows us to integrate the social and ecological variables that impinge on an organisms' well-being today. It allows us to see how

[12] SES diagrams have prompted some social scientists to argue "that the socio-ecological system model represents an effort by natural scientists to 'do social science,' meanwhile ignoring the vast corpus of previous human–environment research in the social sciences, . . . [and] subsuming societies into natural systems and attempting to study them with the tools of natural sciences" (Cairns 2011, 179, quoting an unnamed social scientist in Galápagos).

[13] It may help to remember that the idea of SESs originated with thoughtful social scientists; few have accused Nobel Laureate Ostrom, originator of diagrams like Figure 1.3, of social science "light."

[14] Evolutionary fitness is a surprisingly difficult concept given its central role in evolutionary thought. I like to think of fitness as the "effectiveness of design" of a feature for the survival and reproduction of its individual bearer in a given environment (after Williams 1966; see also Appendix 1). Ideally, we'd use design criteria to measure fitness, but practicalities often require surrogates like energy efficiency or the number of surviving grand-hatchlings. Thoughtfully considered, such surrogates are not likely to present issues in the cases considered here.

the fascinating showcase organisms of Galápagos have been imperiled by an-thropogenic and natural changes in the archipelago. And it allows us to see when and how native organisms have been greatly assisted by recent and continuing conservation efforts.

Hypotheses and Hypothesis Testing

Each of the following chapters builds toward one or more explicit propositions or hypotheses about Galápagos organisms, and then offers at least a preliminary test of each, drawing on my own work and that of other authors. In part, I do this to call attention to one, or a few, central arguments of the chapter; in fact, there are many hypotheses behind each chapter, most of which have been tested and accepted previously. But I have several additional goals in mind with this device.

Goal 1: Formulate Novel Propositions
The hypothesis or hypotheses highlighted in each chapter are, in most cases, not already widely known or established. Labeling them "hypothetical" makes clear that the propositions are recent or new, and thus not as secure or time-tested as other arguments in the volume. They are, however, in most cases the leading contenders from a wider field of working hypotheses that I used in designing the chapters. In some cases, I have included a few of those additional hypotheses to compare, but, most of the time, available evidence allowed me to zoom in on one or several propositions of special promise.

Goal 2: Put Them to a Worthy Empirical Test
Setting up promising arguments as hypotheses subjects them to a useful form of self-discipline: the procedure of testing hypotheses allows pertinent facts and figures to guide our conclusions, not just wishful thinking or theoretical preferences. In my view, these tests are an important step to advance our under-standing: a hypothesis test is only as good as the evidence we bring to bear on it. In each case, I have endeavored to use the most appropriate and reliable data I could find.

Goal 3: Draw Cautious Conclusions
Hypotheses emphasize that the answers we offer today are tentative—the best an author can do with the available evidence. Eventually, better arguments and evidence will emerge. In science, hypothesis testing is how our understandings improve and science itself evolves. Ideally, what we do here offers helpful insights for Galápagos and its conservation today and helps shape and refine the next round of hypotheses and tests for the future. It will be a sign of scientific progress

if some of the hypotheses put forward here are later tested and rejected by better and larger data sets than are available now.

Each chapter thus develops one or several key predictions about the featured organisms. In each, I compare the predictions against the most complete and dependable data that I could find, to see if the hypothesis is supported or not. Sometimes, a conventional statistical test of the hypothesis is appropriate, indicated by a specific test statistic with associated p-value, and, in those cases, I call it a statistical test, following accepted procedure. In other cases, we do not (yet) have an appropriate type of data, adequate amount of data, or both, for conducting a statistical test. Often, this latter effort leads to a different kind of test, showing consistency or inconsistency between available data and one or more hypotheses or showing a greater weight of evidence for or against a particular argument. I call this second type a "preliminary test" of a hypothesis, emphasizing its exploratory and tentative nature. Others might call such efforts "evaluating" or "investigating" a hypothesis, but I use "test" to emphasize that the comparison of expected and observed results can fail to corroborate the hypothesis. This procedure may be unorthodox in some circles, but it has the advantage for conservation work of allowing us to move forward in a cautious, self-correcting manner even when we cannot generate the degree of statistical certainty we would wish. The statistician John Tukey wrote, "There is a place for both 'doing one's best' and 'saying only what is certain,' but it is important to know, in each instance, both which one is being done, and which one ought to be done" (1960, 429). For these reasons I offer a mix of statistical and preliminary tests in these pages, preferring an appropriate statistical test whenever possible, but always "doing my best" for the organisms involved.

Let me now offer a few words about usage. First, I have been fairly liberal in my use of footnotes, hoping to unencumber the text for general readers, whom I invite simply to ignore the notes, or not, as they wish. Second, let me clarify the meaning of a few important words in the text that might otherwise be confusing. For clarity, I speak of the "colonization" of Galápagos by plants and animals, and reserve "settlement" for human occupants. Thus, "pre-settlement Galápagos" refers to the archipelago before 1832 and the first human occupants of Galápagos. I use "endemic" to mean species, past and present, found only in Galápagos, "native" to refer to all species, endemics included, that inhabited Galápagos before 1535 (the first documented human visitors), and "alien," "introduced," or "exotic" species to refer to species that have appeared in Galápagos since 1535 via human influence. When alien species reproduce rapidly and spread widely, I call them "invasive" species, consistent with standard usage. In addition, I frequently discuss "the survival and reproduction" of an organism like the waved albatross, or "its evolutionary fitness," as one way to measure the aggregate impact of surrounding changes on the organism. I use this convention to make it easier to talk about the overall impact of a given change on a population of the organism, all the while recognizing that fitness is best measured or estimated at the level of individual organisms. I use aggregate values to represent an underlying distribution of individual values—a distribution that can always be called into play if and when needed in the analysis. In most cases, it will suffice to think of "survival and reproduction" as representing the average of individual values in a population.[15] Finally, I also distinguish "conservation"—which I take to mean sustainably managed human impact on environment, including sustainable natural resource use —from "preservation," the attempt to eliminate, or minimize, human impact on environment. Conservation is about sustainable use and stewardship, not about preventing use or maintaining "pristine" conditions.[16] As the SES approach makes clear, I do not see Galápagos as pristine, nor do I long to make it so. There is such an enormous amount we can learn from Galápagos—about life, evolution, conservation, the very future of the planet—if, through collective efforts, we can shape the evolving social–ecological systems of the archipelago to where each supports genuinely sustainable use by humans. This book is dedicated to that effort.

How This Book Is Organized

The book has a simple outline. I work through my favorite case studies of Galápagos organisms, their evolutionary history of adjusting to the archipelago, and their contingent conservation challenges today. I include such iconic species as giant tortoises, blue-footed boobies, penguins, marine iguanas, Darwin's finches, flightless cormorants, and waved albatrosses. The discussion also includes several humble, often-neglected species, like the Galápagos rail and little brown sea cucumber. The sample offered here is admittedly selective and biased by my own experience and interests. I tend to favor organisms that are endemic, terrestrial, and zoological. I do not wholly neglect others, such as marine organisms or botanical examples, but my sample is admittedly spotty. Each chapter concludes by looking at the conservation implications of what we have learned about the species at hand.

In Chapter 2, we return to the plight of the waved albatross of Galápagos, because we caught only a glimpse of this fascinating bird in the previous section.

---

[15] This is not to suggest that natural selection commonly works by or through such aggregated measures (on the contrary, consensus since Darwin highlights the importance of selection at the individual level—see Appendix 1). Rather, aggregated measures serve here merely to summarize population-wide impact across the range of individual values in a population. We can readily move to the individual level of detail when needed.

[16] It is important to note that sustainability has an implicit time dimension. I take "sustainable" to mean "capable of supporting indefinitely a given species, or a given set of species."

A closer look also allows us to explore the benefits of an evolutionary perspective for understanding its plight today. How does a bird whose ancestors specialized in long-distance soaring and foraging in strong winds and high waves adjust to Galápagos, where the wind and waves are only half as powerful? Its evolutionary specializations since colonization help us understand the conservation challenges facing the waved albatross today. Chapter 2 also introduces the recurrent climatic pattern so very influential in Galápagos: the El Niño-Southern Oscillation (ENSO). ENSO has far-reaching impact on the waved albatross and many other organisms in the archipelago. The cumulative nature of the greater tale unfolding here makes it worthwhile to read most of the chapters in the order presented.

Chapter 3 begins by stepping back to explore the key factors that give Galápagos its special "creative force," as Darwin put it, to shape the unusual adaptations of albatrosses and many other species. The chapter illustrates the ingredients of this "special recipe" via their role in shaping the diverse tortoise species of Galápagos with their characteristic shell shapes. The islands' geology plays an active role in evolution, often setting up new and diverse "ecological theaters" for the "evolutionary play" taking place above water.[17] Tortoises are a fitting example of adaptive radiation of multiple species from one common ancestor. The chapter also considers the role of tortoises as shapers, metaphorical "engineers," of the ecosystems they live in and the conservation implications of those roles.

Chapter 4 features humans in a more central role by focusing on a small, innocuous ground bird, the Galápagos rail, which continues to be heavily affected by human activity. Many previous works about Galápagos offer cogent summaries of the human history of the archipelago, allowing me the liberty of presenting a succinct thumbnail sketch. I focus on two aspects: how human settlement of the islands since 1832 provided surprising diversity within the emerging social system, including major differences of culture, class, and ethnicity; and how human activity in Galápagos has affected conditions there, with the focal example being the unassuming near-flightless rail. Galápagos rails were drastically affected by a small number of alien species that came to the islands with people—especially goats, rats, blackberry plants, and quinine trees. The "tale of the rail" provides a fitting backdrop for a broader look at how human activity is busily eroding many of the age-old features that made Galápagos special.

Chapter 5 explores the awesome adaptations of marine iguanas to those age-old features—including the iguanas' dietary focus on seaweed and associated ability to snort record concentrations of salt. We will also explore their fine-tuned hormone-mediated stress response, which activates to high levels in the

---

[17] The metaphor of "the ecological theater and the evolutionary play" was first suggested by the ecologist G. Evelyn Hutchinson in his book of essays in 1965 by that name.

face of oil pollution or even tourism. That same response extends appropriately in the opposite direction, helping the iguanas to shut down, and thus to display the astonishing ability mentioned earlier: they shrink during El Niño events and regrow afterward, almost like a concertina. A key conservation issue then comes into focus: with anthropogenic climate change making El Niños stronger, more frequent, and longer lasting, will the marine iguana's concertina be able to keep the beat? These unique iguanas are iconic of both the delightful adaptations and the sobering conservation challenges of Galápagos today.

Chapter 6 provides an example of coadaptation, when two or more species coevolve new properties in tandem. It tests the hypothesis that the tree daisies of Galápagos have coevolved with the group of tree finches, the daisies providing habitat and foraging space for the finches, and the finches providing seed dispersal for the daisies and access to new habitats and islands. This implies that finches and daisies have interconnected population dynamics, and thus will be best served by joint or integrated conservation efforts. The tree finches and daisies are each an example of an "endemic radiation"— the adaptive radiation of species in Galápagos, all descended from a common ancestor. But data suggest that theirs is a form of co-radiation, a fitting example of the novel interactions that can shape organisms that wind up as castaways together on remote oceanic islands.

Chapter 7 looks at a contrasting case of "native radiation," where a group of closely related species probably diversified elsewhere in the tropics, but co-inhabit Galápagos today. Three species of boobies—red, white, and blue—are highlighted in this chapter, with their delightfully divergent characters, much like musical "variations on a theme." The chapter looks at the captivating colored feet of the blue-footed booby as a further example of adaptation—this time, to the visual preferences of mates as an example of sexual selection.[18] We also explore their "facultative siblicide"—the actual killing of siblings—as a "booby trap" driven by a highly variable food supply. The blue-footed boobies' behavior is contrasted with two sibling species in the archipelago —Nazca boobies and red-footed boobies—who nest in overlapping areas of the islands. I suggest that blue-footed boobies originally evolved in adaptation to the Cromwell upwelling of Galapagos, and I offer a provocative hypothesis for the aggressive behavior of adult Nazca and blue-footed boobies, who at times beat fiercely upon unrelated chicks that are minding their own business. Finally, the chapter asks why Nazca and blue-footed populations are plunging in size today and what can be done

---

[18] Sexual selection refers to adaptation to a mating system, and thus to the evolution of features that aid organisms in attracting mates. It can be contrasted with environmental selection, which is adaptation to habitat (see Appendix 1).

about it. Ideas for intervening before siblicide provide interesting pathways for the conservation of these species.

Chapter 8 explores the adaptations of two flightless denizens, Galápagos cormorants and penguins. The chapter compares cormorants and penguins because both prey on small fish in shallow waters of the same upwelling zone in western Galápagos and are thus similarly vulnerable to El Niño. But cormorants evolved flightlessness in Galápagos in response to local conditions, whereas penguins were already flightless at colonization. That difference brings important consequences today to the conservation of both species. The penguin population is slightly more "volatile" in response to ENSO oscillations, and the cormorant population is slightly more stable. Nevertheless, the two species have an impressive underlying similarity: an evolutionary convergence on opportunistic, temperature-sensitive mating strategies in response to the ENSO cycle. The resulting reproductive flexibility provides both species with an impressive resilience in the face of El Niño events. The question is, then, will that El Niño resilience suffice for the decades ahead?

Chapter 9 looks at fishing in the archipelago, with a focus on the flurry of fishing activity in the 1990s and early 2000s in the same upwelling areas where cormorants and penguins live. The chapter focuses on the small but valuable echinoderm—the brown sea cucumber—at the heart of this profitable "gold rush" of exploitation. The chapter analyzes efforts by the Galápagos National Park Directorate (GNPD) to regulate the sea cucumber and lobster fisheries, both hosting a mix of long-time artisanal fishers from Galápagos and hundreds of recent immigrants. In the beginning, with every imposition by the authorities, fishers protested vehemently and sometimes violently, eventually leading to a 12-year experiment with participatory management, which gave fishers, for the first time, a voice in their own regulation. Despite a promising start, the co-management system proved cumbersome and eventually failed. It was replaced in 2015 by a much simpler, more direct form of negotiation between fishers and GNPD. Questions remain about the recovery of the fisheries and options for converting some fishers to tourism careers. For now, direct negotiation seems to be working, guided by a charter of accepted rules for each fishery.

Fishing was not the only activity to lack regulation during the same years: tourism also grew exponentially, reaching over 275,000 visitors in 2018. The direct impact of tourism itself is not as bad as it sounds, even with such big numbers, for the GNPD carefully spreads visitors over diverse visitation sites throughout the archipelago. Evidence suggests that the indirect effects of tourism have caused greater damage than tours themselves, via increased economic activity and immigration to the islands, which also grew exponentially and unsustainably before 1998. Recent laws and enforcement are beginning to help curb immigration and exponential tourism increases, as of course is

COVID-19. But there is still a need to set immediate, enforced limits to land-based tourism (boat-based tourism is already regulated), and to promote ways in which tourism can become a positive force for conservation and the sustainable use and enjoyment of Galápagos.

Chapter 10 closes the book with a look back at the main lessons from the cases of evolution and conservation in Darwin's little world. My fondest hope is that this book will help each reader to understand and appreciate the virtual menagerie of unusual living organisms that call Galápagos home, to gain a deeper understanding of their current plight to survive and to reproduce after many centuries of isolation, and to see some ways each of us can contribute to conservation and sustainability efforts underway. The flora and fauna of the islands are beautifully unique. They are also uniquely imperiled. It is up to us, this book argues—to our generation of humanity, not some future generation—to set a course for sustainable coexistence in the islands.

Let's get started.

2

Tough Times for the Loneliest Albatross

> *An albatross is a great symphony of flesh, perception, bone and feathers, composed of long movements and set to ever-changing rhythms of light, wind, water.*
>
> Carl Safina (*Eye of the Albatross*)[1]

Visitors to Española, the southernmost island of Galápagos, are often thrilled by the sight of the endemic waved albatrosses. They are big birds, weighing up to 4 kg (9 lbs) and sporting wingspans up to 2.5 m (9 ft). They come ashore each year, some 15,000 strong, forming colonies from April to December along the island's southern coast. During that interval, their "beak fencing" courtship display is a wonder to behold. Mated pairs waddle back and forth, "salute" each other with raised beaks and honks, and intermittently smack their large beaks together, making a clacking sound as loud as that made by wooden blocks (Figure 2.1). All this done over and over, with great gusto and speed, makes for a mesmerizing experience.[2] In other albatross species, mates undertake similar, if quieter, displays, uniformly leading to nest-building—carefully shaped mounds of mud or sticks—and egg-laying. Curiously, the waved albatross pairs make no nest, a point to which we'll return, but lay their egg directly on the ground.

Visitors are equally awestruck by the quiet elegance of waved albatrosses in flight, soaring along the windy uplift at the cliff edge of the island (Figure 2.2). Because albatross species generally soar when winds are strong, sailors of yore regarded albatrosses at sea as a good omen for favorable winds and general luck. Indeed, they were widely felt to have prophetic knowledge of the wind, hence the zoological name of the waved albatross, *Phoebastria irrorata*—"Prophetess that bathes."[3]

---

[1] Safina's book *Eye of the Albatross* presents a lyrical rendition of the life of Laysan albatrosses—a species closely related to the albatross of this chapter—and the environmental challenges they face.

[2] For a video of waved albatross beak fencing, see https://www.youtube.com/watch?v=SrfTwkLDMfE. Not all mating occurs with ritualized beak fencing: sometimes females are simply ambushed upon landing and forcibly mated, or sometimes extra-pair copulation happens peacefully. "The outcome of this promiscuous behavior will be that as many as one in four chicks will not have been fathered by the male raising it" (de Roy 2008, 128).

[3] Coleridge nicely captured sailors' respect for albatrosses in his famous poem, "Rime of the Ancient Mariner" (1857): When a sailor killed a soaring albatross with his crossbow, shipmates tied the slain bird around the sailor's neck, superstitiously hoping to shorten what seemed a vengeful slacking of the wind.

Exuberant Life. William H. Durham, Oxford University Press (2021). © Oxford University Press.
DOI: 10.1093/oso/9780197531518.003.0002

Figure 2.1. Beak fencing is a key component of mating rituals in many albatross species, but it is especially lovely—and noisy—in the waved albatross.

Figure 2.2. A waved albatross soars along the cliffs of Española Island, over the pounding surf below. Like a hang-glider, their wings lock in place and require little effort in strong winds.

Figure 2.3. A waved albatross takes off from an "airstrip" free of encroaching vegetation. It takes good effort to wobble to the cliff edge and fall into the updraft, but it's easy from there.

It is also fascinating to see waved albatrosses take off or land at one of the "albatross airstrips" on Española—grassy clearings near cliff edges. So large is their wingspan that they can only land in areas with a minimum of 3 to 4 m (10–13 ft) free of encroaching bushes. Their take-off is equally captivating: on big webbed feet, they wobble to the cliff edge, extend their huge wings, and literally fall into the updraft to fly away (Figure 2.3). Flapping is not their forte. Like albatrosses the world over, they are best at soaring and gliding on strong winds. But Galápagos winds are rarely full albatross strength, and thus the waved albatross seems strangely out of place.

Out of place indeed! No other albatross roosts within 20° latitude of the equator.[4] In typical albatross habitat, the winds average from 8.5 to 12 m/sec (19 to 27 mph), at a common soaring height of 10 m above the ocean, giving these latitudes the nicknames "roaring forties" and "furious fifties" (Figure 2.4). With such strong winds, albatrosses engage in "dynamic soaring," obtaining from the

---

[4] The only other albatross to nest in tropical latitudes is the Laysan albatross, a close relative of the waved albatross, whose main colonies lie in Northwestern Hawaiian Islands. Recently, they have started breeding in northern tropical islands, such as San Benedicto, Mexico, at 19° N (Pitman and Ballance 2002).

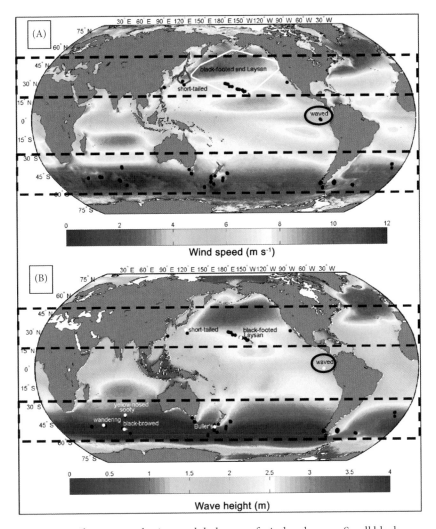

Figure 2.4. Abatrosses colonies on global maps of wind and waves. Small black and white dots represent breeding colonies of various species. *A*, Average wind speed at typical albatross soaring height (10 m, 35 ft). *B*, Average wave height of the world's oceans. Most albatross species live in the "roaring forties" and "furious fifties"—latitudes where wind speed and wave height are global maxima (dashed black rectangles). In contrast, the waved albatross (*black circle at Galápagos*) manages near the equator, where average wind speed and wave height are nearly half the values where most albatrosses live. Polygons in *A* enclose breeding-season foraging areas for three North Pacific albatross species (foraging areas at other times of the year are shown in Figure 2.6 for two of the three species.) *Source*: Reprinted from Suryan, Robert M., David J. Anderson, Scott A. Shaffer, Daniel D. Roby, Yann Tremblay, Daniel P. Costa, Paul R. Sievert et al. "Wind, Waves, and Wing Loading: Morphological Specialization May Limit Range Expansion of Endangered Albatrosses." *PLoS One* 3, no. 12 (2008): e4016. Copyright by Suryan et al. (2008), used by Creative Commons License CC BY 4.0 with minor edits and added circles and rectangles.

wind itself most of the energy needed for sustained flight. With a result much like a wind "pump," they use the wind-speed differential at the top of each wave, repeatedly soaring up from wave troughs into the wind, gaining great energy and speed, then gliding downwind into another trough, repeating the cycle as needed.[5] At high latitudes, strong wind gives this technique great efficiency for albatrosses. In fact, thanks to an evolved shoulder-lock mechanism that holds their wings in the extended, soaring position, albatrosses at high latitudes are able to use little more energy in flight than resting on the ground.[6] In Galápagos, in contrast, the waved albatross heads out to forage in winds averaging no more than half of the albatross norm (just over 4 m/sec or 13 ft/sec, see Figure 2.4), requiring much more flapping and energy expenditure. How does the waved albatross make it work in this equatorial homeland?

To begin to answer this question, let's back up and look at what changed and what stayed the same as waved albatross ancestors took up life in Galápagos, sometime within the last 6 MY (million years), probably about 2 MY.[7] Those ancestors, we may fairly infer, came with the usual toolkit of albatross adaptations. In addition to the shoulder-lock mechanism just described, albatrosses also have a digestive organ shaped like a lab flask, facilitating separation of oil and water from their diet of small fish and squid. This special chamber allows them to save, and not digest as they fly, a rich fish-oil concentrate to bring back to their chicks on Española (shown in Figure 1.2A). Moreover, while traveling, they excrete a concentrated salt solution from nasal glands, allowing them to drink seawater when needed.[8] All albatrosses, including the waved albatross, share this impressive "symphony" of adaptations, in the words of Safina (opening quotation of this chapter), designed by evolution for long-distance foraging. Species living in windy latitudes can travel hundreds of kilometers daily, scavenging low-density prey from the ocean surface, which they spot by flashes of reflected sun or color. An adult female wandering albatross flew with strong winds over 25,000 km in

[5] Richardson (2011, 46) found that albatrosses employ two distinct soaring techniques to gain energy in flight: "wind-shear soaring . . . using the vertical gradient of wind velocity, and wave-slope soaring using updrafts over waves. . . . The energy gain estimated for 'typical conditions' in the Southern Ocean [far south of Galápagos] suggests that wind-shear soaring provides around 80%–90% of the total energy required for sustained soaring. A much smaller percentage is provided by wind shear in light winds and significant swell when wave-slope soaring dominates." The latter is likely the main technique available to the waved albatross in its low-wind latitudes.

[6] This shoulder lock is made of a fan-shaped sheet of tendon extending from the chest (or pectoralis) muscle. When the upper wing bones (humeri) are in a fully forward position, they lock in place, so gliding expends only slightly more energy than resting (Pennycuick 1982).

[7] It is difficult to know the date of colonization by waved albatross ancestors. Most estimates rely on divergence times of mitochondrial DNA (mtDNA), which are notoriously sensitive to small errors of estimation. The generally reliable OneZoom.org website (Rosindell et al. 2017, using data for birds from Jetz et al. 2012) places the waved albatross divergence at 5.7 MY. By comparison with other seabirds in Galápagos, that estimate seems high. Perhaps the ancestors arrived by a circuitous route.

[8] The salt they excrete is guided by grooves in their compound beak down to its tip, so that salty drops do not re-enter the bird's mouth. Some authors proposed that the grooves were an evolved

Figure 2.5. Waved albatrosses regurgitate prey for their months-old chicks. *A*, Two Hood mockingbirds (*Mimus macdonaldi*) wait for a taste. At this age, albatross chicks have beaks long enough to protect them from flying predators (in contrast to Figure 1.2*A*). *B*, The longer beak and bigger mouth accommodate large prey that parents bring from as far as coastal South America, much further than when hatchlings were very young. *Source*: *A*, Copyright Sally Hinton, used with permission.

36 days, averaging 690 km/day (430 miles/day)—an amazing feat (Fitter 2008). But that was done in the winds of the furious fifties. How did the ancestors of the waved albatross make it amid the low winds and waves of Galápagos? And how do today's birds forage adequately with those same equatorial conditions?

In the discussion of the SES framework in Chapter 1, we considered the food needs of new waved albatross hatchlings, and the ability of parents to meet those needs by foraging locally, near Española. As argued there, the fish supply for the brooding parents was influenced by a host of ecological and social variables (Figure 1.3). But the chicks grow rapidly, gaining in size and appetite. Plus, the chicks develop more agility, stronger muscles, and longer beaks to where, around August each year, they can normally defend themselves from aerial predators, mostly hawks and frigatebirds (Figure 2.5). At this stage, and for the next several months (August to November), the parents are freed from the constraints of feeding close to Española. But then not only must they feed themselves over greater periods of exertion, but also they must provide greater rations on Española when they return . . . all of this with record low winds and waves. How do they do it?

---

adaptation whose function was ridding the bird of saline secretions. However, comparative study revealed that a beak of compound plates is actually the ancestral condition for all birds (Hieronymus and Witmer 2010). The grooves predate their function in albatrosses and are thus an "exaptation" (defined and discussed in Appendix 1).

Adaptations to Galápagos

Let's begin with albatross phylogeny (or evolutionary history, from the Greek word, *phylon*, "a group with common ancestry") and compare the waved albatross of Galápagos with its closest relatives living elsewhere. The waved albatross belongs to the genus *Phoebastria*, the genus of largely North Pacific albatrosses.[9] Its closest relative is the short-tailed albatross (or Steller's albatross) of the Izu Islands of Japan, which are volcanic islands like Galápagos. A recent comparative study by Suryan et al. (2008) showed that, after nesting in Galápagos for several million years, the waved albatross, on average, has a 13% larger wing area and 25% less weight than its short-tailed relatives. I infer that natural selection produced in Galápagos a lighter bird with a larger wing surface area, improving its ability to soar on slower winds and smaller waves.[10] With both a function (improved low-wind soaring) and an inferred history of natural selection for that function, waved albatross morphology provides a fitting example of an evolved adaptation.

An adaptive morphology alone would require the waved albatross to forage for much longer times (at its lower wind speeds and wave heights) to cover ocean areas comparable to other albatrosses, and longer absences would eventually put at risk the chicks back on Española, due to lack of energy (for their own defense), dehydration, or both. This is where a second key adaptation comes in: the waved albatross has a behavioral adaptation to a closer, denser food source than any other species in its genus. Satellite tracking data from Anderson et al. (2003), Awkerman et al. (2014), and Birdlife International (2020) show that the waved albatross does not travel anywhere close to "normal" albatross distances in foraging, which can exceed 5,000 to 10,000 km. Instead, as shown in Figure 2.6, it specializes in foraging just over 1,500 km (930 miles) away, a relatively short commute, in the nearby Peruvian upwelling—a coastal zone running the length of Peru that is normally rich with bait fish, especially with the waved albatross's favorite prey, oil-rich anchovies and sardines.[11]

---

[9] Other members of the genus *Phoebastria* include the Laysan albatross, the short-tailed albatross, and the black-footed albatross, all of the Northern Pacific Ocean.

[10] The dimensions of the short-tailed albatross are not necessarily the same as its common ancestor with the waved albatrosses, for which we have no data. The short-tailed albatross may have also evolved new body measures of its own during the millions of years since the short-tailed and waved albatross species diverged. But the short-tailed albatross's roosting latitude is not unusual for the North Pacific group of albatrosses, so it offers the best comparison we have.

[11] The Peruvian upwelling is a stretch of ocean from 4° S to 40° S off the coast of Peru where wind patterns normally cause cold, nutrient-rich water from the deep sea to rise toward the surface. It supports a rich phytoplankton bloom that serves as the basis for one of the world's most productive marine food webs. Although the Peruvian upwelling makes up only 0.02% of the ocean's surface area, it accounts for almost 20% of industrial fish harvest worldwide (Tarazona and Arntz 2001), setting up a major waved albatross conservation issue, as discussed below.

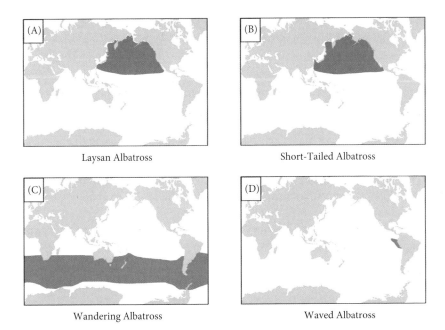

Figure 2.6. Comparison of selected albatross foraging ranges. Most albatross species forage over great distances (*shown in blue*), including two relatives in the same genus as the waved albatross (*A* and *B* at top; their ranges differ mainly in the north) and one of the "great" albatrosses of a different genus, whose range circles the planet (*C*). In sharp contrast, waved albatrosses (*D*) forage over a much smaller range, a behavioral adaptation to the rich Peruvian upwelling. *Source*: Range data from BirdLife International (2020).

Even when brooding and nonbrooding phases are taken together, the waved albatross forages over much less distance than other albatrosses. Thanks to both morphological and behavioral adaptations, waved albatross parents are well fed most years, as are their chicks.[12] For fun, I sometimes use the analogy that "Waved Albatross Airlines" does brisk business on short jaunts with a fleet of sleek regional jets, whereas "Wandering Albatross Airways," and most other intercontinental carriers, depend on large-bodied jumbo jets.

---

[12] One might reasonably ask why other species of albatrosses have not evolved an equivalent specialization on the Peruvian upwelling. At least one other species of albatross *does* frequent the same upwelling zone—Salvin's albatross of New Zealand (Spear et al. 2003). But Salvin's albatross reaches Peruvian latitudes only during the Austral autumn, when waved albatrosses are roosting on Española; hence, there is little competition or interference. When the waved albatross is present, its aggressive behavior in foraging—especially of the males—may well discourage other species. One might also ask, why waved albatrosses do not make greater use of the Cromwell upwelling internal to Galápagos. They certainly do forage in the upwelling area (see Chapter 3), as noted by Merlen (1998). But the density of preferred prey, especially of the Peruvian anchovy (*Engraulis ringens*), is much lower in the Cromwell zone (Grove and Lavenberg 1997, 192), making it much less attractive than coastal Peru.

To this point, the story of the waved albatross is much like that of other Galápagos endemics: we infer that ancestors arrived in Galápagos several million years ago and established a successful breeding colony on a local island—in this case, Española.[13] The population began reproducing and, with the passing of generations, adapted to its new setting. Española had advantages for the albatross colonists: no threats from terrestrial predators, rocky cliffs made take-offs fairly easy, and there was an abundant herbivore population—giant tortoises—keeping the open meadow areas and wide trails that are crucial for wide-wing landings.[14] The avian colonists discovered, or already knew, the bounties of the Peruvian upwelling, and natural selection favored the morphological and behavioral adjustments that were key to accessing that resource. After a few hundred generations, life was probably generally good on Española for the evolving waved albatross, though punctuated by harsh El Niño events. As the warm water phase of ENSO, El Niño stifles the Peruvian upwelling over most of its range, resulting in privation and even starvation for seabirds—conditions that surely have favored the evolution of foraging efficiency in the waved albatross.

Today, the International Union for the Conservation of Nature (IUCN), a body that keeps track of species in trouble by way of its "Red List of Threatened Species,"[15] says the waved albatross "is classified as Critically Endangered because it has an extremely small breeding range, essentially confined to one island, and evidence suggests that it has experienced a substantial recent population decline." The latter is of special concern: according to one estimate, the population declined 42% to 15,500 over the span of 13 years, 1994 to 2007 (Anderson 2008, 190). Given the bird's amazing adaptations, how did this happen?[16]

[13] Alternatively, waved albatross ancestors may have established successful breeding colonies on an earlier island in the same region (that has now subsided below the ocean surface—see Chapter 3) and later moved to Española.

[14] The relative chronology is a bit uncertain in this respect: ancestors of the giant tortoises may have arrived on Española after the ancestors of the waved albatrosses. Eventually, the field of phylogenetics (the use of comparative genetics to reconstruct evolutionary history) may clarify the sequence, but in any case, the wind-swept coast of Española probably had at least some brush-free regions, suitable for albatross take-off and landing.

[15] First published in 1964, the Red List enumerates over 79,800 species, classifying them on a scale from Least Concern to Extinct. The waved albatross can be found online at http://www.iucnredlist.org/details/22698320/0.

[16] There is some uncertainty about the total population of waved albatrosses because thick brush (owing to historical decline in the herbivorous tortoise population) makes censusing difficult. Here I have used the most reliable recent counts (Anderson 2008), rather than looser estimates that range as high as 50,000 to 70,000 individuals. The declining waved albatross population is far from unique. Phillips et al. (2016, 169) found that, of 29 albatross and large petrel species of international concern, "19 (66%) are listed as threatened by IUCN, and 11 (38%) are declining."

Hypotheses

Ironically, the same evolutionary adjustments that made the waved albatross successful in pre-settlement Galápagos—lighter body, larger wings, and a foraging dependence on the Peruvian upwelling—are specializations that spell vulnerability in modern times, especially in the face of growing human impact in and near the archipelago. Here I propose a series of hypotheses regarding waved albatross vulnerability, starting with dependence on the Peruvian upwelling, and then attempt preliminary tests of each, using the best data available. One would, of course, prefer longitudinal (multi-decadal) data from reliable sources, allowing rigorous statistical tests. In this case, the best data we have comes in small chunks from a range of sources. Nevertheless, they point to some worthy lessons from the evolution of the waved albatross for its conservation.

Vulnerability From Behavioral Specialization

First, I propose that behavioral specialization on the Peruvian upwelling has put the waved albatross on a collision course with growing commercial fishing along the Peruvian coast—a course that increasingly threatens adult waved albatross survival.[17] This is *Hypothesis 1*. My reasoning here is simple: since South America's largest fishery works in the same upwelling where adult waved albatrosses feed much of the year, one expects some amount of interference. Competition with fishers is surely a possibility, but a different form of interference, called "bycatch"—the incidental catch or entrapment of foraging albatrosses by fishing gear—almost certainly has the larger impact. By tragic coincidence, the longline technology used in South American fishing since about 1980 generates, from its baited hooks, flashes of color and shiny reflections—the very signaling features that albatrosses use to find their prey.[18] The result: waved albatrosses are fatally attracted to the longlines of mechanized fishery operations.[19] Some of those lines run over 100 km and host 3,500 to 10,000 hooks.

---

[17] Did the waved albatross's foraging specialization during brooding—the short forays to shallow waters near Española, discussed in Chapter 1—also put it on a collision course with the 1990s fishing boom in Galápagos waters? Certainly, there may have been some effect. But as discussed in Chapter 9, the fishing boom of the 1990s focused on lobster and sea cucumbers, which are uncommon in the brood-foraging area of the albatrosses.

[18] In baited longline fishing, a main cable extends from the fishing boat into the sea, punctuated by shorter lines with baited hooks at regular intervals. Longlines were once encouraged as an environmentally friendly fishing method because they were unlikely to damage ocean-bottom habitats and take unwanted fish. Longlines in Peru were specifically encouraged among small-scale fishers in the 1980s and 1990s to reduce the dolphin mortality of gill net fishing (Jahncke et al. 2001).

[19] There are two sides to the attraction of albatrosses to flashes of color and shiny reflection. A recent study shows how closely related Laysan albatrosses (*Phoebastria immutabilis*) are attracted by the flash of surfaced ocean sunfish (*Mola mola*) to clean them of ectoparasitic copepods

Worse still, longlining has grown in popularity in the region, with more than 600 longline fishing boats active in the upwelling.[20] The use of drift gill nets causes a similar problem, where the flashes of shiny bait attract the albatrosses, who then wind up ensnared and drowned.

Conservative estimates of bycatch mortality range from 1% to 5% per year of the waved albatross population, based on the 2008 figure of 15,500 birds.[21] But the tally is even worse than it may seem. Male waved albatrosses are by far the most common victims of bycatch because their competitive behavior in protecting potential prey results in greater exposure to baited hooks and capture.[22] Lower male survival is significant in a species where two parents are essential for feeding and protecting the young: adult male mortality directly reduces the breeding population size.

Fishery effects on the waved albatross offer a good example of the integrative SES perspective outlined in Chapter 1. In this example, depicted in Figure 2.7, waved albatross survival depends as much on the social aspects of longline fishing and baiting as on ecological variables affecting the Peruvian upwelling and their diet species' food web. The framework helps illustrate how the survival and reproduction of organisms like the waved albatross are now influenced as much by human history in Galápagos as by natural history.

As if fishing bycatch were not challenge enough, a second vulnerability is linked to behavioral specialization on the Peruvian upwelling. *Hypothesis 2* holds that El Niño poses another, additive threat to waved albatross survival. A climatological phenomenon of virtually global impact, El Niño—named for the infant Jesus by Peruvian fishers because events often start around Christmas—affects the full coastline of Peru, the Galápagos, and the ocean between, having

(of the genus *Pennella*) in what is apparently a coevolved mutualism (Abe et al. 2012). The other side is dreadful: reports by many scholars (like Young et al. 2009) and photographers (like Chris Jordan, https://www.youtube.com/watch?v=MjK0cvbm20M) show that Laysan and Black-footed Albatrosses, attracted by flashes of color and shiny reflections, bring plastic pollution home from the Pacific gyre to feed to their chicks, who often die with a stomach full.

[20] Alfaro-Shigueto et al. (2010) reported rapid growth in Peruvian longline fishing, and fishing in general, between 1995 and 2005. "Nationwide, the number of fishers grew by 34%, from 28,098 to 37,727, and the number of vessels increased by 54%, from 6,268 to 9,667. At 30 harbors, the number of vessels increased for purse seiners (17.8%) and longliners (357.4%)"—the latter to a total of 637 boats (p. 8). I was unable to locate more recent tallies.

[21] Anderson (2006) and Awkerman et al. (2006) calculated bycatch mortality as 1% of the waved albatross population per year, while Jahncke et al. (2001) estimated between 5% and 13%. I cite the conservative end of their range. Street (2013, Fig. 2.13) also found about a 5% decline in adult survival over roughly the same interval, which correlates well with growth in longline fishing (see Alfaro-Shigueto et al. 2010).

[22] Awkerman et al. (2006) found that 82% of waved albatrosses in local bycatch are males, a finding significantly different from 50/50 (binomial probability < 0.001). In follow-up work, Awkerman et al. (2007) found evidence to suggest that the differential stems from males' hogging the resource via aggressive behavioral exclusion of other albatrosses, including females.

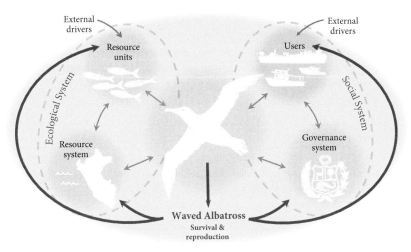

Figure 2.7. SES representation of the impact of commercial fishing in the Peruvian upwelling on waved albatrosses. In response to *external drivers* (e.g., market demand), the resource *users* (Peruvian coastal fishers) working within the rules-in-use of the existing *governance system* (represented by the logo of the Peruvian Ministry of Production) employ fishing technologies—especially longlines and gill nets—that unintentionally catch substantial numbers of waved albatrosses annually. Also affected by *external drivers* like climate, the fishers' harvest of *resource units* (e.g., anchovies and sardines) within the *resource system* of the Peruvian upwelling competes with the albatross's food supply. The net effect of these interactions is to sharply reduce the *survival and reproduction* of waved albatrosses (*red arrow, center bottom*)—a leading reason the species is listed as Critically Endangered by the IUCN. The waved albatross is but one example of the influence of today's social and ecological forces on the evolutionary fitness of Galápagos organisms. Declining waved albatross numbers affect the fishery, especially their prey species (*curved red arrows*). Their population decline also affects the social system, where, if we are wise, changes in the rules-in-use and fishing technology could lead to corrective conservation outcomes. *Source*: Adapted from Ostrom (2009).

far-reaching consequences for albatrosses and other organisms. El Niño causes two main changes in the region: (1) normal trade wind patterns are reversed, bringing unusually heavy rains to Galápagos and coastal Peru, and (2) sea-surface temperature over a wide area rises as much as 4°C to 5°C (7°F–9°F), enough to adversely impact marine flora and fauna, including the upwelling's normal supply of fish.

Let's take a moment here to explore the phases of ENSO, because the wet El Niño and its dry counterpart, La Niña, affect many organisms in Galápagos (as discussed in later chapters). Normally, in the "neutral phase" of ENSO, strong trade winds blow westward along the equator through Galápagos and over to

Borneo, in Southeast Asia's Malay Archipelago, moving water and air that have been warmed by the sun and humidified over the ocean. Enough water is pushed along by the trade winds that the ocean in the western Pacific can be an impressive 60 cm (24 in.) higher than in the east. The warm, wet air rises in the region around Borneo and cools, dropping meters of rainfall—a key input to the tropical rainforests there. The cooled air mass then circulates back east, high over the ocean, and falls with its dry, desiccating air over Galápagos and coastal South America, contributing to the normally dry sea-level conditions in these areas, and rejoining the westward trades. This giant clockwise circulation of air (Figure 2.8A) is called the Walker circulation (or Walker cell), after its discoverer, Gilbert Walker. Because the Walker circulation normally drags the surface water of the ocean along with it, water from deep in the ocean rises to replace that water. Along the Peruvian coast, that upsurge, rich in nutrients, contributes to the coastal upwelling, and Galápagos gets an additional boost from a local upsurge, the Cromwell upwelling (see Chapter 3). The Walker cell is thus mirrored by ocean circulation, flowing in the same direction at the ocean's surface (counterclockwise when viewed from the south side) most of the time. When atmospheric conditions promote a stronger-than-normal Walker circulation, the result is unusually cool sea-surface temperatures from increased nutrient-rich upwelling, a condition called La Niña. La Niña boosts productivity in marine ecosystems around Galápagos to extra-high levels, normally triggering booms of fertility and survival among local seabirds and marine mammals.

In contrast, during an El Niño event, the normal trade winds taper off, the Walker cell reverses, at least partly, and there follow manifold consequences. Since the wind over the ocean now flows eastward, warm, moist air flows toward Galápagos and coastal South America, rising and cooling in the eastern Pacific, and dropping Borneo-worthy amounts of rain on Galápagos and coastal South America (Figure 2.8B). In Galápagos, this means as much as 3 m (10 ft) of rainfall in lowland areas that generally receive one tenth as much. Meanwhile, without the usual trade winds pushing west, massive amounts of warm water slosh eastward, replacing the normally cold surface waters between Galápagos and coastal South America, including the Peruvian upwelling, for the duration of the counterflow.

The severity of El Niño can be accurately measured by the change in sea-surface temperature (SST) in the equatorial Pacific (Figure 2.9). The warm water mass coming west across the Pacific kills off the usual algae and phytoplankton at the base of the food chain, stifling marine life from small to large and inducing hardships for all resident seabirds and marine mammals. Concurrently, low-elevation habitats of the islands, normally qualified as "arid zones," are awash with meters of rainfall, improving conditions for many terrestrial dwellers, including land birds, reptiles, and many plant species.

(A)

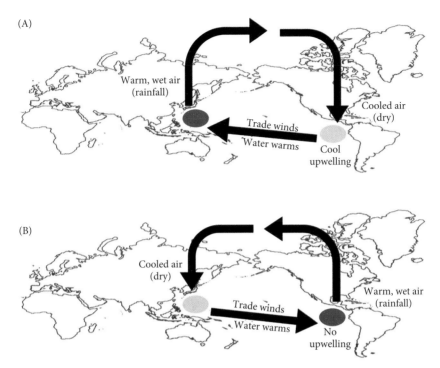

(B)

Figure 2.8. Equatorial winds during different phases of El Nino-Southern Oscillation (ENSO). *A,* Wind patterns, known as the "Walker circulation," during ENSO neutral conditions. Strong trade winds blow westward through Galápagos, moving water and air warmed by the sun toward Southeast Asia. The warm, wet air rises there and cools, dropping meters of rainfall. The cooled air mass then circulates back to the east, high over the ocean, and falls as dry, desiccating air over Galápagos and coastal South America, contributing to dry sea-level conditions. The air circulation, clockwise from this perspective, is normally matched by a counterclockwise oceanic circulation (not shown) that helps fuel the cold, rich Peruvian upwelling. During La Niña, trade winds promote a stronger-than-normal Walker circulation, resulting in an extra-rich upwelling near the western islands, unusually cool sea-surface temperatures, and very dry conditions on land across the islands. *B,* Wind patterns during El Niño. Trade winds weaken or reverse and warm water sloshes east, reversing Walker circulation and stifling upwellings in the eastern Pacific, raising the sea-surface temperature. Warm, moisture-laden air rises in the Galápagos region, bringing record levels of rainfall. During such an event, many land organisms thrive; many marine organisms migrate away or starve.

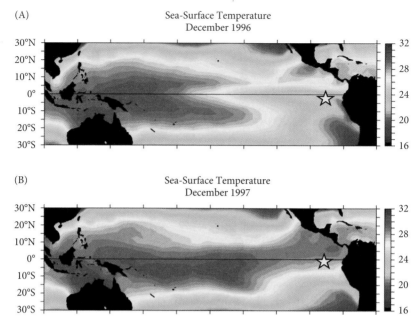

(A)   Sea-Surface Temperature
December 1996

(B)   Sea-Surface Temperature
December 1997

Figure 2.9. Pacific sea-surface temperatures (SSTs) during ENSO phases in 1996 and 1997 (*gold star* marks Galápagos). *A,* Typical SSTs during an ENSO-neutral period, with cool water all around the archipelago. *B,* Much warmer SSTs in the eastern Pacific, including Galápagos, during the El Niño period 1 year later. *Source:* McPhaden (1999), public-domain image from NOAA, originally published in *Science* 283: 951.

The waved albatross is one of the organisms negatively affected by the warm-phase changes. One might even argue that El Niño's impact on waved albatrosses is especially severe, because of their dependence on the Peruvian upwelling. With virtually all their year-round diet coming from the waters of Galápagos and coastal Peru, their entire feeding range is "ground zero" of El Niño. Other Galápagos seabirds, like the three booby species (Chapter 7), are affected by El Niño, too. But when times are bad, they commonly take off and fly away.[23] Such a strategy has not evolved among waved albatrosses, quite possibly because of the energy demand of flapping for added distance with such large wingspans. In any case, one expects a big impact of El Niño on the world's only tropical albatross because of its specialized feeding range.

---

[23] During El Niño events, for example, Galápagos blue-footed boobies often flee the islands and head for the Bay of Panama and the west coast of Central America.

Sure enough, careful measurements by biologists David Anderson, Jill Awkerman, and colleagues demonstrate that El Niño takes a big toll on waved albatross survival. Life is already rough for waved albatross chicks: in ENSO-neutral "normal" years, not even a quarter (23%) of mated pairs produce a surviving chick. But in an El Niño year, on average, only one chick survives per 12 nests (7.9%, Awkerman et al. 2006). Moreover, breeding-age waved albatrosses also suffer ENSO hardships: whereas their normal yearly survival rate runs 92%, it drops to 86% during El Niños, and is only that high because adult albatrosses have learned to forage in temporary "cells" of cool (and fish-rich) upwelling that break through warm El Niño waters along the coast (Jahncke 2007, 6). Alas, evidence indicates Peruvian fishers use the same cells during El Niño events, thus combining the processes behind Hypotheses 1 and 2, inevitably making more trouble for waved albatrosses. Even worse, episodes of El Niño are becoming more intense and frequent, likely due to anthropogenic influences.

Having seen that behavioral adaptations of the waved albatross present vulnerabilities today in light of local and regional changes, let's consider parallel liabilities arising from its morphological adaptations.

Vulnerability From Morphological Specialization

Two consequences come from being a relatively light-bodied albatross with large wings. First, large wings make it more difficult and energy intensive to generate lift by flapping, especially for fledglings. Hence, it is greatly advantageous to waved albatrosses to roost close to windward cliffs, where they may routinely take flight without prolonged flapping. Many fledglings practice for some days over dry land, but then walk to the cliff face and fall into the wind to fly—or, sadly, sometimes just fall—in their inaugural flight. Proximity to windward cliffs, in an area without terrestrial predators, and as close as possible to the Peruvian upwelling, is a recipe that leaves few potential nesting areas. *Hypothesis 3* proposes that, given their specializations, waved albatrosses experience a scarcity of high-quality nesting locations. Supporting this hypothesis is the fact that waved albatrosses nest only on the windward sides of three islands: Española, where over 99% of the pairs nest, Genovesa, a northern Galápagos island where a few pairs are sometimes found, and Isla La Plata near the Ecuadorian coast, with another 10 to 20 pairs (Awkerman et al. 2014). In the Española population, as many as 40% of birds do not breed in a given year, suggesting that nesting activities may sometimes approach "habitat saturation" (Anderson et al. 2002). The topic warrants further study, but the very fact that the albatrosses are considered a "point population" (nearly all from one small island) is consistent with Hypothesis 3.

The second morphological vulnerability is simply that long wings require a wide area cleared of brush for take-off and landing, lest the bird become entangled or crash—again the Waved Albatross Airline metaphor is apt. This argument is *Hypothesis 4*. In pre-settlement Galápagos, meadows on Española were maintained by herbivorous giant tortoises, providing important clearings for the albatrosses. With the advent of buccaneers and whalers (see Chapter 4), for whom those tortoises were a convenient and durable food resource, tortoises were overharvested to where bushy regrowth made a tangle of former albatross airstrips. One study found that dozens of waved albatrosses died by hanging and starvation when they became ensnared in the encroaching brush (Anderson et al. 2002), a clear confirmation of this hypothesis.

Vulnerability of Waved Albatross Fertility

In contrast to earlier hypotheses focused on survival, *Hypothesis 5* holds that waved albatross fertility is also seriously impacted by El Niño events. A major cause of fertility loss is egg abandonment, in which albatross parents simply depart, leaving their egg "marooned." Because waved albatrosses do not construct nests, while all other albatrosses do, a marooned egg simply sits on the ground, out in the open or near a bush. Eggs appear to be marooned for two related reasons: (1) when El Niño undermines the food supply, forcing hungry parents to abandon reproductive efforts in favor of their own survival; (2) when mosquito attacks become insufferable, especially during El Niño years. An investigation during a moderate El Niño event in 1986 found that egg abandonment frequency on Española was related to the biting of endemic black salt marsh mosquitoes (*Aedes taeniorhynchus*), the females of which take ravenous blood meals from Galápagos birds and mammals.[24] They then lay eggs in moist, even salty, soil, and the eggs hatch when flooded by rain, as during El Niño, producing dense swarms that are a nuisance for animals . . . including people.[25] In April and May, 1986, for example,

> The only areas of the island consistently free from mosquitoes were strips of coastline facing the prevailing southwesterly winds. . . . On April 29 [after

[24] Egg rolling and abandonment could also be related to overheating: "On Española, breeders minimize the thermal challenge by nesting during the coolest part of the annual cycle, but incubating birds and chicks still thermoregulate laboriously by panting" (Anderson et al. 2002, 68).

[25] The mosquitoes may well be disease vectors, a problem potentially even more serious than their ability to annoy. A recent paper reports avian pox and pox-elevated mortality among waved albatross nestlings (Tompkins et al. 2017). The vector has not yet been identified, but mosquitoes are a definite candidate. Avian pox has infected Galápagos bird fauna since it was introduced by human settlers in the late 1890s (Parker et al. 2011).

rains and amid a mosquito outbreak,] we first noted unattended waved albatross eggs. . . . Over the next 4 days increasing numbers of eggs were abandoned throughout the eastern region of the island until approximately half of all eggs were unattended in some places by 2 May. We never saw adults return to an egg once it was neglected, and neglected eggs soon became partially covered by the muddy substrate, and many were [eaten] by the [local species of] mockingbird. . . . Egg neglect appeared to be associated with geographical variation in mosquito density that we perceived as we travelled about the island. We suspected that desertion was a response to mosquito bites when we saw swarms of mosquitoes taking blood meals from the facial skin of incubating adults in some areas. We tested [and confirmed statistically] this hypothesis by sampling mosquito population density in four subcolonies that differed dramatically in numbers of neglected eggs. (Anderson and Fortner 1988, 727)[26]

In areas where the tally averaged as high as 63 female mosquitoes per minute on the chosen sampling surface (the exposed arms of investigators!), an impressive 47.3% of all waved albatross eggs were abandoned, significantly more than in areas with fewer mosquitoes. Earlier observations elsewhere on Española, for the interval 1961 to 1972, revealed "seven good years, four bad years, and one intermediate." The bad years (1965, 1968, 1969, 1972) were each associated with El Niño rains, and each produced mass desertions of eggs: "Such marked failures of breeding have not been recorded in any other species of albatross" (Harris 1973, 496). El Niño brings the only tropical albatross a dual squeeze of both increased mortality and reduced fertility.

Finally, consider waved albatross egg-rolling, a behavior rare among birds, and the related absence of nest-building. These are not at all rare or aberrant acts within this species: Anderson (2008, 161) reported that, in his experience, roughly 93% of all waved albatross eggs are moved. Before him, Harris (1973, 493) found in two colonies he watched that 510 eggs of 970 laid (52.6%) failed to hatch and noted that "80% of the failures were due directly to the habit of moving the eggs." Harris also noted that eggs were moved as much as 40 m (131 ft) from the initial site. Anderson (2008, 161) considered a range of possible explanations—to seek shade, to have eggs temporarily adopted by other adults, general dissatisfaction with nesting sites, and more—only to reject most alternatives. Two of the remaining hypotheses are worthy of our attention here. Let me call the first *Hypothesis 6A:* egg-rolling and nest omission in the waved albatross are adaptations to living in close proximity with Galápagos giant

---

[26] Peterson (1967) suggested, with convincing photographs, that thick swarms of mosquitoes drove the incubating waved albatrosses from their eggs in 1965. The mosquitoes were a result of heavy El Niño rains that year.

tortoises. In Anderson's words, "tortoises might plough ahead like a bulldozer, crushing an egg that parents don't move. This idea could account for the initial evolution of egg-moving behavior, but fails to explain any of the movement that we have documented, because no tortoises were present during our observations of egg movement." But if tortoise bulldozing had been a prominent selection pressure over the course of waved albatross evolution in Galápagos, then nest omission and egg-rolling could be innate in them, and thus performed without practice or specific provocation. The behaviors could well continue today without the stimulus of advancing tortoises, or with fewer instances. Second, the two behaviors might, instead, be additional products of the selection pressure reviewed above for marooning. *Hypothesis 6B* holds that egg-rolling and nest absence are additional adaptations to the menace of the black salt marsh mosquito, alone or in combination with other insect pests like ticks or flies. How can we put 6A and 6B to a test?

Headway can be made from the follow-up study of egg-rolling by Awkerman et al. (2005b), if we may also assume that most meadowlands along the southern coast of Española had, for millennia, been kept in their open state by hungry and heavy resident tortoises of the island.[27] On that assumption, egg-rolling could not be a tortoise avoidance mechanism, countering Hypothesis 6A. Why not? In a careful study of egg-rolling, Awkerman et al. (2005b, 14) found that "incubating birds nesting far from the coast tended to move their eggs farther and more frequently than did birds close to the coast, and that they oriented movement toward the coast, especially if their egg-laying site was sparsely populated." For the behavior to have evolved for the benefit of avoiding tortoises, the innate response would generally have to be directed *away* from areas of high tortoise activity. The same observations, however, support Hypothesis 6B, since the windy coastline would consistently have had lower mosquito densities than some distance inland.

The topic warrants additional investigation, but the preliminary test offered here supports the argument that egg-rolling and associated lack of nest-building are evolved behavioral adaptations to mosquito avoidance, plus perhaps other insects. There is one further implication of this argument: today's mosquito densities may also be linked to tourism growth on Española. Unsuspecting visitors are an easy alternative blood-meal source (personal observation) during the annual period, roughly December to March, when albatrosses are off at sea. Even a small food subsidy from tourists would temper what was formerly—before the tourism boom—a food bottleneck for female mosquitoes. Additional research is needed to disentangle various interactions and their magnitudes, but this much

---

[27] This assumption is also behind the GNPD decision in 2015 to clear woody growth from waved albatross airstrips at various locations in the south of Española. See Clue 4 below.

is clear: mosquitoes and egg-rolling are associated with fertility decline in the waved albatross, even as the population also undergoes survival decline, all because of its evolved prolonged dependence on the Peruvian upwelling.

In summary, we have seen how the original adaptations—behavioral and morphological—of the waved albatross to life in a tropical region, with low winds, small waves, and no terrestrial predators or people, have become key vulnerabilities today. The net result is a decline in species' fertility at roughly the same time as an ongoing increase in mortality. Little wonder that the IUCN lists the species as Critically Endangered.

Conservation Implications

How can we help this threatened albatross? Looking back, one can see that each of the waved albatross's specific adaptations to Galápagos is not only a key source of vulnerability today, but also an important clue for its successful conservation. Let us look at each clue and some ways it can be, and is being, used in conservation efforts.

Clue 1: *Adaptation to the Peruvian Upwelling*

The evolved feeding specialization of the waved albatross has run headlong into confrontation with fishing technology in the Peruvian upwelling. There is little or no problem during the brooding season each year, when albatross parents generally feed within the limits of the Galápagos Marine Reserve, where fishing is now tightly regulated (see Chapter 9). But on the coast of Peru, where waved albatrosses feed the rest of the year, fishing was, for a long time, "catch as catch can" using both longlines and gill nets, with inevitable albatross bycatch. In 2001, a group of scientists and policymakers drafted the international "Agreement on the Conservation of Albatrosses and Petrels" (ACAP), and a nongovernmental organization by the same name, ACAP, began having a meeting of its members every 3 years to promote conservation of albatrosses and petrels, including the Galápagos species. By 2010, ACAP had grown to 13 member nations—not including the United States, where membership still awaits Senate action—with an action plan to coordinate international activity mitigating bycatch and other threats to albatrosses. Ecuador and Peru have both ratified the Agreement, and many fishers are cooperating, adding scarecrow devices (e.g., dangling lines called "tori streamers" or albatross-perplexing colorful cones) to longlines, and taking other measures to reduce accidental catch of albatrosses. It's important for the U.S. Senate to approve with no further delay the ACAP Treaty Doc 110-22, which would bolster scientific and economic resources available to the ACAP initiative. U.S. funding would help with monitoring fertility and survival rates of waved albatrosses on Española, and for further corrective action in fisheries throughout their range.

Clue 2: *Adaptation to a Niche That Is El Niño's "Ground Zero"*

Unwittingly, waved albatross ancestors settled in the very geographic center of El Niño's effects in the eastern Pacific. We infer from their historical numbers that the ancestors fared well most years, deriving sustenance from the upwelling sea. But every few years the bottom fell out: the trade winds slackened, warm water moved in, and the life-giving upwelling shut down. Today, evidence is growing— some of it *from* Galápagos (e.g., Thompson et al. 2017)—that El Niños are becoming longer lasting, more frequent, and of higher magnitude (i.e., with greater rise in SST) than in the past, at least partly in response to climate change. These changes can be seen in simple interdecade plots of the multivariate ENSO index called "MEI" (Figure 2.10), and in integrative paleoclimatic reconstructions going back to the year 1500 (Figure 2.11).[28] These El Niño trends are surely the most serious of all issues facing the waved albatross today, because they have the potential to drive the species to extinction in a matter of decades. If one is concerned about waved albatrosses and other El Niño-sensitive species, it is important to do what one can, individually and in groups and organizations, to curb growing greenhouse gas emissions in the United States and worldwide.[29]

Clue 3: *Adaptation to Low Equatorial Wind and Waves*

Evolutionary changes in morphology endowed the waved albatross with a lighter body and greater wing area, the better to navigate its challenging tropical homelands. But it still needs generous wind for efficient flight, especially for take-off—hence the importance of Española's southern cliffs and south-facing slopes. Recognizing that importance, the Galápagos National Park Directorate

---

[28] Techniques more advanced than visual comparison confirm these same conclusions (see Cai et al. 2015a, 2015b, 2018; Liu et al. 2017). A particularly detailed historical data set from the Bay of Bengal revealed a cycle of El Niños in the region roughly every 6.5 years from 1893 to 1940, and then the cycle shrank to every 5.0 years between 1980 and 2001, both tragically matched by a cycle in cholera incidence. That's an almost 25% increase in El Niño frequency in less than a century (Rodo et al. 2002).

[29] For those wanting to help curb climate change, a few prominent organizations include: 350 (https://350.org/), Union of Concerned Scientists (http://www.ucsusa.org/), Sierra Club (http://www.sierraclub.org/), and Greenpeace (http://www.greenpeace.org/usa/). Other suggestions for action can be found in Online Appendix 2.

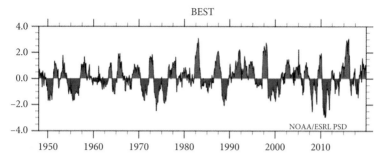

Figure 2.10. Time-series graph of the ENSO index known as the Bivariate El Niño-Southern Oscillation Index (BEST), which handily integrates both sea-surface temperature changes and atmospheric pressure differences in the equatorial region around Galápagos where ENSO is most pronounced. Warm El Niño events and cool La Niña events show as peaks in red and blue, respectively. Starting with the 1983 El Niño event (roughly midpoint on *x* axis), El Niño events are noticeably more frequent, stronger in magnitude, and longer in duration than they were before 1983. *Source*: Public domain image from NOAA Earth Systems Research Laboratory, Physical Sciences Division, Boulder, Colorado (from http://www.esrl.noaa.gov/psd/).

(GNPD) has taken care to protect the full southern coast of Española. Tourists, for instance, may visit only Punta Suarez, the westernmost edge of Española that has no more than 1% to 2% of the cliff-edge length of the island. Moreover, by Park regulations, visitors may not cross important albatross "runway" areas, nor get within 2 m (7 ft) of mating and nesting sites (unless spatially unavoidable from marked visitors' paths). Since waved albatrosses also roost in small numbers on Isla La Plata near the Ecuadorian coast, one recommendation would be to enhance protection for that small colony.[30] The colony of La Plata could then help "mitigate the extinction risk" of the Española population over the long term (Awkerman et al. 2014, 376).

Clue 4: *Dependence on Cleared Terrestrial Airstrips*

Before human impact in Galápagos, the abundant giant tortoise population on Española kept large open areas, or tortoise meadows, free of brush and woody species, as noted earlier. The tortoise population and its herbivory changed

---

[30] As noted by Awkerman et al. (2014), waved albatrosses living on La Plata have the advantage of shorter travel to foraging areas than those on Española. But on La Plata there is a need for greater protection from human activity and its byproducts. While the situation has improved since Nowak (1986, 17) reported a waved albatross killed by feather hunters, there remains a serious threat of egg predation by introduced fauna.

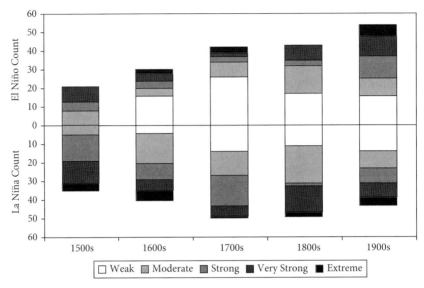

Figure 2.11. Paleoclimatic reconstruction of ENSO patterns from 1500s through 1900s. Pooling a wide array of ENSO proxies—tree ring histories, coral records, drought records, ice cores, and more—this plot of centennial trends is one of the most complete historical ENSO compilations to date. Strength indicators are based on percentiles of estimated magnitude: extreme (> 90th percentile), very strong (70th–90th percentile), strong (50th–70th percentile), moderate (30th–50th percentile), and weak events (< 30th percentile). According to the authors, the 20th century witnessed 43% of all extreme and 28% of all protracted (3 years or more) ENSO events. One would expect 20% per century if events were evenly distributed across the centuries studied. *Source*: Reprinted by permission from Springer Nature and J. Gergis, from Gergis, Joelle L., and Anthony M. Fowler. 2009. "A History of ENSO Events Since A.D. 1525: Implications for Future Climate Change." *Climatic Change 92, nos.* 3–4: 343–87.

drastically with the arrival of buccaneers, whalers, and explorers, who discovered that live tortoises could be stored for food in ships' hulls—an inadvertent blow to the waved albatrosses, who had evolved extra-large wings back when tortoises maintained coastal meadows that doubled as brush-free airstrips. The Española tortoise population dwindled down to tiny numbers and, eventually, had to be translocated to the Darwin Research Station for a protected breeding program. Meanwhile, alien goats were also eliminated from the island and so woody brush grew into impenetrable thickets. Hundreds of tortoises have been repatriated from the breeding program since 1975, but the thickets are too much for the toothless herbivore. In 2015, with new reports of waved albatrosses entangled in brush or starved from the inability to take flight, the GNPD sent a team to clear a series of albatross airstrips on Española, and more may soon be needed[31]—a fitting interim measure until the endemic tortoise population once again approaches historic densities and vegetation control. Meanwhile, albatrosses use the pathways of repatriated tortoises for intra-island movements (Gibbs and Woltz 2010). Close monitoring is underway and will be needed for years to come to ensure that waved albatrosses have ample access to brush-free airstrips.

Clue 5: *Adaptation to Seasonal Pest Problems*

The egg-rolling and nest-free behavior of waved albatrosses may well have evolved for the advantages of mobility during seasonal and El Niño-related mosquito swarms. Available data are consistent with this hypothesis, though not yet fully convincing (since ticks, lice, and flies may also be involved). Also tentative is the suggestion that increased year-round tourism to Española contributes to the mosquito problem for albatrosses. Further investigation is sorely needed. Given the severity of waved albatross population decline, however, it seems prudent to try conservation measures based on the hints we have.[32] One recommendation is for the GNPD to add one or more alternate tourist trails far from Punta Suarez, perhaps near Punta Cevallos at the eastern end of Española, to be used every 2 to 3 years or in some other alternating pattern. This alternative would allow continued waved albatross viewing—important for the educational and motivational value of visitors' experience—but mitigate any localized byproducts, including mosquito buildup, from continuous tourism. Another recommendation is for tourists visiting Española to wear an approved mosquito repellent during time onshore in the wet season. In any case, monitoring of *Aedes* and other mosquitoes is called for, because some of them have considerable range and could possibly track alternate tourist routes. The overall goal is to reduce buildup of mosquitoes just inshore from the southern cliffs of Española, with the aim of abating the egg-rolling and egg abandonment.

Clue 6: *Reducing Egg Abandonment*

The unique egg-rolling of waved albatross parents may be an adaptation to avoiding mosquitoes and/or other pests, as argued here, but it also has detrimental

---

[31] On artificial albatross airstrips, see Gibbs (2015) at https://www.Galápagos.org/blog/preparing-for-Española-2015/

[32] One could indeed argue for stricter limits on numbers of tourists visiting Española or, for that matter, all of Galápagos. This is a complex topic that is returned to in Chapter 9.

side effects when eggs get stuck. Jill Awkerman and her research team demonstrated the benefit of manually repositioning the marooned waved albatross eggs to places very close by, so that breeding birds might resume incubation. Marooning occurred in 14.9% of the monitored nests in 2004 (132 of 886); half of the eggs were repositioned and half were left in place. The researchers found that five times more eggs hatched in the relocated, and thus incubated, half than in the control (unmoved) half. Human repositioning of marooned eggs "may not be enough to fully compensate for losses through fishery interactions and will require sustained human intervention, but it does provide an effective, low-cost management strategy for increasing reproductive success with minimal disturbance" (Awkerman et al. 2005b, 15). Perhaps Galápagos naturalist guides and GNPD rangers could be coached on this technique to help boost declining waved albatross fertility. The strategy also offers teachable moments to share with visiting tourists.

Figure 2.12 summarizes the specific adaptations, modern vulnerabilities, and proposed conservation solutions for the waved albatross. It also illustrates an

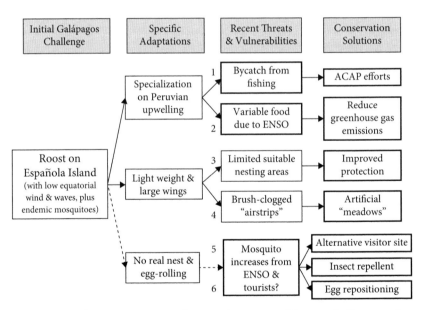

Figure 2.12. Schematic summary of major adaptations of the waved albatross, plus recent threats to the species and their conservation implications. *Solid arrows* are links with mounting empirical support; *dashed arrows* are tentative links; boxes caused or affected by human activity are *bolded*. Small numbers 1 to 6 indicate the main hypotheses of this chapter. Some key implications for conservation action, actual or potential, appear in the right-hand column, ordered by the threat to which they relate. For simplicity, the schema does not include the seasonal specialization of waved albatrosses for feeding new hatchlings from short forays near Española.

important point that is returned to in chapters ahead: the world of the waved al-batross today can no longer be considered in the domain of "natural history;" it is now in a hybrid domain of natural and human history, in which many outcomes are simultaneously influenced by human activity (bolded boxes in the figure) and natural processes. The waved albatross provides a good example of the value of the SES perspective. Even in this little world in the middle of the Pacific Ocean, human activity is now an integral part of the daily life and demography of what were once isolated endemic organisms. Indeed, we have now seen how the fertility and mortality of the largest bird of Galápagos are directly affected by human activity.

Last, Figure 2.12 also suggests there is great potential to promote a rebound in the waved albatross population via conservation measures like those shown in the figure. Notice that many are simple, low-technology actions—like an appro-priate mosquito repellent, an alternate visitor site at the other side of the island, supporting ACAP in its efforts to work with Peruvian fishers, and reducing our own carbon footprints. Time and again, what seems at first to be a problem vast, far away, and even esoteric—to help the Galápagos waved albatross—turns out to be comprehensible, accessible, and valuable for all the world. Humanity can make a positive difference here, if we're willing to try. But until our efforts suffice to reverse the decline of this threatened species, we shall all have an albatross hanging around our necks: the beautiful waved albatross of Galápagos.

3

The "Secret Recipe" of Galápagos

Hideous and disgusting as is their appearance, no animal can possibly afford a more wholesome, luscious, and delicate food than [do Galápagos tortoises. . . . After one exposure,] every other animal food fell greatly in our estimation. . . . But what seems most extraordinary in this animal is the length of time it can exist without food [or water, and suffer] no diminution in fatness or excellence.

Captain David Porter
(*Journal of a Cruise Made to the Pacific Ocean*)

The compact geographical scale of Galápagos makes it all the more surprising that the terrestrial plant and animal communities differ on each of the 15 major islands. This feature makes visiting the islands much like visiting different continents or subcontinents in the rest of the world. It especially surprised young Charles Darwin on his famous visit in 1835. "It never occurred to me," he later wrote (Darwin 1839a, 474), "that the productions of islands only a few miles apart, and placed under the same physical conditions, would be dissimilar. I therefore did not attempt to make a series of specimens from the separate islands." He even failed to note the specific island source for many of his collected finch specimens, as he would later regret. However, Darwin received one unambiguous "heads up" about inter-island differences during his visit, from a resident, Nicholas Lawson, who Darwin met on recently colonized Floreana:

It was confidently asserted, that the tortoises coming from different islands in the archipelago were slightly different in form; and that in certain islands they attained a larger average size than in others. Mr. Lawson maintained that he could at once tell from which island any one was brought. Unfortunately, the specimens which came home in the Beagle were too [young and too few] to institute any certain comparison. (Darwin 1839a, 465)

As the largest terrestrial animals in the archipelago, tortoises have weights up to about 250 kg (550 lbs) for males and half that for females, so differences among island populations are relatively easy to see (Figure 3.1).

Exuberant Life. William H. Durham, Oxford University Press (2021). © Oxford University Press.
DOI: 10.1093/oso/9780197531518.003.0003

Figure 3.1. Three species of Galápagos giant tortoises, illustrating their distinctive shell shapes. *A*, The dome-shaped shell of *Chelonoidis porteri* of Santa Cruz Island—named for Captain Porter in the epigraph for this chapter—extends down close to its neck. *B*, Also dome-shaped, the shell of *C. donfaustoi*, also of Santa Cruz—named

Figure 3.1. Continued

for park ranger "Don Fausto" Llerena, who ran the first tortoise breeding center for more than 40 years—is smaller, lower in front, and often features plates (scutes) with visible points. This species was only fully distinguished from *C. porteri* in 2015 (Poulakakis et al. 2015), showing that species differences are still being identified today. *C,* The saddleback shell of this still smaller species, *C. chatamensis* from San Cristóbal, leaves open a big gap above the tortoise's neck. So prominent were these and related species in the 1600s that early mapmakers named the archipelago after the Spanish word for "tortoises."

Comparative study since the *HMS Beagle* voyage has confirmed what Lawson first reported, as illustrated in Figure 3.2, which summarizes the variation among adult carapaces (shells) from different islands.[1] At the time of Darwin's visit, unbeknownst to him, there were 15 different tortoise species in Galápagos, all assigned today to the genus *Chelonoidis* (from the Greek *chelone,* "tortoise"). Indeed, there was *a different species on each major volcano* of the archipelago, and two on the especially large volcano of Santa Cruz. Today's roughly 20,000 tortoises are survivors of what were once over 200,000, their numbers diminished by loss to buccaneers, whalers, sailors, and collectors, who took the animals for food, oil, and museum specimens.[2] Even the *HMS Beagle's* crew consumed 48 tortoises from San Cristóbal (*C. chatamensis,* Figure 3.1*C*) on the way back to England (Olson 2017).

New tools for reconstructing tortoise phylogeny on the basis of genetic similarity (see especially Poulakakis et al. 2012, 2015) allow us to infer that the 15 Galápagos species were all related by descent to a single ancestral colonization. It's a fitting example of adaptive radiation: a group of species diversified over a relatively short period from one common ancestor. Such radiations resemble Darwin's full theory in miniature, forming a metaphorical tree

---

[1] Carapaces consist of ribs fused together with dermal bone, covered with plates (scutes) of keratin. Major differences in the shape of adult tortoise carapaces trace to genetic differences in underlying bone structure. For full details and discussion of morphology and most other tortoise topics, see Gibbs et al. 2020.

[2] Working from ship's logs, Townsend (1925, 57) counted over 13,000 giant tortoises taken by a sample of 79 U.S. whaling vessels between 1831 and 1900, at a time when there were over 700 whalers in the American fleet. Thus, one could reasonably "credit American whalers with taking not less than 100,000 tortoises subsequent to 1830" (p. 70), not to mention whalers of other countries and so-called seal hunters who also killed two endemic species of sea lions. Following Townsend, Chambers (2006, 104) proposed that another 100,000 tortoises were taken "in the few decades prior to 1830." A majority of the estimated total (200,000) were inevitably female, because they are generally smaller, are more often found close to the coast for egg laying, and are easier to carry. Tortoises were also taken earlier by buccaneers, but we have no good way to estimate the numbers. Estimates of the total pre-human tortoise population run as high as 336,000 (Gibbs 2017, Gibbs et al. 2020, 408); the number I use here, 200,000, is deliberately conservative.

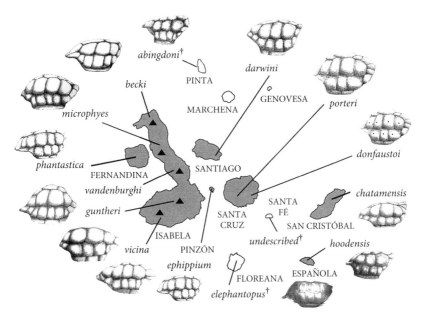

Figure 3.2. Distribution of giant tortoises, all of the genus *Chelonoidis*, among the islands of the archipelago (*shaded* islands have living populations). Fifteen species are recognized today, each with a different shell shape as shown (all facing to the right), including three species now extinct (marked with *daggers*). Note that each island has a different species except for Isabela, which has five (one on each intact volcano, marked by *triangles*), and Santa Cruz in the center, which has two (as in Figure 3.1). The species on Santa Fé is marked "undescribed" because no complete carapace exists, only a few bones. The islands of Marchena and Genovesa (*near top*), plus Darwin and Wolf to the northwest (not shown) did not have tortoises. *Sources*: Map from Wikimedia Commons, Public Domain. Carapace drawings by Martin Withers (from Fitter et al. 2016) by permission of Princeton University Press, except *C. nigra,* which is from Gunther (1902, plate XVII); *C. phantastica,* which is adapted from the Withers sketch of *C. ephippium*; and *C. donfaustoi,* which is adapted from *C. porteri.*

whose "branches" are species shaped largely by natural selection, producing adaptations to diverse local environmental conditions. But 15 is a lot of species from one ancestor in a short period of time, especially given the long life span—over 100 years—of tortoises. How did that happen? What conditions are responsible for the unusual "creative force" of the archipelago, as Darwin put it (1845, 398), such that tortoise species differ from island to island, even volcano to volcano?

The Secret Recipe of Galápagos

Three ingredients are key to what I call the "secret recipe" that made Galápagos so productive of new terrestrial species: (1) active volcanic origins and clustering of the islands, (2) isolation on the equator, amid a convergence of ocean currents, and (3) unusually demanding terrestrial ecological conditions. The pertinent time frame for this recipe is pre-settlement Galápagos, because many aspects were later modified by human activity.

Consider the first ingredient: the islands have been, and are being, formed by sequential volcanic eruptions of a Pacific Ocean "hot spot" (Harpp et al. 2014; Toulkeridis 2011). The hot spot is located near the Galápagos Spreading Center (GSC), where two oceanic plates of the Earth's crust, the Cocos and Nazca Plates, spread apart as new plate material is formed (Figure 3.3).[3] The Cocos Plate moves northeast at about 83 mm/year (83 ± 3 km per million years, or roughly 52 miles per million years), and the Nazca Plate moves southeast at about 59 mm/year (59 ± 1 km per million years, or roughly 37 miles per million years).[4] The leading edges of both plates suffer a similar fate: they subduct when they meet continental plates, forming trenches like the Peru–Chile Trench, in a region of great seismic activity. With heat and pressure, the mass of descending oceanic plates is eventually recycled back into the pool of molten rock from which it came.

At the hot spot, according to a leading model (Courtillot et al. 2003), a mantle plume from the Earth's interior rises toward the surface and forms a magma dome of molten rock just below the upper mantle (Figure 3.4). Molten material from the dome feeds a magma chamber that periodically melts through the crust in eruptions, whose flowing lavas form relatively shallow shield volcanoes (resembling a warrior's shield viewed from the side) on either or both of the plates. These volcanoes vary in location, size, and shape as the plates move along in their respective directions. Over time, they erode and subside, eventually falling below sea level to become seamounts along the ocean floor. As the plates move along over millennia, the seamounts contribute to ocean-floor ridges that represent "volcanic traces" of the Galápagos hot spot. The existing Galápagos islands are simply the most recent peaks of this activity, rising from a volcanic platform (360–900 m or 1,200–3,000 ft below sea level) that was itself formed from submarine eruptions from the same plume (Geist et al. 2008).

---

[3] The Earth's outer crust is comprised of seven major tectonic plates, each roughly 50 to 100 km thick, that move on the order of 50 to 100 mm each year. The movements of these plates against and away from one another constitute continental drift, giving rise to much seismic action, volcanic activity, and mountain range formation.

[4] Plate velocity estimates from O'Connor et al. (2007, 344). For comparison, these rates are roughly 1.5 to 2 times the growth rate (41.6 mm/year) of the average young adult fingernail (Yaemsiri et al. 2010).

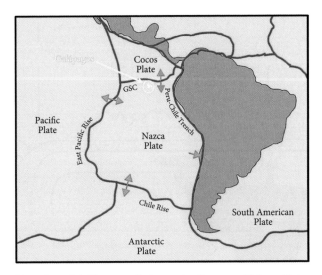

Figure 3.3. Location of Galápagos relative to the Cocos and Nazca Plates and the Galápagos Spreading Center (GSC) between them. As oceanic plates spread apart at such centers (*diverging orange arrows*), mid-oceanic ridges are formed (e.g., East Pacific Rise), which count among the most volcanically active areas of the Earth. Where plates collide (*converging orange arrows*), the denser plate subsides, forming a deep trench (e.g., Peru–Chile Trench) where friction and heat fuel additional volcanoes. Virtually surrounded by volcanic activity, Galápagos was, and is, formed by volcanic activity from another source: a magma plume (hot spot) that intermittently erupts through the Nazca Plate on the west side of the archipelago. *Source*: Public domain illustration by Paula Keener, NOAA Office of Ocean Exploration and Research, Chile Margin 2012 Expedition, with slight modifications.

This mid-ocean hot spot is exceptional in several respects. First, it lies near the GSC, so that volcanoes sometimes formed on both plates, providing fault lines and cracks that allow magma to escape various places within plates. As a result, the hot spot has formed clusters of islands, rather than a linear chain, as happens when a single plate passes over a hot spot (e.g., the Hawaiian–Emperor seamount chain).[5] Compared to chains, a cluster has a lower average distance between islands, providing myriad pathways for life to move among them. I especially like the analogy offered by Geist et al. (2014, 159): "In areas like Galápagos, where the islands do not form a linear chain, colonization and dispersal [of organisms] are

[5] The Hawaiian–Emperor seamount chain extends from the middle of the Pacific Ocean (20° N) to the Kamchatka Peninsula (50° N), spanning over 5,000 km, with seamounts dating back 70 MYA. With a life span four times greater than Galápagos, and a current land area more than twice as large, Hawai'i also had a greater number of endemic species. But Hawai'i has sustained more than a thousand years of human impact (see, for example, Grigg 2012).

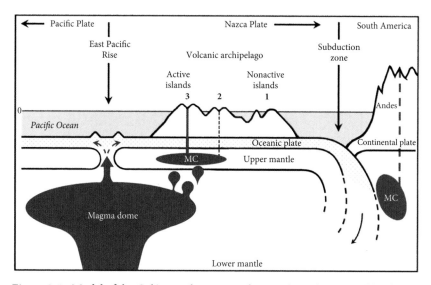

Figure 3.4. Model of the Galápagos hot spot and its products (not to scale). The magma chamber (MC) of the hot spot is fed from a magma dome of molten rock coming up from the Earth's core. A strong plume supply fuels the active shield volcanos of islands 2 and 3 (*broken red line* indicates less frequent eruption). As the Nazca Plate moves east, volcanoes move away from the hot spot, become inactive (for example, island 1), and eventually erode and sink below the ocean surface to become seamounts. On its eastern edge, the Nazca Plate sinks under the South American Plate in the subduction zone, generating the uplift that created the Andes and friction that feeds a separate magma chamber and intermittent Andean volcanic activity. *Source*: Modified and used by permission of Princeton University Press from Kricher, J. (2006), *Galápagos: A Natural History*, Princeton University Press, with revisions based on Courtillot et al. (2003), Kelly et al. (2019), and Cleary et al. (2020).

not so much like a series of stepping stones as they are a game of checkers on a board whose geometry changes every 100,000 years."

This quote points to a second unusual feature of the Galápagos hot spot: "the number of simultaneously active volcanos" (White et al. 1993, 549). The hot spot fuels a restless archipelago in which the number, size, and configuration of the islands are constantly changing in time. What we see today as a static image on a map is but a single frame of a long, geologically dynamic movie in which different volcanic islands come in, speak their part, and exit below sea level. The process means that the Galápagos islands of five million years ago, for example, were completely different in properties and configuration from those of today. Because those earlier islands remain as seamounts, one can "read backward" from ocean ridges to reconstruct basic features of earlier archipelagoes, and even relate those features to the evolutionary history of particular species. The radiation of tortoises offers a good example of such reconstructions.

A third novel feature of the Galápagos hot spot is its increasing distance over time from the GSC. Before six million years ago, "the hotspot lay directly beneath the GSC and contributed magmas concurrently to the production of the Cocos and Carnegie ridges, on the separating Cocos and Nazca Plates" (White et al. 1993, 534). At roughly two to three million years, the spreading center had moved to as much as 100 km (62 miles) away, but still close enough to the hot spot to receive most of the magma flow from the plume. Only a limited supply was then available for forming volcanoes directly over the hot spot; thus, the relatively smaller and structurally simpler islands of the eastern archipelago were formed. The GSC continued to move away, and beginning one to two million years ago, there was "a fundamental change in the intensity of the magma supply and consequently [in] the constructional mechanisms responsible for Galápagos volcanoes" (Cleary et al. 2020, 9). As a result, "the western Galápagos volcanoes formed with robust magma supplies, sufficient to [build taller volcanoes with] summit calderas" (p. 10). Hence the western islands tend to be larger, higher, and structurally more complex.

Why does it matter? Altitude makes a big difference in Galápagos, because islands reaching 300 m or more have a special feature: a humid (or moist) zone, owing to the interaction of altitude and climate. These "high islands" receive both more rainfall in their upland areas and additional moisture from cloud water interception. They therefore support denser, more diverse communities of organisms than lower-elevation arid zones. High islands also have temperature gradients, producing an array of habitat types—littoral, dry, transition, humid, and on high enough islands, high-altitude dry habitat—all within a short distance of one another (Figure 3.5). Low islands may have only two or three climate–vegetation zones, typically a littoral zone, an arid zone, and sometimes a transitional zone with slightly more precipitation.

Putting it all together, the islands are each a unique, single-volcano petri dish, temporarily poking up from the sea, offering different conditions for life. The exceptions are Isabela, where six major volcanoes fused together forming one island with craggy breaks between them, making each volcano its own separate "dish," and San Cristóbal, where two volcanoes coalesced.[6] Analysis of seamounts along the Carnegie Ridge from Galápagos to South America, and along the Cocos Ridge from Galápagos to Costa Rica, confirms that a cluster of

---

[6] It's easy to miss the sixth volcano of Isabela. Given the island's overall shape like a seahorse, volcano Ecuador is the mouth of the seahorse (see Figure 1.1). It looks like a mouth because it is partially collapsed on the western side: one can look into its caldera from a boat at sea level. On San Cristobal, see "Galápagos Geology on the Web." http://www.geo.cornell.edu/geology/GalápagosWWW/Cristobal.html

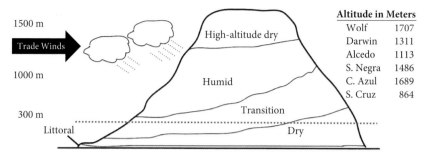

Figure 3.5. The climate–vegetation zones of an idealized high island, generated by temperature and precipitation changes with elevation. Islands reaching 300 m or more have a humid zone and, if tall enough, a high-altitude dry zone. Islands not exceeding 300 m (*dotted line*) receive much less rainfall and have only littoral, dry, and perhaps transition zones. A rain-shadow affects the downwind side of high islands (*right side as shown*), making dry and transition zones more extensive there. For comparison, the inset table shows the altitude of the tallest volcanos of the archipelago. *Source*: Used with permission of A. Tye, from Tye, Alan, and Javier Francisco-Ortega. 2011. "Origins and Evolution of Galápagos Endemic Vascular Plants." In *The Biology of Island Floras*, edited by David Bramwell and Juli Caujapé-Castells, 89–153. Reproduced with permission of Cambridge University Press through PLSclear.

islands—of variable number and configuration—has existed continuously in the same general area for at least 17 million years (Werner and Hoernle 2003). In short, volcanic origins have made the changing Galápagos islands a special long-term incubator for life—the first key ingredient of the recipe.

The second key ingredient is *isolation on the equator amid a convergence of currents*. The islands are isolated by 1,000 km (600 miles) of salt water and intense equatorial sun from mainland Ecuador. But they are also isolated from the mainland, and from one another, because they sit in the center of a major convergence of ocean currents in the eastern Pacific (Figure 3.6A). Two big currents come from the coast of South America, propelled by the equatorial trade winds. First, the famously cold Humboldt Current from the south is propelled out to the archipelago by the southeasterly trade winds. Called by Darwin, "the great southern Polar current" (1845, 373), it brings a wide plume of cold water all the way from southern Chile to Galápagos. Second, the much warmer Panama Current, also called El Niño Flow (Figure 3.6A), comes from the north, pushed along by the northeast trade winds. The two converge near the equator and integrate into the South Equatorial Current (SEC), which flows westward at velocities as high as 70 cm/sec (2.3 ft/sec). Such currents make it easy to move within the archipelago in some directions (i.e., with them), and very difficult to move other directions (i.e., against them), as every snorkeler learns.

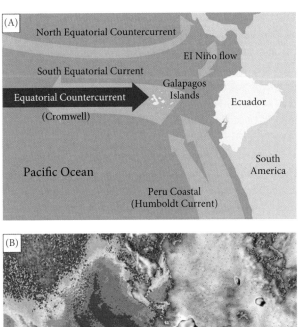

Figure 3.6. *A,* The convergence of ocean currents near Galápagos. The Peruvian Coastal (or Humboldt) Current from the south and the El Niño Flow (or Panama Current) from the north join the massive, westward South Equatorial Current (SEC). Much of that water is replaced by the Equatorial Countercurrent (commonly called the Cromwell Current or Equatorial Undercurrent, EUC), flowing eastward, below the surface, toward Galápagos. *B,* As the Cromwell Current meets the volcanic platform and islands, it produces a major, nutrient-rich upwelling, fueling an extremely productive marine ecosystem (here orange and red, representing the high chlorophyll concentrations of a phytoplankton bloom; the upwelling normally flows around Isabela into the western archipelago, though covered here by clouds). This unique mid-Pacific upwelling is a key reason Galápagos supports such a rich and diverse biota. Note the different spatial scales of *A* and *B. Sources: A,* D'Ozouville et al. (2010, 172) in Galápagos Report 2009–2010. Puerto Ayora, Galápagos; public domain; *B,* Satellite image of May 23, 1998, from NASA Scientific Visualization Studio, public domain.

The currents also affect climate. The Humboldt brings abundant cold water to Galápagos, dominating during the cool and dry season (June to December).[7] This period is also known as the *garúa*, or fog season, when most days bring moisture-laden cloud cover. In these months, daily Galápagos temperatures are unusually temperate for the equator—cool enough for sea lions and densely coated fur sea lions normally found at much higher latitudes.[8] Conditions are also cool enough to sustain coffee cultivation at 300 m (1,000 ft) and above (elsewhere in the tropics, coffee grows only over 1,500 m). In contrast, the Panama Current flows southwest toward the equator, dominating during the warm and wet season (January to May). Even then, seasonal sea-level precipitation can only support desert habitat in low-lying regions of the archipelago.

Perhaps most important of all, water propelled westward by the SEC is partly replaced by a giant Equatorial Undercurrent (EUC), commonly known as the Cromwell Current (named for the oceanographer who documented it in 1952). This current is about 30 km wide (almost 20 miles; see Figure 3.6A) and flows eastward, below the surface, at a depth of 40 to 400 m (130–1,300 ft). At roughly 1,000 times the volume of the Mississippi River, and moving more than twice as fast as the westward surface current (150 cm/sec, 5 ft/sec), the Cromwell produces a large, cold, nutrient-rich upwelling as it comes up the western side of the Galápagos platform, producing the phytoplankton bloom of Figure 3.6B. Part of its richness comes from iron it picks up along the volcanic escarpment west of Isabela and Fernandina, an element notoriously scarce in other mid-ocean marine ecosystems.[9] Sardines and other small bait fish normally abound in this little corner of the world, fittingly home to two flightless seabirds of Galápagos—the Galápagos penguin and cormorant, revisited in Chapter 8. Wings literally could not carry these two birds to any better place for food. The cold, nutrient-rich waters are deflected around Isabela at its northern and southern ends, reaching

[7] Seasons in Galápagos change with the movement of the Intertropical Convergence Zone (ITCZ), where southeast and northeast trade winds meet. In the cool and dry season, the ITCZ is north of the equator, the southeast trade winds predominate over Galápagos, and sea-surface temperature (SST) falls. In the warm, wet season (January–May), the ITCZ migrates south, the northeast trade winds increase, and with them precipitation and SST increase.

[8] Galápagos fur sea lions (erroneously also called "fur seals") and sea lions, both of the family Otariidae ("eared seals," from the Greek *otarion*, "little ear"), came to the islands from opposite high latitudes. The bear-faced Galápagos fur sea lion (*Arctocephalus galapagoensis*) came up from Tierra del Fuego, propelled along the South American coast by the Humboldt Current, about 1.3 MYA (Yonezawa et al. 2009; Trillmich 2016). Assisted by clockwise ocean currents, the Galápagos sea lion (*Zalophus wollebaeki*) diverged from the larger California sea lion about 2.3 MYA (Wolf et al. 2007). Each species is the smallest of its genus as befits its new equatorial home.

[9] Iron has low solubility in salt water and is therefore scarce in ocean regions far from iron-containing continental margins. "Martin's iron hypothesis" posits that this lack of iron limits the productivity, in terms of phytoplankton and chlorophyll production, in iron-deprived ecosystems (Martin et al. 1991). As Figure 3.5B shows, this is not normally an issue in the area of the Cromwell upwelling.

well into the archipelago (largely hidden by clouds in Figure 3.6B), adding enormously to the churn of currents and contributing to one of the most productive marine ecosystems of the world.

This combination of currents in Galápagos makes movement between islands difficult and unpredictable—for sailors and other organisms. The 19th-century author Herman Melville put it well: "nigh a month has been spent by a ship going from one isle to another, though but ninety miles between" (1854, in Melville and Michelsohn 2011, 19). The churn contributes to the isolation of the islands from one another, adding a special wrinkle to the floral and faunal distributions: often, a given plant or animal's closest relatives are not found on the closest islands as normally measured, but on the closest islands *downstream* along the prevailing current, sometimes at considerable distance. The currents have added to the archipelago's reputation as "enchanted islands"—thwarted by the power and speed of the sea, early sailors imagined the islands carried a curse making them appear to move.

The third ingredient of the secret recipe of Galápagos is *unusually demanding ecological conditions on land,* conditions that seem almost otherworldly. A colonizing organism of Galápagos had to survive days of ocean crossing with salt water and sun, only to arrive on the soil-scarce rocky desert shore of the islands. There, fresh water is notoriously scarce, often with less than 400 mm (16 in.) of annual precipitation, much of it from mist. Only the very toughest organisms can survive in such a place, like cactus plants of the genus *Opuntia*, the prickly pear cacti (well known from dry areas of the Americas). As Darwin noted, even insects are scarce here (Keynes 2001, 297). The low productivity of low-elevation terrestrial Galápagos stands in striking contrast to the normally high productivity of its marine habitats just offshore. One can visit a barren volcanic point, like Punta Espinoza on Fernandina, for example, and marvel at the abundance of marine life—sea lions, marine birds (including pelicans, penguins, and cormorants), and untold numbers of marine iguanas—lying on barren lava or in the shade of a saltbush or mangrove. And at the same time, land birds prove to be few and elusive, preferring to hide in shade, and there is only the occasional lava lizard—maybe 1/100th the size of big male marine iguanas—or a pencil-sized racer snake. The lowlands of Galápagos are generally very harsh.

But the desert-adapted lowland flora and fauna face another challenge. El Niño arrives every few years (as described in Chapter 2), bringing warm ocean temperatures, 10 times normal rainfall at sea level, and even more rain in the upland areas. The rain and warm temperatures mean renewed vegetation and food, increasing the vitality and fertility of most terrestrial wildlife, including Darwin's finches and giant tortoises (if the rain doesn't wash the latter out to sea). El Niño gives land birds something to celebrate: finches sometimes manage three clutches per year during the rains and lush vegetation. Much of the terrestrial vegetation thrives on the extra rain, but some species—notably large arborescent *Opuntia*—can topple in the rain and wind. Furthermore, mounting evidence suggests global climate change is adding to the magnitude, duration, and frequency of El Niño events (also described in Chapter 2), jeopardizing some of the organisms so well adapted to this difficult place.

But if El Niño brings temporary terrestrial abundance with its rainfall, the opposite El Niño-Southern Oscillation (ENSO) phase—dry La Niña—makes up for it. During La Niña, terrestrial organisms face their stiffest challenges, often going without precipitation for many months, sometimes a year. Even the normally moist highlands can dry out, leaving land birds, reptiles, and insects facing severe food scarcity. Since there is nowhere to go to escape these effects within 1,000 km or 600 miles, mortality is often at its highest levels during La Nina. To make matters worse, Galápagos lies in the very heart of ENSO: no land on Earth is more affected by this cycle, which aptly illustrates the hugely demanding and variable terrestrial conditions of the archipelago.

These three basic ingredients, together with everyday ecological processes, endow pre-settlement Galápagos with three more special properties. First, they create *highly simplified terrestrial ecosystems.* Because of the hardships of getting there and getting established, "in oceanic islands," as Darwin noted, "the number of kinds of inhabitants is scanty" (1859, 244). The plants that survive in Galápagos, for example, can be called "waif" species (see Carlquist 1967)—and they are mostly bird-carried plants and drifters, with a notable lack of bush and tree species compared to mainland South America. A few tree species make it in the lowlands, including the deciduous incense tree (*Bursera graveolens*), known as *palo santo* (holy tree) in Spanish, famous for the scent of its sap. But common mainland tree species are noticeably lacking on the islands, including the moist highland areas of taller islands (except for trees brought in by human settlers).

Darwin's early observation that island inhabitants are "scanty" is an aspect of what we call today "island biogeography" (after MacArthur and Wilson 1967). We now recognize that island biotas can be thought of as a product of two variables: (1) immigration rates for new species, which vary by distance from the mainland (closer islands have higher rates), and (2) extinction rates of established species, which vary by island size (small islands have higher rates). Hypothetically, after many generations, the rates of arrival and extinction balance at an equilibrium number of species for any given island size and distance from the mainland. Because Galápagos islands are, on a global scale, remote *and* small, it's no wonder their terrestrial ecosystems are diverse but scanty.

Another aspect of simplified ecosystems that warrants recognition is the *absence of terrestrial mammalian predators or competitors* during the millennia before human presence. Often a thrilling discovery for today's visitors, the terrestrial fauna is completely dominated by reptiles and birds—the endemic

Galápagos hawk (*Buteo galapagoensis*) being the main apex predator, plus the native barn owl (*Tyto alba)* and native short-eared owl (*Asio flammeus*; we return to these organisms later on). Often the landscape feels like the set for a science fiction prehistory movie: on many islands, one can spend whole days surrounded by reptiles, some gigantic, and birds, some flightless, without once seeing a terrestrial mammal other than people. As Darwin wrote, "These islands appear to be paradises for the whole family of reptiles" (Keynes 2001, 353). The native mammals are limited to a few "rice rats" (mouse-sized endemic rodents) and one bat.[10] It's a terrestrial community *very* different from rural continental areas.

Third, in the absence of terrestrial predators and competitors, but faced with otherwise harsh ecological conditions, Galápagos wildlife generally had *time to evolve a high degree of endemism* before human settlement. Many local organisms are found nowhere else on Earth, and sometimes even those that look alike from island to island are actually different species. This is true, for example, of the mockingbirds of Galápagos that so intrigued Darwin in 1835. At first glance, they all look alike. But to Darwin's surprise, he noticed consistent, small, morphological differences—the number of subtle stripes on the wings, for example—among the mockingbirds of the first three islands he visited, with one repeat on the fourth island. Reflecting on the island-by-island differences after leaving Galápagos, he wrote in a now-famous passage in his notebook: "I must suspect that they are only varieties," for if they were indeed different species, "such facts would undermine the stability of Species" (Darwin 1836 [1963], 262).[11] Months later, back in England, British ornithologist John Gould poured over Darwin's collection and confirmed that the mockingbirds were, indeed, three distinct species—otherwise "the experience of all the best ornithologists must be given up" (Darwin 1839b, 64). Gould's pronouncement, based on those subtle differences, was a definitive turning point in young Darwin's thinking (see discussion in Durham 2012). He had no choice but to wonder how some "creative force" in the islands, whatever its source, had shaped such differences. This thinking, plus Thomas Robert Malthus' book (1798), *An Essay on the Principle of Population*, soon led Darwin to the idea of evolution by natural selection.

Today we recognize a common pattern in the evolution of new, endemic species in Galápagos. The pattern simplifies to four processes: colonization, separation into reproductively isolated populations, genetic differentiation of the

---

[10] There were once more than a half-dozen endemic species of "rice rats" of the genera *Nesoryzomys* and *Aegialomys* distributed over various islands (see Castañeda-Rico et al. 2019).

[11] The inter-island differences Darwin found among Galápagos mockingbirds were unexpected by the special creation paradigm prevalent at the time. Other mockingbird species he had seen on mainland South America covered huge ranges—nearly the size of Chile, for example. By extension, one would expect all of Galápagos to have a single species. Darwin realized that if the mockingbirds of different small islands in the Pacific were not varieties of a single species, this would suggest something other than their special creation.

isolated populations, and, with time, the emergence of breeding barriers indicative of species formation. The processes are described more fully in Appendix 1; here we need only discuss two of the four. First, consider colonization, where the likely pathways are: fly, swim, or drift; raft on another object; or be blown by wind. Because of proximity to the Americas, plus the convergence of winds and currents of the region, it is unsurprising that the vast majority of colonists of the Galápagos came from the Americas, including North America (Grehan 2001).[12] Colonists from elsewhere were extra-hardy hitchhikers, including twenty-plus species of insects, mostly mites, and two species of land snails that came from the western Pacific near Asia, probably via birds or rafts along the North Equatorial Countercurrent (see Figure 3.6A; Peck 2001, 60).

Second, we recognize that separation of successful Galápagos colonists into reproductively isolated populations occurred by one of three distinct processes, or some combination. The first is *dispersal*—movement of a small number of individuals to a new island or to a new and separate habitat within an island. The second is *vicariance* (from Latin *vicarius*, "substitute" or "proxy")—splitting of an initial population into two or more "sister" populations by some external force or process. The subsidence of previously connected island volcanos, for example, is one clear way for vicariance to divide one Galápagos species into two populations. The third process is *hybridization*, or cross-breeding of genetically different organisms. Hybridization often reduces diversity, but exceptions occur when accidental, complete duplication of parental genes occurs in progeny (as happens in plants), or when striking new, heritable features are produced by a crossing that reproductively isolates the hybrid offspring.[13]

With enough time, the result is the formation of new, endemic species in the archipelago, many of them among the Earth's most interesting and unusual. But their specializations also spell vulnerability in times of change. Indications of this vulnerability are seen among the waved albatrosses of Chapter 2, aptly symbolized by their building no nest, in contrast to albatrosses everywhere else. Their disarming naiveté can inspire awe and wonder in us, but it brings with it nearly complete helplessness in the face of introduced cats and rats. Life in isolation has left Galápagos organisms with amazing specializations . . . and a general susceptibility to the human-induced pressures of today.

---

[12] A thorough evaluation of Galápagos flowering plants (angiosperms) estimated that they derived from at least 117 colonization events, of which at least 110 (94%) came from source regions in the Americas (Tye and Francisco-Ortega 2011, Table 5.3). Of those, at least 64 (58%) came from South America, 28 (25%) from Central America and the Caribbean, and 7 (6%) from southwestern North America—other origins being more obscure. By washing durable seeds out to sea, El Niño may well have played a substantial role in plant colonization (see Hamann 1984).

[13] This last process has been highlighted by Lamichhaney et al. (2018). As discussed in Chapter 6, the successful mating on Daphne Major of two distinct morphospecies of Darwin's finches produced a lineage of fertile hybrids with features that promote mating only with one another.

Fortunately, with enough time, Galápagos endemics do evolve a measure of *resilience* that helps offset some of their vulnerability ("resilience," for our purposes, is simply the ability of individual organisms to withstand perturbations or to rebound quickly after them).[14] It is helpful to distinguish two forms of resilience: (1) the general resilience that came with initial colonists to Galápagos—that is, their ability to rebound via previously evolved mechanisms for handling challenges; and (2) the specific resilience added during evolution in Galápagos in response to local perturbations and selection pressures. These latter perturbations and pressures run the gamut from repeated volcanic eruptions, extensive flows of molten lava, and island subsidence, all the way to El Niño rainstorms, La Niña droughts, recurrent floods, fires, and new colonizing species. While it is conceivable that an organism could arrive in Galápagos already "fully equipped" for all or most of its challenges—coming from another tropical archipelago, for example—it is much more likely that natural selection would go to work, shaping some capacity for Galápagos-specific resilience.[15] Such resilience will be a product of the evolving adaptations of endemics to Galápagos and can be thought of as the resources of morphology, physiology, or behavior that help individual organisms meet a given challenge. They may well come in handy, serendipitously, when endemics face today's human-induced challenges, but only with gaps and imperfections. In short, enough time for selection pressures to promote endemism with specific resilience is the sixth and final ingredient of the "recipe" as shown in Figure 3.7.

Alas, with human-induced changes underway in the archipelago today, the lovely naiveté of endemic organisms also makes them seem at times like "babes in the woods." The latter expression has proved an apt descriptor even for Galápagos giant tortoises—hardly one's image of youthful innocents—in the face of anthropogenic change, so let us now return to them.

---

[14] A word of caution: the term "resilience," like "adaptation," is used today with very different meanings across diverse fields and disciplines, from psychology to ecology to climate science. It is also used in reference to widely different levels of organization, from individuals to populations to ecosystems. The usage here is specific to evolutionary analysis at the level of individual organisms.

[15] One native species came close: the Galápagos flamingo (*Phoenicopterus ruber glyphorhynchus,* meaning roughly "red Phoenix-winged bird with an engraved beak"), whose ancestors were lagoon-loving Caribbean flamingos (same genus and species, also called American flamingos) who flew out sometime in the last 350,000 years. By virtue of a wide-ranging diet and energy-efficient behaviors (such as standing on one leg), their general resilience resulted in very few genetic changes in Galápagos (Frias-Soler et al. 2014). Even El Niños have little impact: when threatened, the flamingos simply fly to better conditions among dozens of lagoons in the archipelago (Vargas et al. 2008). More research is needed, but current evidence suggests just a single specific adaptation: the Galápagos flamingos have a 5% to 10% smaller body size, and associated smaller eggs and legs (Tindle et al. 2016), in response to limited food supply in their new home.

The Secret Recipe of Galápagos
(What makes its terrestrial habitats so creative of new species?)

(1) Active volcanic origins & clustering of islands

(2) Isolation on the equator amid converging currents

(3) Demanding terrestrial ecological conditions

(4) Highly simplified terrestrial ecosystems

(5) Absence of terrestrial competitors & predators

(6) Enough time for the evolution of endemism & resilience

Figure 3.7. The secret recipe of Galápagos includes six ingredients responsible for what Darwin called its unique creative force.

The Tortoise Example

One of the best examples of the creative force of the islands is the radiation of the giant tortoises from one common ancestor. The radiation shows how each island or volcano offered unique conditions for living organisms, allowing evolution to shape tortoises' divergent morphologies. And it shows how each volcano is part of an active landscape—rising, falling, sometimes merging, sometimes separating into pieces—and how each is surrounded by strong currents that can either propel recurrent colonists to the next island downstream or send them out to sea. It is a fascinating evolutionary history with enough details today from genetic and geologic analysis that one can appreciate the role of each ingredient in the special recipe.

The story begins on the mainland in South America, where comparative genetic studies among Galápagos tortoises and their much smaller, contemporary mainland relatives provide a convenient starting point. Analysis by Poulakakis et al. (2012) indicates that Galápagos tortoises diverged from their last common ancestor with mainland tortoises about 3.2 MYA (million years ago). It was a time when giant tortoises still roamed South America and North America as well, all of which have since gone extinct, often under human influence.[16] The closest living mainland relative today, the much smaller Chaco Tortoise of Argentina (also called the Patagonian tortoise), whose adult length is roughly 25

[16] Tortoise authority Peter Pritchard (1996, 17–18) put it well: "Huge terrestrial chelonians used to be almost everywhere. . . . But the spread of mankind over the face of the globe, more than any other factor, has plunged the giants into ignominious retreat [on remote islands]. . . . But finally these island survivors too were found. . . . Whole islands were stripped of their last tortoise so that [mariners] . . . might be victualled with something better than salt pork. . . . Only in Galápagos did some diversity of giant tortoise life survive until mankind finally developed a conscience. . . . [We] extirpated them on the continents, so we now have to go to Galápagos to see them."

to 40 cm (10–16 in.), is descended from the same ancestor, with evident selection for small body size. Although Galápagos tortoises are a terrestrial species (not to be confused with sea turtles), they float and paddle with their heads above water when need be, as we infer their ancestors did.[17]

As floaters, they were poised to benefit from the convergence of strong equatorial currents. Elsewhere on Earth, islands 1,000 km or more from the nearest mainland do not have tortoises of any kind, as noted by Gorman (1979), nor do they have rodents, small carnivores, large herbivorous mammals, or freshwater fish, and thus are highly simplified ecosystems, as expected from island biogeography. Gorman's findings hold true in Galápagos for most terrestrial animals, but obviously *not* for tortoises, nor for rodents, as noted earlier. While the simplified ecosystems rule remains accurate, Galápagos tortoises and rodents are exceptions—exceptions whose presence testifies to the strong east-to-west equatorial currents that drastically reduced transit times and made unusual colonizations possible.

It is likely that the first tortoise immigrants—perhaps even a single gravid female (sperm remain viable within females for long periods)—were dislodged from coastal South America in a storm, perhaps an El Niño-related flash flood in the Andes (for recent dislodgings of the kind, see Marquez et al. 2008, 10; Cayot 2016). Or possibly they floated off in search of more abundant food, as do other tortoises (Ernst and Lovich 2009). With help from the strong westerly currents, they floated all the way to an island on the eastern edge of the archipelago. Geological reconstructions of the period (Geist et al. 2014, Karnauskas et al. 2017) suggest that landfall probably occurred at the "proto-island(s)" that later became Española and San Cristóbal.[18]

When added to those reconstructions, the genetic analyses of Poulakakis et al. (2012, 2015, 2020) allow us to sketch a tentative "phylogeographic history" of Galápagos tortoises.[19] As techniques improve, so will the details, but current

---

[17] As noted by Austin et al. (2003, 1415), "Large tortoises have colonized many oceanic islands by transmarine migration. . . . Their buoyancy predisposes them to such journeys, as [does] the position of the lungs near the top of the shell, which makes them self-righting in water, and their long necks, which permit them to keep their heads well above the water surface and so breathe easily." The tortoise family Testudinidae, including the Galápagos genus *Chelonoidis*, is a marvelous case in point: genetic analysis traces their ancestry to Africa, confirming that ancestors of *Chelonoidis* crossed the Atlantic, then crossed South America, eventually arriving in Galápagos (Le et al. 2006, 528).

[18] Islands reconstructed from estimated plate movements, plate flexure, island subsidence, and volcanic activity are called "proto islands." The articles sited here provide two views of the eastern archipelago at the time of tortoise colonization, 3.2 MYA. In Geist et al. (2014), San Cristóbal and Española were separate but close proto-islands; in Karnauskas et al. (2017), they formed one large proto-island that later split in two. In both cases, say the genetic data, ancestors to today's Española tortoises arrived from San Cristóbal in separate and much later dispersal event, 0.7 to 0.8 MYA. That means no tortoises survived on Española when it split off, ruling out vicariance in the second scenario.

[19] Hennessy (2015, 87) is critical of "the molecular turn" in such studies because she says they opens up "new technologies for imagining pristine nature." Hennessy is always stimulating, but she overstates the "quest" of conservationists in Galápagos for the "previously 'pristine' sanctuary of

information suggests that the sequence began with dispersals, as tortoises floated, or possibly walked in some cases, between separate island areas. Later, vicariance and extinction also played a key role, as volcanic islands with tortoises split up and moved apart, freeing subpopulations to evolve differently in response to local conditions (for details, see Online Appendix B).

In short, a history of dispersal, extinction, and vicariance gave each major island—plus the five complete volcanos of Isabela—a tortoise population of its own that then evolved into new species. Major islands lacking tortoises were either far north and small (Darwin and Wolf), or else not directly "downstream" on prevailing currents from islands with tortoises (Genovesa).[20] The resulting order of colonizations was from southeast to northwest (Figure 3.8), paralleling the order in which the islands initially formed over the hot spot, and also roughly in the direction of the Humboldt Current, which surely facilitated dispersal phases. This pattern, known from other radiations as well (Parent et al. 2008), is called the "progression rule": sequential colonizations often mirrored the historical emergence of existing islands.

Figure 3.8 makes a second point: much later, human-induced tortoise dispersals (dashed arrows) did not follow the Humboldt Current. In this case, whalers and buccaneers carried them by ship to remote areas (e.g., Banks Bay, northwestern Isabela) for food or oil, and dropped them in the sea or left them ashore as space or weight required (for a memorable example, see Litt 2018, 49). Many years later, these castaways would play a key role in breeding and repatriation efforts for tortoise populations pushed to extinction on their home islands.

Tortoise Adaptations

Following the geographic distribution of tortoises by dispersal and vicariance, their differentiation proceeded over millennia, we infer, through island-specific combinations of founder's effect, drift, inbreeding, and natural selection, causing carapaces, body size, and other features to change. Key among the processes was Darwin's favorite, natural selection, owing to the demanding ecological

---

Darwinian nature" (Hennessy and McCleary 2011, 132). Conservation is not preservation: the goal is *not* to return to a mythical "pristine" baseline, but to constrain human impact so that other species can survive indefinitely. Genetic analysis and geological history combine to show that evolution produced 15 species of tortoises. That tally is useful for assessing adverse human impact in the past, and for taking steps to ensure the survival of remaining species going forward.

[20] Marchena is an exception for which I know of no convincing explanation (maybe just chance; possibly extinction). Rábida is another apparent exception, given the boney carapace fragments and pieces of tortoise eggshell, dating to 8540 BP, found by Steadman et al. (1991) in a sediment-filled volcanic cone on the island. Further DNA analysis of this material is needed.

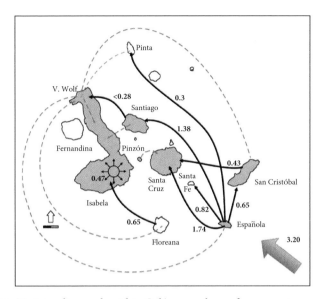

Figure 3.8. Tortoise dispersals within Galápagos, shown for convenience against the present configuration of islands. Natural dispersals (*solid black arrows*, with adjacent dating in MYA), ranging from 1.74 to 0.28 MYA, roughly followed the flow of the Humboldt Current (*large arrow*). In this reconstruction, Española was the site of initial colonization, remaining a major source of dispersals for many years; Floreana and Santiago were secondary sources. Not shown are vicariance events, related to the breakup of larger islands, during the same interval (see Online Appendix B). In contrast, recent human dispersals of tortoises (*dashed green arrows*)—occurring, for example, when mariners carried tortoises to remote camping sites—often do not follow the flow of the dominant current and sometimes cross it. Here Fernandina is not shaded (indicating no living tortoise population) pending genetic analysis of a female found there in 2019. *Source*: Republished with slight modifications by permission of N. Poulakakis from N. Poulakakis et al. 2012. "Unravelling the Peculiarities of Island Life: Vicariance, Dispersal and the Diversification of the Extinct and Extant Giant Galápagos Tortoises." *Molecular Ecology* 21, no. 1: 160–73. Permission of John Wiley and Sons conveyed by Copyright Clearance Center, Inc.

conditions especially on low, dry islands (or islands that became that way via subsidence as the tortoises evolved). Today, one can "read" the nature of the demanding ecological conditions right from the carapaces.

The tortoises were first systematically analyzed by herpetologist John Van Denburgh (1914) of the 1905–1906 California Academy of Sciences expedition, who identified four categories of tortoises by shape. Two are classic forms or "morphs," as in Figure 3.1: "domed" (or dome-shaped) and "saddleback" (with the carapace turned up behind the neck). The two others are "intermediate"

Figure 3.9. Fieldworkers checking a saddleback tortoise on Española, amid a cluster of arborescent *Opuntia*. On low, dry islands, evolution of the upturned carapace of saddlebacks allowed additional reach for scarce cactus pads and fruits. The cacti of such islands, meanwhile, evolved added height through the advantage of evading the tortoises' ever-longer reach. The coevolution of these species was thus an evolutionary "arms race." *Source*: "Española Galápagos tortoises by Christian Zeigler," copyrighted, exposure adjusted. Used under Creative Commons License 2.0 (from https://www. flickr.com/photos/gianttortoise/9518921884/in/dateposted/).

(between domed and saddleback) and "unknown" (the Sante Fé tortoises, of which only a few bones remain). We know today that saddleback carapaces evolved on dry, lower islands, where vegetation is scarce and where, we infer, a higher reach for eating, from the upturned carapace "collar," conferred a survival advantage.[21] Also, the limbs and necks of saddlebacks are longer than expected for a tortoise of their size (Fritts 1983, 109), adding further to their reach. On these same islands, the tortoises' preferred food—pads and fruits of the prickly pear *Opuntia*—have a surprisingly tall, treelike form (see Figure 3.9). It is no coincidence. As argued by Fritts (1984) and summarized by Gibbs et al. (2010, 208):

---

[21] Saddleback carapaces may well have a second function in their superior self-righting capacity on land (Chiari et al. 2017). Nothing requires adaptations to come with a single function, especially in organisms with long life spans like tortoises, wherein several different benefits to a trait may have additive net fitness effects.

Grazing pressure by tortoises and, to a lesser extent, land iguanas on *Opuntia* [was] likely an important evolutionary driver of the substantial variation observable in growth form of the cactus among islands. More specifically, islands lacking tortoises typically support *Opuntia* with a recumbent [horizontal] growth form, whereas islands with either tortoises, land iguanas, or both more typically harbor *Opuntia* with an arboreal form, which raises pads above the [average] grazing height of iguanas and tortoises.

Moreover, *Opuntia* cacti have sharp and penetrating spines on islands with tortoises and land iguanas, and only fuzzy bristles on islands lacking the two herbivores (e.g., Genovesa). This correlation is further evidence of a long-standing coevolutionary relationship between saddleback tortoises and *Opuntia* cacti.

The simplicity of arid ecosystems was part of the harsh conditions shaping saddleback tortoises. Given the paucity of native food plants, we infer that saddlebacks evolved their smaller overall size, compared to domed counterparts, as they adapted to the food scarcity of low islands.[22] In addition, competition was surely intense.[23] As noted by Fritts (1983), competitive interactions contribute to the selective advantage of long necks and limbs, the better to intimidate conspecifics and aggressively secure limited food items. These benefits, moreover, appear to have been advantageous enough on low islands that the saddle-shaped carapace evolved independently a number of times (see Marlow and Patton 1981). The saddleback tortoises of Pinzón came from domed ancestors, much like those of Española.

The absence of terrestrial predators and competitors also played a role in the evolution of the saddleback carapace. We know this because the saddlebacks cannot provide the same protection as domed carapaces. When a domed tortoise is threatened, it simply retracts its extremities inside the carapace and covers the openings with its thick-scaled legs. When a saddleback similarly retracts, it inevitably leaves large areas exposed and vulnerable. This shape could only have had a net advantage, we infer, in the absence of terrestrial predators.

Last, two prominent behavioral differences among tortoises correlate with their shell shapes. First, on higher islands with more moisture and vegetation close to the ground, dome-shaped tortoises evolved to be mostly grazers, feeding on low-lying grass and herbaceous plants. Because grasses and herbs are often

---

[22] Similarly, dome-shaped tortoises almost certainly evolved to be larger and more rounded than their continental ancestors. These traits are advantageous for retaining heat in the cool highlands, for storing calories and water for seasons of want, and for moving through dense brush in their habitats.

[23] This inference is indirectly supported by the research of Sulloway (2015) on another low island, South Plazas. Sulloway found that the demand for cactus pads and fruits from another herbivore, the Galápagos land iguana (*Conolophus subcristatus*), has prevented the successful growth of any juvenile *Opuntia* cacti since the 1950s.

abundant, especially in humid zones, domed tortoises are relatively gregarious, often gathering peacefully around water holes. In contrast, saddleback tortoises evolved to be mostly browsers, feeding on sparser vegetation above ground level, as facilitated by their greater reach and competitiveness. The benefits of competitive ability in these landscapes favored a generally solitary lifestyle among saddlebacks; only during El Niño events is there enough water for a temporary lowland water hole.

The second behavioral difference has been recognized since at least the 1800s. Back then, whalers noted, while stocking up on tortoises for food, that domed tortoises migrate down from the highlands for part of the year, especially females, who lay their eggs in the lowlands (Porter 1823, 66; also Heller 1903, 50). The trek is long—sometimes exceeding 10 km each way—and arduous, requiring climbs from sea level to 400 m or higher along age-old tortoise trails, dryly known as "tortoise highways." Migration studies in recent years by Steven Blake and colleagues (2012, 2013, 2015) reveal a complex migration pattern with the following characteristics:

1. Following hatching, small domed tortoises remain in the lowlands year-round, because they lack the energy reserves required for migration until they reach a large size.
2. Most of the largest domed tortoises, male and female, migrate to the highlands in the cool and dry months of the year (July to December), when forage is generally reliable.
3. Many of the largest tortoises then return to the lowlands during warm and wet months (January to June), when the lowland vegetation offers tender shoots and warm soil for nesting females. During the deluges of El Niño, a majority of the tortoises migrate to the lowlands.
4. Adult females lay their eggs in the lowlands during the warm and wet months, and the cycle begins anew. The incubation temperature determines the sex of progeny: eggs incubated over 29°C (84°F) develop into females; eggs incubated at or below 28°C (82.4°F) develop into males. (There are obviously profound implications here for the warming associated with climate change.)

In contrast, saddleback tortoises do not migrate, even when experimentally introduced to an island with a humid highland zone (Hunter et al. 2013). If we reasonably assume a fitness advantage to seasonal feeding on tender shoots in the lowlands, then migration is one more domed tortoise adaptation, having both a function and a history of selection for that function.

Tortoises of both morphologies exhibit the two forms of resilience described above: the older, general resilience that came with the ancestors at colonization, and new, specific resilience added during evolution in Galápagos. Their general resilience is already an impressive baseline. Their phenomenal adult size, their super-durable carapace (of whatever shape), and the strength that comes with these features all help with a range of everyday challenges, including plowing through dense vegetation. In addition, tortoises can hold and resorb from their bladders vast amounts of water, up to 40% of their body weight (Jørgensen 1998). This capacity allows them to "tank up" in times of plenty for later use in times of want—advantageous for the annual cycle of warm–wet and cool–dry seasons, and even more so for ENSO cycles. Moreover, tortoises store superabundant calories as fat, drawing upon those reserves later when browse and graze falter. All this "standard equipment" at the time of colonization allows tortoises today to survive for months without food and water, as summarized in the opening quote of the chapter. To these age-old capacities, evolution added the novel Galápagos-specific adaptations, including the migratory response of domed tortoises, and the special capacities of saddlebacks for reaching and defending *Opuntia* pads and fruits.

Thanks to the whole bundle of adaptations, forms of resilience, and absence of terrestrial predators and competitors, one can easily see how tortoises thrived for millennia in their island hideouts, reaching aggregate numbers of 200,000 or more before the advent of humans. As summarized in Figure 3.10, each of the "secret ingredients" played a role in shaping the tortoises to this place. With 15 unique species found in a tiny archipelago just larger than Delaware, they dramatically make the point about endemism. But Galápagos tortoises offer an additional key lesson, which leads now to the main hypothesis of this chapter.

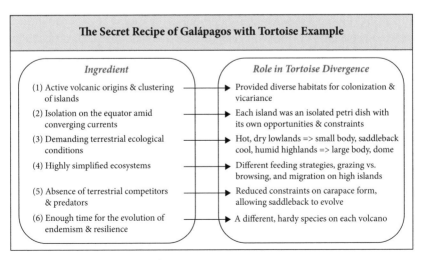

Figure 3.10. The secret recipe of Galápagos as it influenced the divergence of tortoise species.

Hypothesis

Much as tortoises have been shaped by the creative force of Galápagos environments, so may tortoises have played a role in shaping that environment, owing to their size, numbers, diet, and mobility. Elsewhere on Earth, other organisms have undertaken efforts at "niche construction" (Odling-Smee et al. 2003) via behaviors that improve habitats for their use, thereby enhancing their own survival and reproduction. Beavers are a familiar example, building dams of toppled vegetation and creating protective lakes as moats around their lodges. Tortoises, the largest herbivores in Galápagos, with a pre-settlement population roughly 10 times what it is today, qualify as the local megafauna, with a potentially "dramatic role in ecosystem function through herbivory, trampling, seed dispersal, and nutrient cycling" (Blake et al. 2012, 1961). In a simplified island ecosystem, going about their daily activities and constructing niches for themselves, what impact have tortoises had in shaping the broader communities and habitats of Galápagos? Building on work of other researchers, especially James Gibbs and colleagues (2008, 2010, 2014), here I offer a preliminary test of the hypothesis that *Galápagos tortoises are impactful "ecosystem engineers" through their wider, incidental effects on the distribution of resources and habitats in areas where they have lived.*[24]

Let's begin with the seed dispersal, a service Galápagos tortoises perform as a byproduct their herbivory and frugivory. Their capacity as seed dispersers for *Opuntia* cactus species was inferred years ago by Dawson (1966), and tested later by Gibbs and colleagues. One study (Gibbs et al. 2008) found that tortoises on Española inhibit regeneration near adult cacti, where they act mostly as consumers, but promote regeneration via droppings in areas away from adult cacti and woody plant cover (which impedes their movement). In a later study, Gibbs et al. (2010) found the cactus population to be increasing on Española, as a consequence of tortoise dispersals. They concluded that tortoise droppings play a crucial role, promoting a "scattered distribution in *Opuntia* cactus and its associated biota in the Galápagos Islands" (p. 208).

But do tortoises provide dispersal services for more than *Opuntia*? Stephen Blake et al. (2012) took up this question and found, in a painstaking analysis of seeds in 120 dung piles, that tortoises distribute the seeds of more than 45 plant species. No other endemic species in the archipelago disperses more than 13—the number carried by the small ground finch of the Darwin finch group (Heleno et al. 2011; see Chapter 6). Blake and colleagues also confirmed what Gibbs had found earlier, that tortoises carry seeds good distances—on regular jaunts of 100

[24] Ecosystem engineers are "organisms that directly or indirectly modulate the availability of resources to other species. . . . In so doing, they modify, maintain and create habitats" (Jones et al. 1996).

m (328 ft) or more from the parent plant and sometimes as far as 4,000 m (13,400 ft). "The dispersal distances we obtained dwarf by one to three orders of magnitude those for herbs dispersed by wind, ballistic, and non-specialized means" in similar habitats (Blake et al. 2012, 1969).

Galápagos tortoises' seed services have earned them the nickname, "gardeners of Galápagos." Moreover, an earlier study by Rick and Bowman (1961) suggested that tortoises, during digestion, might even facilitate germination of the seeds they carry. In experimental trials with the endemic Galápagos tomato, Rick and Bowman found tortoises to provide both an enzymatic treatment that breaks tomato seed dormancy during transit through the gut, and the obvious pulse of nutrients and moisture at the deposition site.[25] Alas, tortoise germination trials by Blake et al. (2012) on five other plant species, including endemic cactus species *Opuntia echios,* failed to show the same effect. So, tortoises may not be full-service cultivators, but they do transport seeds better than any other species, and they do provide soil conditioner as part of the "package deal."

Tortoises have also been shown to influence vegetation at a landscape scale, by trampling and eating back brush in favor of open pathways and "tortoise meadows" like those on Española Island discussed earlier. Pinta, too, formerly supported enough tortoises to feature an enduring mosaic of tortoise meadows and forest patches. But Pinta tortoise numbers were so reduced over the years by hunting and collecting that, by 1972, only the adult male called "Lonesome George" remained. He was removed to the Darwin Station for safekeeping and possible breeding (see Nicholls 2006)—alas, efforts encouraging George to mate did not succeed before he died in 2012, the last of the Pinta species.

Without tortoises, Pinta vegetation changed drastically, with uncontrolled expansion of woody plants, such that meadows all but disappeared. In 2010, the Galápagos National Park Service initiated an "ecological replacement" experiment: they repopulated the island with a diverse collection of 39 sterile tortoises of different origins, domed and saddleback, in an effort to recoup tortoise ecological functions. Using data from the vegetation-reduction action of these herbivores, Hunter and Gibbs (2014) simulated their long-term vegetation effects. In a series of 50-year simulations, at a modest density of 0.7 tortoises per hectare (0.28 per acre), they found that tortoise activity reduced woody

[25] Rick and Bowman (1961, 415) went on to suggest that "the giant tortoises . . . might play a role in the germination and dispersal of native tomato species" in Galápagos, but cautiously noted that their experimental feeding trials carried out with a pet tortoise in California "do not prove the case." After months of investigation, a literature review and interviews in Galápagos, I found no evidence supporting their suggestion that free-ranging tortoises are effective "in spreading the native tomatoes within islands" (p. 425). Mockingbirds eat tomato fruits (Gibson et al. 2020) and surely distribute seeds within islands, and rarely between them. But Rick and Bowman claim that "the period of bird digestion in general is too short to be effective" at promoting germination (p. 415). Further research is needed. Although there is no evidence for a tortoise–tomato mutualism in Galápagos, there is solid evidence for such a mutualism involving ebony and giant tortoises in Mauritius (Moolna 2007).

vegetation density enough to restore a meadow/forest mosaic pattern like the one that had been there before. Replacement tortoises, they concluded, could indeed perform the ecosystem engineering needed to maintain meadows on the island.

But we have yet to consider the most pervasive example of tortoise niche construction: wallows, which dot the landscape of high islands' humid zones (Figure 3.11A). A product of domed-tortoise activity, wallows are truly emblematic of their engineering role, as they not only create long-lasting water holes by compaction but also shape surrounding vegetation, often to the tortoises' advantage. For example, the fruits of the poison apple tree (*Hippomane mancinella;* in Spanish, *manzanillo,* "little apple") whose milky sap is toxic to humans, are an important food source for tortoises, traveling with them for days after ingestion. "Normally an inhabitant of the coastal zone and arid lowlands" (McMullen 1999, 79), the tree also occurs in humid highlands in "dense stands" around many of the wallows (Hamann 1979, 70)—a 'tortoise-engineered' food supply, perpetually close at hand.

A related finding, with an impressive lesson about tortoise engineering, comes from recent analysis of sediment cores from the highlands of Santa Cruz (Froyd et al. 2014; Nogué et al. 2017, 6).[26] Today, nobody remembers tortoises grazing or even passing through at the 700 m (2,300 ft) level on Santa Cruz. Blake's aforementioned studies of tortoise migration track them only to around 400 m (1,300 ft). But analysis of cores from highland peat bogs reveals "dung-affiliated" fungal spores in the 700 m range of altitude, confirming a dense pre-settlement tortoise presence there. Moreover, the cores reveal that today's sphagnum-filled bogs—see Figure 3.11B for an example—once had high densities of open-water aquatic plants, including several that are presently rare or extinct. The aquatic plants persisted, cores indicate, until several hundred years ago, when human hunting began taking its toll. As tortoise numbers dwindled, the freshwater pools at altitude converted into the sphagnum bogs seen today. The implication is that tortoises were keepers of many open, shallow, freshwater pools over a great range of elevations—at least as high as 700 m—much in the way that they keep wallows of open water at lower elevations today.

This section concludes with a final suggestion about the muddy niches tortoises make for themselves. Wallowing in pools at all elevations has advantages for *Chelonoidis* tortoises in terms of avoiding skin parasites, mosquito and fly evasion, and thermoregulation, as also reported for Aldabra giant tortoises of the Seychelles Islands (Grubb 1971). But there are special benefits to highland wallows during dry periods in Galápagos, beginning with the obvious

---

[26] Not surprisingly, there used to be highland peat bogs on other high islands—for example, Pinta—but they have been destroyed by goats (Coffey et al. 2012, 61).

Figure 3.11. Niche construction by Galápagos tortoises. *A*, Tortoises in a wallow at about 800 m altitude (2,625 ft) in the caldera of Alcedo Volcano, Isabela. These shallow watering holes are scattered over the humid zones of islands with domed tortoises. In pre-settlement Galápagos, when tortoises were at least 10 times more abundant than today, tortoises maintained wallows over a much wider range of altitudes. Wallows are often rimmed by trees, including the manzanillo or "poison

Figure 3.11. Continued
apple" tree, a common food for tortoises (but toxic to people). *B,* Psidium Bog at roughly the same altitude on Santa Cruz Island, one of the former wallows, shown here during a dry spell (whitish area and path at center). This bog provided core evidence for open-water wallows at altitude when tortoises were at pre-settlement abundance on Santa Cruz. The bog is flanked by a dying forest of the invasive red quinine tree, victim of a fungal infection (see Chapter 4). *Sources: A,* Tui de Roy/ Minden Pictures; *B,* Photo by Edward A. D. Mitchell, University of Neuchâtel, from Fournier et al. (2016, 534), used by permission.

---

ones, food and drink. Added to these is an additional advantage in this restless volcanic archipelago: wallows afford fire and heat protection, especially under La Niña conditions when fires easily spread. In the shadows of active volcanos, there were untold numbers of lava flows and natural fires before human settlement. In one noteworthy case, lava flows appear to have "covered most of the prime tortoise habitat with meters of hot pumice, [and] caused a dramatic population reduction" of *C. vandenburghi* on Alcedo Volcano of Isabela Island. By the genetic evidence, today's Alcedo tortoises are derived from one lineage that survived (Beheregaray et al. 2003, 75).

Of course, the frequency and scope of such events are hard to estimate today,[27] and it is impossible to know that wallowing saved the day. The point is simply that fire and heat add an additional potential fitness benefit to wallows and their "biocompaction" (technical term for aggregating and trampling), especially during the droughts of La Nina events. This added benefit could explain the observation that "even when dry, [wallows] appear to have some special significance for the tortoises" (Rodhouse et al. 1975, 302). It's possible that wallows were part of a domed tortoise adaptation to ENSO from the start, possibly with migration being another part, although the annual nature of the trek today doesn't look that way. But it seems highly likely that wallows offered protection during La Niña— from drought, heat, and fire—as an added benefit, making them at least an ENSO *exaptation.* Tortoises are clearly not the only species to show candidate ENSO adaptations and exaptations (see Chapters 5 to 7; also Curran and Leighton 2000), but they are certainly one of the more picturesque. In their wallows, if I am right, tortoises really are slow but sure.

Taken as a whole, the processes of this section add up to a remarkable case: in the course of everyday behavior, Galápagos tortoises reshape the habitats they live in—especially, we infer, at their pre-settlement densities. One group of

---

[27] But some idea comes from the analysis of bog cores in the Santa Cruz highlands by Coffey et al. (2012). Carefully sifting through core debris and small fossils, they found charcoal evidence of at least six fires before the year 1550 (first record of human visitors was 1535—see Chapter 4), including three between 1100 and 1550, and three with evidence of concurrent volcanic activity.

tortoise researchers put it well: "Because disruptions to ecosystem interactions can be so detrimental to biological diversity, the restoration of [the tortoises] and their associated functions should be a conservation priority" today (Hunter et al. 2013, 702), so let us turn to that topic.

Conservation Implications

Tortoise adaptations to the diverse habitats of Galápagos bring us back to a central theme from Chapter 2: human actions can quickly turn adaptations into vulnerabilities. Vulnerability was perhaps inevitable in the case of tortoises, because so many of their features made them a prize capture for Galápagos visitors over the centuries. The muscles ("meat") that provide locomotion for such large organisms, the fatty tissue ("oil") that prepares them for lean times, and the enormous bladder with storage capacity ("water") were all alimentary delights to early mariners, who harvested tortoises by the score. Better still, from the mariners' perspective, tortoises in the lowlands were relatively small and easy to carry, either because they were saddlebacks or because they were immature domed tortoises or females during migration. When tortoises were not needed for food or drink, the oil from their fat lit the lamps of Quito and other mainland towns in the 1800s. Figure 3.12A provides a ghastly glimpse of a harvest, probably for oil, on Isabela.

But these were just the first of the vulnerabilities that came with human activity. Tortoises also existed in greatest numbers and highest densities on islands with extensive humid zones, the very areas most conducive to human settlement and agricultural land use. No doubt, many were hunted during the early years by farmers or by humans' escaped pets. Later on, farming created other problems including, fences, hedgerows, walls, and other obstacles. Untold numbers of tortoises have thus been blocked from migration, being fenced out and/or trapped for long periods in human-modified landscapes. Figure 3.12B offers the example of a male tortoise, monitored by the migration project, who was blockaded for nearly a year by fences. Blake et al. (2015, 148–49) conclude their initial analysis of tortoise migration difficulties with three recommendations:

1. "Removal and/or reduction of barriers to migration," including clearing pathways through impenetrable thickets of invasive blackberry and elephant grass.
2. "Mitigation of impacts due to roads," and preventing further road construction in highland areas.
3. "Maintenance of high-quality habitat at both ends of the migration," including crop and pasture options that allow tortoise passage.

Figure 3.12. Human impact on Galápagos tortoises, then and now. *A,* Remains of hundreds of slaughtered tortoises on Sierra Negra Volcano, Isabela, ca. 1903. Photo by R.H. Beck; original caption read "Work of the oil hunters" (Pritchard 1996, 76). *B,* Adult male Santa Cruz tortoise, *C. porteri,* "trapped inside an upland farm for a year by nearly impregnable fencing and was thus unable to complete his seasonal migration to the lowlands" (Blake et al. 2015, 148). *Sources*: *A,* Public domain, Library of Congress, https://www.loc.gov/item/99472326/ *B,* Photo courtesy of Stephen Blake, Galápagos Tortoise Movement Ecology Programme.

Yet, more barriers to tortoise migration have been documented, partly in response to farmers' perceptions of tortoise-induced damage to their fields (Benitez-Capistros et al. 2018).

The SES perspective is again helpful in framing these aspects of human impact. Tortoises went from *being* the resource, during the days of exploitation by mariners, to *being impacted by use of the resource,* namely the use of agricultural land by farmers with a damaging byproduct in introduced species. External drivers, especially market demand for tortoise oil, were key to the transition between these phases.[28] The great decline in tortoise numbers was thus a product of two changing SES relationships: not until very recent conservation efforts did tortoises get a breather. As a result, noted Cayot (2008, 40), "Giant tortoises were among the most devastated of all species. . . . Only the endemic rice rats (Tribe Oryzomyini) were hit harder, with the majority of species now extinct."

Today the tortoise is iconic, but not simply of endemism produced by the creative force of Galápagos. It is iconic of human impact on endemic species, and of the associated need for human intervention to restore and maintain their populations (see Hennessy 2019). Time and again, people have had to step in to keep alive the varied tortoise species of the archipelago. Here are some of the main restoration initiatives, spanning decades, run by the Galápagos National Park Service (GNPS) and its collaborators:

1. *The Pinzón "head start" program* began in 1965, when only mature tortoises existed on that island, invasive rats having killed all hatchlings for decades. Eggs were transferred to the tortoise breeding center on Santa Cruz, incubated, and reared to age five, when they were considered rat proof. With this head start, the first 20 were repatriated to Pinzón in 1970. Rats were eradicated in the 1980s and now over 500 tortoises have been repatriated.

2. *Española repatriation program,* "one of the most successful, but least heralded, species reintroduction efforts ever attempted" (Milinkovitch et al. 2004, 341), began in the mid-1960s when the surviving 14 *C. hoodensis* tortoises (two males and twelve females) were transferred to the tortoise breeding center on Santa Cruz. The first hatchlings emerged in 1971, were repatriated in 1975, and by 1978 invasive goats were eradicated from Española by GNPS. In 1977, the famously prolific male, "Diego" from the San Diego Zoo, was added to the breeding center population, and in a few years nearly 2,000 tortoises were repatriated. A sequence of assessments confirmed the continuing success of the effort (for example, Gibbs et al. 2014), and showed that reproduction was again successful in free-ranging populations. In 2020, the Galápagos National Park Directorate (GNPD) announced the successful completion of the project, and returned all the remaining tortoises—including Diego—to Española.

3. *Pinta repatriation program:* As noted earlier, in 2010, 39 sterilized tortoises of various origins (domed and saddleback) were released on Pinta, where

---

[28] A turning point was the discovery in 1854 that an alternative, kerosene, could be made from coal and then, in the 1860s, from petroleum. Because kerosene burned with a cleaner, brighter flame, the demand for tortoise oil plummeted, as did the demand for oil from whale blubber. Whaling near Galápagos (and elsewhere) declined sharply thereafter, greatly reducing the hunting of tortoises.

no tortoises had existed since 1971, to serve as ecosystem engineers (Hunter et al. 2013). Their ecological impact is believed sufficient to restore the mosaic of habitats on the island and to provide seed dispersal services for *Opuntias* and other plants.

These efforts are impressive for the amount of care and dedication they have taken to keep tortoise species alive (Pinzón and Española) and to preserve habitats and ecosystems that tortoises maintained for millennia (Pinta). The Santa Cruz breeding center, and more recent satellite centers on San Cristóbal and Isabela, were and are crucial to the restoration of tortoises and their eco-system services on impacted islands. They also embody the argument that the very survival and reproduction of endemic organisms are now directly affected by human activity.

A bigger, more ambitious program is the Giant Tortoise Restoration Initiative (GTRI; see Tapia et al. 2017). Organized in 2014 by the GNPD in collaboration with Galápagos Conservancy and other organizations, GTRI is a 10-year ef-fort to restore tortoise populations across Galápagos and to "repair and balance ecosystems through the beneficial actions of these apex herbivores" (Galápagos Conservancy 2018). By 2020, the initiative already reached an impressive number of benchmarks, including (a) repopulation of Santa Fé with almost 400 juvenile tortoises of Española stock (closest relatives of the long-extinct Santa Fé tortoise); (b) a captive breeding program for the newly described eastern Santa Cruz species, *C. donfaustoi* (Figure 3.1); (c) repatriation of 163 juvenile *C. guntheri* from captive breeding to their home range on Sierra Negra Volcano; and (d) expeditions to Wolf Volcano looking for descendants of cast-off tortoises that mariners had captured elsewhere (as in Figure 3.7).[29]

Amazingly, in the last of these efforts, researchers found surviving descendants of Pinta and Floreana tortoises on the slopes of Wolf Volcano, quickly prompting breeding efforts to "re-produce" close genetic relatives of their respective spe-cies. These efforts, beyond our collective imagination even just a few years ago, have already produced the first hatchings of largely Floreana ancestry. Project managers hope the effort will eventually allow repopulation of Floreana with tortoises of a "mostly Floreana genome" (Poulakakis et al. 2008; Miller et al. 2017). GTRI also envisions a captive breeding program to "re-produce" Pinta tortoises, of which Lonesome George was the last survivor. Pinta may well take longer than the Floreana effort, given that only 17 hybrids were found on Wolf Volcano (Edwards et al. 2013), but efforts continue to identify more breeding stock on Wolf and elsewhere.

---

[29] For more on all aspects of GTRI, see https://www.Galápagos.org/conservation/our-work/tortoise-restoration/

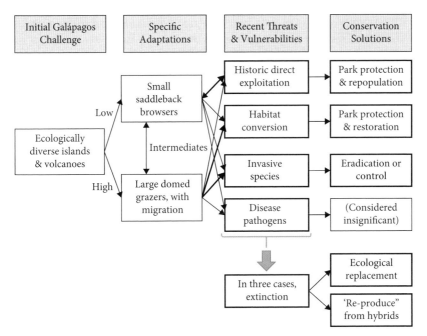

Figure 3.13. Summary of adaptations, threats, and vulnerabilities affecting the giant tortoises of Galápagos, with key conservation efforts, recent or currently underway. Especially important links are shown as bold arrows; boxes caused or affected by human activity also have bold edges. Three major threats are shown: direct harvest by hunters and mariners, which especially affected smaller, more portable saddlebacks; habitat conversion by early farmers and ranchers, which heavily impacted the migration routes and highland areas used by domed tortoises; and the impact of invasive species, especially mammals (e.g., goats, dogs, and rats) as predators and competitors. A fourth—disease pathogens—is not yet a major threat but could easily become one via introductions from outside. The cumulative effect of the threats and vulnerabilities (*bracket and blue arrow*) has produced the extinction of three species since 1535. Experiments in ecological replacement are underway for the extinct populations on Santa Fé and Pinta Islands, as are efforts to "re-produce" Floreana tortoises through controlled breeding of translocated hybrids from Isabela Island.

The GTRI efforts and their predecessors, included in this chapter's summary (Figure 3.13), are impressive conservation milestones, celebrated in Galápagos and beyond. Some of them are held up today as examples of efforts to "resurrect" lost species (Poulakakis et al. 2008, 15467), and to allow genes of the late Lonesome George to "live on" (Miller et al. 2017; see critique in Hennessy 2015). But the key point is that the Galápagos giant tortoises, long iconic of the unique biodiversity of the archipelago, have recently become iconic of human impact—an impact that now reaches even to their reproductive success.

The overall goal is to build back independent tortoise populations in the islands, free of reliance on people, but are we sure we can now prevent the destructive forces, old and new, that might again menace these populations? Can we avert a vicious cycle of renewed threats and never-ending interventions? The giant tortoises may well be the "gardeners" of Galápagos, but we must guard against setting ourselves up to become their keepers. The goal must be a self-sustaining future for the species that gave Galápagos its name.

4

Galápagos Derailed

What havoc the introduction of any new beast of prey must cause in a country, before the instincts of the aborigines become adapted to the stranger's craft or power.

Charles Darwin (*Narrative of the Surveying Voyages of His Majesty's Ships Adventure and Beagle*)[1]

Nearly opposite in scale to the lumbering giant tortoise, but no less important for the story it tells, is a nimble, diminutive ground bird: the Galápagos rail, also called Darwin's rail, and in Spanish, *el pachay* (Figure 4.1). This species of rail, *Laterallus spilonota*—from Greek, aptly, for rail with "back spots on the side"—is endemic to Galápagos. Only 15 cm (6 in.) long, it has been reported from all eight islands that have a highland humid zone, and, in the pre-settlement period, it was reported from coastal mangrove forests of many of the same islands. The rail's preferred habitat is moist, dense, and dark herbaceous cover, so it thrives in highland areas where native ferns and sedges form a thicket. Sadly, rails are no longer found in mangrove areas, probably because of nearby human activity, especially fishing, and the dogs, cats, and rats humans have introduced.

Like many rail species, the Galápagos rail is a diurnal insectivore, complete with a long probing beak. Spending most of its life on the ground and nesting there, it is also territorial, such that mated pairs defend a foraging patch against others of their species. Like its mainland relatives, it remains secretive and wary of its predators, the Galápagos hawk, the short-eared owl, and the barn owl. Since all three are aerial foragers, *el pachay* is understandably elusive when approached from above and is especially good at hiding in vegetation and darkness. Researchers on their hands and knees, however—at a terrestrial predator's height—find Galápagos rails to be "exceptionally tame" and easy to follow along their well-traveled "runways" in the vegetation (Franklin 1979, 202). Before human settlement, the rail had no terrestrial predators.

---

[1] Note the use of "adapted" in this quotation, in a fully evolutionary sense. The quotation was written after Darwin had discovered evolution but had kept it mostly secret, 20 years before the publication of *On the Origin of Species*.

Exuberant Life. William H. Durham, Oxford University Press (2021). © Oxford University Press.
DOI: 10.1093/oso/9780197531518.003.0004

Figure 4.1. Galápagos rail of Santa Cruz Island. With its big feet, this "water hen" occurred in good numbers in highland Floreana and Santiago, Darwin noted during his visit (1835 in Barlow 1963, 264). Rails have lived on all eight islands with a humid zone; today they are feared extinct on San Cristóbal and Floreana. *Source*: Photo courtesy of Yanick Saindon.

Galápagos rails are members of a global bird family, the Rallidae, which has over 130 species of rails, crakes, coots, and gallinules. However, like nearly all other species of the family, Galápagos rails are reluctant flyers. Ripley's synthesis of "rails of the world" (1976, 630–31) put it well:

> All observers note the difficulty with which the whole rail family flies. . . . It is quite true that rails prefer not to take wing if they can avoid it, running off instead. If they do fly for a short distance, the legs are usually dangled downward. . . . However, once on the wing, as in migration, rails tuck their legs out evenly behind . . . and seem able to fly as well as any other bird. It is this anomaly of being a poor, reluctant flier in day-to-day activity and yet being able to fly well in migratory or dispersal flight that is one of the most provocative aspects of the behavior of the entire rail family.

True to form, Galápagos rails fly when startled, with the legs hanging down, and their flight appears weak and "labored" (Franklin et al. 1979, 202). They readily get themselves to a fence post or low branch, but they clearly prefer just to run.[2] Not surprisingly, full-grown adult male wings average just 68 mm in length (2.6 in.; Livezey 2003, 624), comparable to the wings of an adult yellow warbler, which is a tiny Galápagos denizen of less than one-third the weight of the rails.

With respect to wing length, the Galápagos rail is intermediate between two other closely related members of the genus *Laterallus* (using average sizes for adult male birds, from Livezey 2003, 624). First, there's the super-secretive, migratory black rail of eastern North America (*Laterallus jamaicensis,* subspecies *jamaicensis*), a smaller, sparrow-sized bird (10–12 cm long, 4–5 in.) that weighs only 35 g (1.2 oz), 17% less than *el pachay*, and has a wing length 8.8% longer (74 mm, 2.9 in.). This little rail is otherwise so similar in appearance that Leck (1980), among others, proposed that Galápagos rails originated as errant black rails from migration to South America (where other black rail subspecies live today) or the Caribbean. Second, there's the Inaccessible Island rail (*Laterallus rogersi*) of the same overall size (13–15 cm, 5–6 in.) and weight (42 g; 1.5 oz) as the Galápagos rail, but with wings 17.6% shorter (56 mm, 2.2 in.; see Figure 4.2). With wings that small, plus plumage "decomposed and hairlike" (Taylor and Sharpe 2020), it is a completely *flightless* rail, having adapted to its isolated mid-Atlantic island home within the last 1.5 MY (Stervander et al. 2019). The feathers of the Galápagos rail are sturdy, if small. Taken together, the data suggest that the Galápagos rail qualifies as an "intermediate flyer," with a flight capability somewhere between the long-distance migrant and the flightless relative.

Evolution of the Galápagos Rail

So, how far can *el pachay* fly? The question has importance well beyond its theoretical implications, tracing back to the origin of the species. García-Ramirez et al. (2014) showed that the Rallidae are all related, in one "deep global evolutionary radiation" that runs back 30 to 40 MY. Their radiation has the same general form as that of tortoises, branching from one common source, but the Rallidae radiation is ancient and universal, involving dozens of species across all major continents. Second, Rosindell et al. (2017) estimated that ancestors of the Galápagos rails separated from their closest mainland relatives around 3.27 MYA, when we might surmise that some number of them flew, with help from trade winds, out to the islands of that time. Using again the reconstruction of

---

[2] On Santa Cruz, I've seen them fly, when startled, 120 feet or so (37 m) to a nearby branch. In 150 days of study, Franklin et al. (1979) reported "13 observations of [rails] flying short distances; 6 of these were of birds flushed by the observer in open meadows. These individuals flew an average of 8 m (1–30 m)."

Figure 4.2. Piddly wings of *Laterallus rogersi*, the flightless rail of Inaccessible Island (a mid-Atlantic overseas territory of the United Kingdom). The wings of this species are the shortest of its genus, averaging 56 mm (2.2 in.) in adult males (Livezey 2003, 624). The Galápagos rail's wings are intermediate, averaging 68 mm in adult males, just 6 mm (0.24 in.) shorter than those of the migratory black rail of North America (*Laterallus jamaicensis*). *Source*: "Inaccessible Island Flightless Rail with Piddly Wings" copyright by Brian Gratwicke, cropping added, used under Creative Commons License 2.0 (from https://www.flickr.com/photos/briangratwicke/6954261774).

Poulakakis et al. (2012 and 2020; see Online Appendix B), the large, centrally located proto-island (that later became Española and San Cristóbal) likely offered the first humid zone for colonizers. Thus colonists flew in, evolution went to work, and many generations later the result was . . . the same species of intermediate flyer on all eight islands with extensive humid zones (San Cristóbal, Santa Cruz, Santiago, Floreana, Isabela, Fernandina, Pinta, and Pinzón).

The Galápagos rail's evolution is all the more interesting because similar evolutionary chronologies have been proposed for rails and giant tortoises (3.27 MY vs. 3.20 MY for tortoises), making available a similar sequence of emerging islands for colonization, whether by dispersal or by vicariance. Yet the outcome is very different: 15 species with different carapaces in the case of the tortoises, and just one species and not even any discernable subspecies in the case of the rail (e.g., Hill 2020).[3] It's even more curious because generation times, which

---

[3] Here and in later chapters the term "subspecies" is taken to refer to populations within a species that are recognizably distinct from place to place—genetically, morphologically, or both—but still capable of successful, fertile mating.

inversely affect the rate of evolutionary change, are so very different: conservatively estimated at 60 years for giant tortoises (meaning slow change) and 3.7 years for the endemic rail (IUCN 2018a). It all adds up to another puzzling question: why the same rail species everywhere it occurs, and no subspecies?

There are several possibilities. First, the Galápagos rail has never been subject to a systematic phylogenetic analysis[4]—effectively, we are working from visual similarities between populations. Could it be that there is simply more inter-island differentiation than meets the eye? Or could it be that the rail's humid zone habitat is so similar from island to island that evolution has favored similarity, not difference, between locations? Both of these suggestions seem unlikely in an archipelago known, since Darwin, for a strong "creative force" that typically shapes visible, and often visibly different, adaptations on each island. Second, could the estimated date of the rail's separation from mainland relatives (3.27 MY) be too early, overstating the time evolution has had to produce inter-island differences like those in tortoises? A recent review across the Rallidae suggests such a possibility (García-Ramírez et al. 2020). But even if the date is off by a million years or more, the harsh conditions of the islands plus their famous isolation by distance, wind, and currents would surely be enough for evolution to shape discernable differences among them. As is described later, evolution in other land birds has worked wonders in a million years.

Speaking of isolation, it provides a third possible explanation for one rail species and no subspecies across the archipelago. Could it be that, for rails, the islands are not as isolated as they look? Isolation, after all, is a fundamental requirement for diversification as discussed in Chapter 3. Despite a reputation for "reduced flight ability" (Stervander 2019, 96), maybe rails fly at least occasionally among different islands? Are they "loath to fly, [but] difficult to stop once in the air," as Ripley (1976, 628) put it, until they land upon a different suitable surface? Despite appearances, Galápagos rails may best be described as "near flightless"— a term helpful here because "near" indicates that the birds do still fly; they are not flightless, at least not yet. All considered, this explanation is the most likely of the alternatives: Galápagos rails remain one species because flight keeps island populations from persistent isolation. This matter certainly warrants more research, including GPS dataloggers, birdcall monitoring, and other tracking devices.

Meanwhile, one can fairly ask, if rails do make inter-island flights at times, why has no one observed and reported them? There may be a simple answer: other species of closely related *Laterallus,* including the diurnal eastern black rail, have been recorded flying substantial distances, but only at night, as

---

[4] Happily, this situation changed as this book was in production with the phylogeny from comparative genetics described by Chaves et al. (2020). We'll consider its implications below.

reflects their general predator wariness (Evans and Rosenberg 2000, 155). For that matter, aerial predator avoidance may also explain the rails' overall reluctance to undertake daytime flight as part of the same pattern.[5] But if Galápagos rails do fly at consistent low frequencies between islands, probably in small numbers, we might expect to see sudden, new appearances at times, or reappearances on islands with local extinction. Data are few, but we do have examples of each. For example, there is the first-ever record of Galápagos rails in the humid zone of Pinzón, following a recent rat eradication effort there.[6] And as is shown below, rails have also made sudden reappearances after they appear extinct from habitat loss. It's early to say for sure, but it does fit with Ripley's general pattern: rails are reluctant to take to the air, but, once airborne, are capable of substantial flight.[7] Time will tell, but repeated inter-island flight would succinctly explain the rails' remaining a single species since colonization.

There's one further relevant difference between tortoise and rail evolution in the archipelago. Tortoises evolved the ability to live on low islands that lack a humid zone, and rails never did. Crucial to the difference is the fact that tortoises arrived in Galápagos with unusually great general resilience, including their large water- and fat-storage capacities, which enabled them to survive not only in moist highlands, like the rail, but also in the harshest of low island environments, where they went on to evolve specific dry zone adaptations. On their arrival, the diminutive rail ancestors certainly had much less dry zone resilience: they evidently survived then—and now—only in humid zones. Low elevations have a depauperate insect fauna, much less dense vegetation, and ready surveillance by aerial predators, including the Galápagos hawk.

So, no wonder the Galápagos rail shows "close association with the dense moist zone vegetation" (Franklin et al. 1979, 203), where its insect prey abounds. No wonder it has the lifestyle of furtive "skulking" (keeping out of sight) in deep cover, while protecting its foraging area. And no wonder the rail is found today only on islands with an intact layer of deep cover. Rails are highly vulnerable to aerial predation: dense ground vegetation is their first line of defense.

---

[5] The rail's former use of mangrove habitat, which is impervious to large aerial predators, may have been part of the same pattern. It's even possible that the use of mangrove habitat may largely have been diurnal, with rails flying in from highlands to forage.

[6] With a maximum height of 458 m, little Pinzón Island (only 18.2 square km or 7.0 square miles) also has a small humid zone (roughly 0.5 square km, 0.2 square miles), where Galápagos rails have been seen since the eradication of invasive rats in 2013. This is the first report of rails on the island (Rueda et al. 2019, 54; James Gibbs email of August 16, 2020). See also https://www.Galápagos.org/conservation/our-work/ecosystem-restoration/rat-eradication/

[7] Ripley (1976, 634) described a remarkably parallel, but recent, rail colonization: "The little-known paint-billed crake, *Porzana erythrops*, of mainland South America . . . is found in reedbeds, swampy areas, humid woodland and on farm edges. . . . It is a poor flier that has to be forcibly flushed, fluttering in flight on rapidly beating wings, legs dangling. . . . [Nevertheless, in 1962 ornithologist R. Bowman was first to spot] the paint-billed crake breeding on the farmland edges of Santa Cruz Island in the Galápagos," where it co-occurs to this day with *el pachay*.

Before proceeding, a brief update is warranted, prompted by the timely phylogeny of Chaves at al. (2020) as this book was in production. From detailed analysis of modern and century-old genetic samples (the latter from the 1905–06 California Academy Expedition to Galápagos), Chaves and co-workers found (1) that the Galápagos rail's closest living relative is indeed the black rail, which confirms our understanding of their similarities; (2) that both of these *Laterallus* species, plus the Inaccessible Island rail, are part of the same small clade that diverged in the last 6.5 MY or so, confirming that the latter is the most different of the three; and (3) that *el pachay* split off from its common ancestor with the black rail about 1.2 MYA, presumably coming directly out to Galápagos from South America, meaning that its arrival was indeed more recent than 3.27 MYA as discussed earlier. It is not so surprising, then, that the rails have not differentiated into as many species as the giant tortoises, and the finding means that Darwin's finches are a more apt comparison, as we'll see in a later chapter (also Chaves et al. 2020, 10). But for us, it's especially interesting that with a full 1.2 MY to differentiate, the genetic diversity among rails from different islands is so low that "there was no appreciable phylogenetic signal" between them (in mitochondrial DNA; Chaves et al. 2020, 5). As they note (pp. 10–11),

> "These results were contrary to our prediction that [the] Galápagos rail should show higher levels of genetic structure not only based on its limited dispersal ability, but also given its habitat specialization. . . It is possible then that the rails performed seasonal (or random) elevational migrations toward the coast and thus increase the chance for between-island connectivity. . . A degree of swimming capacity has been reported for this species [citing Franklin et al. 1979], which could be relevant for movements between islands, but alternative means of dispersal (i.e., rafting and nocturnal flights) remain speculative." (Brackets added)

I suggest here that nocturnal flights are not speculative, but that they are both (a) expected from comparison with the closely related black rail species—a confirmed nocturnal migrant—and (b) fully testable. Despite the reputation of the rails, the hypothesis that best fits our current data, and that requires future testing, is that Galápagos rails continue to fly by night on a local scale. . . between islands.[8]

---

[8] The rails' low genetic diversity surely reflects an initial founder effect (i.e., colonization from a few individuals) plus continuing movement between islands, or gene flow (found earlier for tree finches). In addition, as Chaves et al. (2010, 11) note, there will be further reduction of diversity from repeated population bottlenecks. When rails die back on a given island from anthropogenic influences (especially from invasive species as we'll see below), any population rebounds will be from a small number of individuals and thus a fraction of the local ancestral diversity. Reduced genetic diversity can hamper the long-term viability of a population, and is thus itself a serious conservation issue (as we'll see again in penguins, for example).

Anthropogenic Vulnerability

As other analysts have noted, several characteristics beyond wing size set up the whole family of rails—including the Galápagos rail—for the repeated evolution of near- or full-flightlessness on islands. First, the birds spend most of their lives on the ground, including foraging, resting, and nesting activities. Second, in many species, "The young are covered with down [not feathers] and are capable of running one day after hatching, begin to feed themselves on the third day, and are self-sufficient in picking up food by the seventh day, and yet require weeks to develop substantial wings [and] flight feathers" (Steadman 2006, 298). Indeed, the sternum (breast bone where flight muscles attach) does not fully ossify for weeks, such that all rails, even continental ones, are flightless from hatching to near-adult size.[9] There are also advantages to a large adult body size, related to territorial defense of foraging area and probably also to thermoregulation in cool and damp habitats as on high islands in Galápagos. When evolution favors both at once—reduced flight apparatus *and* a large body—in a ground-foraging bird, it is easy to see how flightless and near-flightless lifestyles emerge. As a result, "more species have become flightless in rails than in any other family of birds" (Steadman 2006, 296). Island rails are thus considered a classic case of *convergent evolution*: the independent evolutionary emergence of parallel features in ecologically similar but geographically dispersed locations. Steadman (2006) reviewed evidence that as many as 1,500 rail species on Pacific islands had parallel adaptations.

Flightlessness entails additional vulnerability when humans come on the scene. "Flightless rails were often the largest, most conspicuous animals on remote oceanic isles and they succumbed quickly to predation by hungry humans and the dogs, pigs, and rats that accompanied them (Steadman 2006). Losses were greatest in the Pacific, where flightless rail species may have numbered in the hundreds" (Kirchman 2012, 57). These rails were part of an "unprecedented avian extinction event," where between 1,000 and 2,000 species, "most of them rails, were lost on tropical Pacific islands after the arrival of humans" (Donlan et al. 2007, 521). But what of the Galápagos rail? It's not flightless but near-flightless, with a heavier body and shorter wings than frequent flyers. Plus, it's ground nesting, insectivorous, and similarly vulnerable to "hungry humans and the dogs, pigs, and rats that accompanied them." Yet the rail lives on in Galápagos. So, what's different?

---

[9] Rail flightlessness on islands is likely a case of "neoteny"—the evolutionary retention of juvenile features in adulthood. Absent terrestrial predators, there is little benefit to flight and, instead, advantages to saving energy with prolonged chicklike features: reductions in flight muscles, wing size, and flight-related skeleton (Livezey 2003).

It's surely pertinent that Galápagos has only recently been settled by humans and their invasive companions—less than 200 years, compared to 2,000+ years for many other Pacific islands. And surely their hypothetical inter-island flight capacity, if confirmed, must be part of the answer. But we also know from the two islands where rails are extinct, or nearly so—Floreana (Rosenberg 1990) and San Cristóbal (IUCN 2018a)[10]—that Galápagos rails have been seriously threatened these last 200 years. How have they held on in most of the islands? An historical perspective will help us answer this question.

Human History of Galápagos

Human history in Galápagos, like the natural history of many species, has presented its fair share of challenges and adjustments as the islands have become more integrated into the "world system" of commerce, trade, and human activity. Fortunately for our purposes, the grand tour of that history has already been conducted by several authors,[11] allowing this text to focus on a selected sample of topics. The pertinent background is readily summarized:

1. The first recorded human visit to Galápagos was accidental: in 1535, on a papal mission to mediate a colonial dispute in Peru, the Bishop of Panama, Tomás de Berlanga, and his shipmates were becalmed in the Pacific and drifted out to Galápagos. There, he reported, "It is as though God had sometime showered stones; and the earth there is worthless, because it has not even the power of raising a little grass." (from his 1535 letter reprinted in Benz 2000, 31.) Desperate for water, they explored islands that Woram (2005, 9) surmised were, first, Española—a low, dry island—and later Floreana, where they did find a limited amount of fresh water, enabling them eventually to reach Peru.

2. 1600 to 1790 was the "Buccaneer Period." One such buccaneer was Ambrose Cowley, whose 1684 map gave major islands English names that were adopted for hundreds of years (e.g., Chatham, Charles, James, Albemarle, Narborough, etc.). Captain John Cook also visited "one of the islands the Spanish call enchanted islands, saying they are but shadows and

---

[10] Happily, the rail survives on San Cristóbal, in certain patches (personal observation, November 2019).

[11] For more detailed historical accounts of Galápagos, see Hickman (1985), Lundh (1999), Grenier (2000, 2012), Idrovo (2005), Woram (2005), Ospina (2006), Watkins and Oxford (2009), Epler (2013), and Maldonado and Llerena (2018). Reports of early explorers are given in Van Denburgh (1914, specific to tortoises), Rose (1924), and Larson (2001).

no real islands. Here being great plenty of provisions, as fish, sea and [land] tortoises, some of which weighed at least 200 pound weight, which are excellent good food." (from a 1684 journal entry reprinted in Benz 2000, 32.)

3. 1790 to 1870 marked the "Whaling Period," when an estimated 700 whaling ships came to Galápagos. Whaling started when the British Royal Navy lent an experienced Captain, James Colnett, to a private whaling company to explore the Pacific, because the Atlantic had been overharvested. The company was not disappointed: Colnett observed great lines of whales moving westward "as if they were in haste to reach the Galápagos." (from a 1792 journal entry reprinted in Benz 2000, 33.)

4. British whalers worked the seas around Galápagos until the War of 1812, when half of them were captured in 1813 by the surprise tactics of U.S. Captain David Porter and the *USS Essex*. Porter found great numbers of stored tortoises on the captured whalers (leading to the quote opening Chapter 3), along with water and other supplies (Porter 1823). As discussed in Chapter 3, a total of at least 200,000 tortoises were taken by whalers by 1870, plus untold numbers of whales, sea lions, and fur sea lions. Porter and his crew contributed to the impact on Santiago Island: goats they set out to graze on the island escaped into the interior (Porter 1823, 68). Their descendants, joined periodically by other escapees and releases, were not eradicated until a major campaign, "Project Isabela," concluded in 2006. On another expedition in 1820, a crewmember of the *Essex,* attempting a prank on his tortoise-hunting mates, set a fire that spread rapidly over most of Charles Island (now Floreana), killing hundreds of animals and contributing to the demise of both the endemic tortoise (*Chelonoidis elephantopus*) and the Charles mockingbird (*Mimus trifasciatus*) on the island.

5. In 1832, the islands were officially annexed as the "Archipelago of Ecuador," with General José Villamil as the first Governor, presiding over a colony of about 120 people on Floreana. The colony had substantial environmental impact: by the time of Darwin's visit, there were already enough feral goats and pigs to support sport hunting as a pastime (Keynes 1988, 356). The Ecuadorian government sent criminals to the colony as well; disputes multiplied until a poor and disillusioned Villamil returned to the mainland after just 5 years in Galápagos. Between 1830 and 1890, other colonization efforts followed, most dependent on cattle and cultivation of crops like sugar cane, coffee, and potatoes, and the harvest of dyer's moss, orchil,[12]

---

[12] Orchil (*Rocella gracilis*), or dyer's moss, is an epiphyte (a plant that grows on other plants, in this case on branches of the Palo Santo tree) that yields a red or violet dye.

in the humid zones of San Cristóbal, Isabela, and Santa Cruz. Settlements were inevitably difficult to maintain and most ended in collapse.

6. In 1892, on the 400th anniversary of the voyage of Columbus, the Ecuadorian government changed the islands' name to "Archipelago of Columbus" (*Archipel de Colón*)—although he never visited—with related Spanish names for the islands, like Fernandina, Isabela, Pinta, Santa Maria, and San Cristóbal. These names remain in use today, although the English names are still used for some species (e.g., Charles mockingbird). In 1885, the town of Puerto Baquerizo Moreno was founded on San Cristóbal, and in 1893 the town of Puerto Villamil was established on Isabela. The late 1800s and early 1900s were also the "golden age" of collecting expeditions, including those by the Smithsonian, Stanford University, and the California Academy of Sciences (see Larson 2001; James 2017).

7. Biologist-explorer William Beebe's book *World's End* ([1924] 2012) painted an appealing picture of Galápagos and set off a second wave of immigration. Alas, Beebe's description was exaggerated in Europe by unscrupulous profiteers, promising mines of gold and diamonds for those who would sell their possessions and move to the islands. In 1926, a group of susceptible Norwegians and Germans colonized several islands; when they found no riches, they tried farming in the humid zones and fishing along the coast. Life was hard, and many gave up and returned home, dis-enchanted. A few hardy souls and their descendants have endured to the present, some by harvesting the local sailfin grouper (*bacalao)*, turning the fish into a dried, salted export called *seco salado*. Others succeeded in making the highlands moderately productive with coffee cultivation or cattle. The impact of crops and cattle was often enough to extirpate the native vegetation (e.g., Bush et al. 2014).

8. Disturbing misadventures also befell German and Austrian colonists who arrived from 1929 to 1932, as in the well-traveled story of Floreana, where again, few survived (as in Goldfine and Geller 2013).[13] Survival is perhaps the most underestimated challenge in every history of Galápagos, human or otherwise. In 1941, after negotiations with Ecuador, the U.S. Army Air Corps built the Baltra island airbase, nicknamed "The Rock," to protect the Panama Canal. As many as 2,400 servicemen lived there in short stints. After World War II, the U.S. building materials were distributed far and

---

[13] In the 1930s, a series of European eccentrics moved to the uninhabited island of Floreana, starting with the misanthropic German dentist Friedrich Ritter and his mistress. Another of these newcomers, an Austrian baroness, ultimately disappeared in 1934 from the island (along with one of her multiple lovers) amid circumstances Icelandic settler Walter Finsen deemed "passing strange." Another inhabitant nearly died in childbirth, and Ritter succumbed to food poisoning (Finsen n.d.). Key events are still shrouded in mystery (Treherne 1983; Wittmer 2013).

wide, contributing to growing settlements like Puerto Ayora and Bellavista on Santa Cruz. With a new runway, Baltra is still used today for flights.

9. The impetus for conservation in Galápagos came in two historical waves, both on significant Darwin anniversaries. The first began in 1934–1935 in the months leading up to the centenary of the *Beagle*'s arrival in Galápagos. Despite good intentions and some initial declarations, little was done in practice.[14] The second wave was more effective: as the centenary of Darwin's "big book" approached, an international group of scientists renewed the lobby for a genuine park, and in 1959 Galápagos National Park—Ecuador's first national park—was established with jurisdiction over 97% of the terrestrial area of the archipelago.[15] The Galápagos National Park Service (GNPS) began operations in 1968. In 1959, Julian Huxley, grandson of British biologist Thomas Huxley ("Darwin's bulldog"), together with other leading scientists of the time, organized a UNESCO effort to establish the Charles Darwin Foundation (CDF) and its Charles Darwin Research Station (CDRS) in Puerto Ayora, Santa Cruz. Since its dedication in 1964, CDRS has played a crucial role in research and restoration, working in partnership with the GNPS. In 1978, Galápagos was among the first World Heritage Sites declared by UNESCO,[16] and became an official Biosphere Reserve in 1984. In 1986, the Galápagos Marine Resources Reserve was established to provide parallel protection for the surrounding marine environments. The Reserve was expanded and converted to the Galápagos Marine Reserve (GMR) in 1998, and itself became a World Heritage Site in 2012.

10. Two further events of significance warrant introduction here. The first organized tour group arrived in 1969, opening the islands to new forms of commerce. Initially, tourism in the archipelago grew slowly, being limited by flights and infrastructure, but it began a dramatic increase in the 1980s and 1990s, stimulating parallel immigration to the islands. The island economy grew at near-record rates, with as much as 68% of the growth explained by tourism; however, immigration absorbed much of the increase (Taylor et al. 2009). At the same time, drastic change came to Galápagos

---

[14] Citing U.S. parks as a model, Ecuadorian diplomats lobbied for protection of endemic species, a research station in the archipelago, and a national park. As noted in Lundh (1999, Ch. 15) as well as in Moore and Cameron (2019, 128), a 1935 statute, "Regulations for Fisheries and Marine Game," placed most Galápagos fauna under protection. But the government had few means for enforcement, and the onset of World War II soon put on hold any idea of doing more.

[15] The other 3% of terrestrial Galápagos was set aside for human occupation, and it is the area of human use in agriculture, hospitality, and urbanism to this day.

[16] Galápagos was the first site nominated for listing as a World Heritage Site, but it shared the first date of listing with 11 other sites named by UNESCO in September, 1978 (Moore and Cameron 2019, 94).

fisheries. Collapse of the mainland's coastal sea cucumber fishery in 1992 triggered another wave of immigration and a sea cucumber "gold rush" in the mid-1990s, with millions of sea cucumbers harvested from Galápagos waters in peak years. As discussed in Chapter 9, the political and social turmoil that followed attempts to regulate sea cucumber and lobster fisheries are illustrative of further lessons learned from evolving social–ecological systems of the archipelago.

11. Local response to the changes in tourism and fisheries led to a sweeping legislative reform in 1998 called the "Special Law of Galápagos." It outlawed industrial fishing in an expanded Marine Reserve, permitting only artisanal fishing by registered local members of fishing cooperatives, and established a novel, if complex, system of participatory fisheries management. It began an effective curb on immigration to the islands, restricting permanent residents to persons with at least 5 previous continuous years living in Galápagos (1993 to 1998). It offered state subsidies for residents' income and travel costs and stipulated that all future tour operation licenses be reserved for permanent residents. Its many far-reaching implications were not fully realized, however, owing to problems of complexity and enforcement. In 2015, the Special Law was revised by the national government to simplify the governance structure of key fisheries, also changing some subsidies to less favorable terms in the view of locals. More recently, debate over those terms has fueled demands for yet another revision of the Special Law as the cycle continues.

12. As a legacy of this history, Galápagos today "confronts in microcosm the full complexity of balancing economic growth and ecological sustainability" (Gardener and Grenier 2011, 75). The archipelago faces enduring social and ecological challenges ranging from tourism growth and fisheries management to the perennial problems of introduced species and pollution—challenges that seem to grow more serious with each passing year. Consider tourism, for example. Already the largest business in the islands, tourism grew at alarming rates before the COVID-19 pandemic, surpassing 275,000 visitors per year in 2018, almost eight times the size of the resident population (35,000).[17] Implications abound from this trajectory, from impact on visitor sites and their fragile flora and fauna (see Vora 2018) to outbreaks of dengue and other infectious diseases related to the growing influx of visitors (Ryan et al. 2019). Even before COVID-19, there was a growing sense of urgency in the islands for finding a pathway

---

[17] Boat-based tourism declined from its peak of 90,000 visitors in 2008 to 73,000 in 2015. More popular today is land-based tourism, or "island hopping," which has grown from 83,000 visitors in 2008 to 152,000 in 2015. Before the pandemic, the latter was predicted (Izurieta 2017, 85) to increase to a whopping 206,000 land-based visitors.

to sustainable tourism, and sustainability generally. It remains to be seen when and how tourism will rebound from the pandemic, and what changes will be implemented to curtail its uncontrolled growth. Chapter 9 returns to these important topics.

As this brief recap suggests, the human history of Galápagos is one of increasing connectivity to the greater world system. Despite its relative isolation, Galápagos has certainly not escaped globalization and its implications, a point duly emphasized by the speedy arrival in March 2020 of the new coronavirus. Looking back, a few themes stand out. First, because the archipelago had no indigenous inhabitants, no peoples were colonized or displaced when Europeans arrived. Instead, the islands served as a base for exploration, piracy, and extractive activity by temporary visitors who gleaned raw materials from their flora and fauna. The consumption to extinction of several endemic tortoise species is powerfully symbolic of this theme. Second, consider the growing impact of human introductions to the islands, sometimes even their deliberate seeding with destructive mammalian herbivores, like goats and pigs, for the improved food supply of people in extractive industries. That settlers on Floreana already hunted feral goats and pigs at the time of Darwin's visit in 1835 is certainly emblematic of this issue. Another theme is the exceptional challenge of Galápagos for human settlement before the age of fossil fuel, owing to its isolation and lack of fresh water. Symbolic of this theme were the many tragic failures of colonization efforts, including appalling accounts of Norwegian settlers who sold everything to come to Galápagos in response to phony claims of gold and diamonds awaiting them in a tropical Eden.

A fourth theme stands out: despite its demanding conditions, Galápagos eventually came to be inhabited by an impressively diverse array of human settlers. From prisoners to coffee growers to would-be emperors to indigenous *Salasacas* of the Ecuadorian Andes, they came in fits and starts and from diverse lands—Ecuador, Germany, Austria, and Norway—and later from just about anywhere. The mélange put an indelible stamp of cultural diversity on local society. Initially, most settlers came from farmer, fisher, herder, and other working-class backgrounds. Later, especially with the advent of the Park, the Darwin Station, and conservation NGOs, a greater proportion of immigrants were of educated middle- and upper-class origins, giving rise to a full set of differences by class, race, ethnicity, and religion—and associated differences in income, wealth, and education—in a small society in a small place. Such social and cultural diversity has resulted today in a panoply of stakeholder categories, complicating decision making and local politics (this topic is revisited in Chapter 9). In this way, too, Galápagos has become a "little world within itself," facing many of the social issues of our larger world.

Fifth, conservation started late in Galápagos history, more than four centuries after its discovery, eventually coming in the form of a proposal for "protected area" conservation, following the general model of national parks in the United States. But, for its time, the implementation of the model in the archipelago was more progressive than usual: unlike so many contexts, where the model was used to dispossess and remove local inhabitants, here most settlers and settlements were allowed to remain. "Wildlife reserves" (which became Galápagos National Park, GNP) were largely fitted around existing human communities. For example, settlements and agricultural areas in the eastern half of Santa Cruz were allowed to remain, a bit like "gateway communities" near the U.S. parks, while the western half was set aside as part of the new GNP.[18] Unfortunately, agricultural areas also served as gateways for introduced organisms, especially feral farm animals, producing tensions between settlers, GNP, and the conservation community. Other tensions emerged from fishing, especially when the rush for sea cucumbers overtly challenged the Park's protected coastlines and their vulnerable coastal species. Key to reducing tensions in the future will be finding more and better ways for local people to derive tangible benefits from the protected areas (the Park and the Marine Reserve) and from their efforts toward biodiversity conservation.

A final theme is this: 200 years is precious little time for a diversity of human beings to learn how to live and work comfortably in this demanding environment and with each other in the pursuit of ways to live indefinitely in this unique place. The challenge has been especially great because human use is restricted to but 3% of the land area and, at the same time, because the islands experience ever-increasing links with the outside world. The latter include "boom and bust" cycles for exploitable natural resources, like lobster and sea cucumber, and exponentially increasing tourism—an industry that changes in both goals and format as the years pass. Consequently, human activity in Galápagos has increased in major ways over time, modifying many of the same conditions that, in prehistory, gave Galápagos its special creative force. Let's now turn to the hypotheses of this chapter regarding these changes and their consequences.

---

[18] But as often happens when protected areas are established, there were boundary issues. Notes Hennessy (2018, 495), "But in practice delineating the park was not so straightforward. . . [it] involved more than a decade of negotiation, territorial disputes and legal cases over the resettlement of families who did not want to leave the land they had recently claimed."

Hypothesis 1

Chapter 3 lays out the secret recipe of Galápagos and how it was able to reshape an ancestral giant tortoise lineage into 15 different species. Here, we draw on that recipe to understand the impact of human activity in the archipelago, and especially how it has brought on the sustainability urgency of today.

The first hypothesis is that human activity is undermining all but one of the six key ingredients in the recipe: only the archipelago's volcanic origins have not been significantly changed in recent years by human activity (Figure 4.3). Current conservation and livelihood issues in the archipelago stem largely from erosion of the other five ingredients that originally made Galápagos so evolutionarily creative and biodiverse.

To begin, the *isolation* of Galápagos has diminished at an accelerating rate. A hundred years ago, one could cross from coastal Ecuador to Galápagos in 10 to 12 days, depending on the ship and currents. Today, with jet service, one can cross in under 2 hours, that is, one twelfth of one day. Moreover, before the COVID pandemic, one had a choice of over 107 commercial flights a week, 15 per day, to the two jetports (Toral-Granda et al. 2017), and if we include Ecuadorian military and naval operations, there may be as many as 18 to 20 daily flights each way. A 2007 study found that 52% (24 of 46) of inspected aircraft were carrying stowaway species, including 32 insect species that would have been new species for

---

The Secret Recipe of Galápagos
(What makes its terrestrial habitats so creative?)

Before Human Impact	*Situation Today*
(1) Volcanic origins & clustering of islands	(1) No change
(2) Isolation on the equator amid converging currents	(2) Fully interconnected within the world system
(3) Demanding terrestrial ecological conditions	(3) Human-altered landscapes & habitats
(4) Highly simplified ecosystems	(4) Hundreds of introduced species
(5) Absence of terrestrial competitors & predators	(5) Over 20 species of mammalian competitors & predators
(6) Enough time for the evolution of endemism with resilience	(6) Striking vulnerability of endemic species

Figure 4.3. The secret recipe of Galápagos and its erosion by human activity. All key special features of Galápagos except its volcanic origins have been at least partly undermined by human impact and globalization.

Galápagos (Causton 2007).[19] In addition, several supply ships go to Galápagos each week, mostly from the port city of Guayaquil. It is believed that these ships historically introduced, in routine shipments of fruits and produce, many of the damaging exotics (e.g., fire ants, predatory wasps, cottony cushion scale) and diseases (malaria, avian pox, dengue) affecting Galápagos natives.

Second, *demanding terrestrial conditions* are no longer so harsh, especially around settled areas. Today, more than 35,000 people live and work in Galápagos, inhabiting five of the islands, with towns, farms, airports, roads, and even two highways. Humans have created enough urban habitat with water collectors and sewers that Galápagos now supports its first amphibian, Fowler's snouted tree frog (*Scinax quinquefasciatus*), introduced during the 1997–1998 El Niño. Such a frog could never have survived the trip and colonization without human assistance. Humans have also inadvertently supported the frog's reproductive cycle via climate change and the increasing frequency of El Niño deluges, creating larger and more frequent freshwater pools across the islands. Today, human impact is readily appreciated in the humid zones of inhabited islands, where, as of 2002, 93% of the humid habitat had been converted to agricultural pursuits on San Cristóbal, and 74% on Santa Cruz (Bensted-Smith 2002)—all with introduced cultivars, of course. Newcomers are often stunned to learn that, as of 2017, introduced terrestrial plant species outnumber natives by 821 to 560 (Toral-Granda et al. 2017; Watkins and Cruz 2007).[20]

But introduced plants themselves change the *highly simplified ecosystems* of yore, as do the hundreds of introduced invertebrates—at last count, 643 species (545 of them insects)—plus 50 introduced vertebrates, of which 21 are mammals. Added together, human introductions of the last 500 years number over 1,500 species, not including microorganisms (for thorough analysis, see Toral-Granda et al. 2017; Causton et al. 2017). Only two islands—Fernandina and Genovesa—have no resident exotic mammals today. Among the most invasive of all introductions are two tiny organisms of major impact on Darwin's finches: the avian pox that impacts the survivorship of adult finches, and the pernicious parasitic fly *Philornis downsi*, whose offspring attack finch nestlings, greatly reducing their evolutionary fitness (Figure 4.4).[21]

---

[19] Mandatory baggage inspection and light bug spray before landing have been implemented to stem the tide.

[20] Toral-Granda et al. (2017, 12) noted that "For the first 440 years (1535–1975), [alien species] arrived at an average of less than one species per year. Since tourism accelerated in the 1970s, nearly 30 new alien species have been recorded annually." But it's not so much that tourists are bringing in the introductions. Rather, tourism has boosted the local economy, which has prompted immigration of mainlanders looking for work and bringing organisms with them, including pets and ornamental plants for their house sites.

[21] In the last 60 years, flesh-eating larvae of *Philornis downsi* have become the leading cause of nestling mortality in the critically endangered Darwin's medium tree finch (O'Connor et al.

Figure 4.4. Small ground finch nestlings killed by *Philornis downsi*, a parasitic fly introduced from mainland Ecuador, whose larvae feed on the chicks of finches and other birds and are a major cause of nestling mortality (Fessl et al. 2018). *Source*: Courtesy of J. O'Connor and S. Kleindorfer.

But the most widespread impact came from the legions of introduced *terrestrial predators and competitors*, like cats, dogs, rats, pigs, cattle, donkeys, and goats, absent in Galápagos before human activity. In the 1990s, observers were shocked by the numbers of feral goats, pigs, and donkeys that burst forth on several of the larger islands, at the expense of native flora and fauna. Alcedo Volcano on Isabela was mowed clear by the army of goats there, leaving only stubble for the giant tortoises. Because they bring with them superior speed and competitive

2010) and the second most serious threat to mangrove finch nestlings, after rat predation (Fessl et al. 2010). What's more, any surviving nestlings are left with larvae-enlarged nostrils that alter the song of mature males, affecting their fitness in the mating period of the life span (Kleindorfer et al. 2019). Happily, control efforts focused on the fly are showing considerable success, and hand-raised finch chicks are being reintroduced to their native habitat with success (see reviews in Fessl et al. 2018; McNew and Clayton 2018). Another hopeful indicator, reported by Cimadom et al. (2016), is that several species of finches have learned to apply a natural insect repellent, derived from an endemic tree, to their feathers.

ability, the impact of these invasives on habitats and native populations has been enormous, the very "havoc" of which Darwin spoke in the quote at the beginning of this chapter.

That threat was especially challenging for the products of *endemism* in Galápagos, the organisms like the rail that evolved into new species over generations following their initial colonization. Freed from the pressures of continental predators and competitors, they evolved new features suited instead to the isolated archipelago. These new features, although effective in the harsh terrestrial conditions of the islands, generally spell calamity in the face of increasing human impact. Just as in the waved albatross and tortoise examples, colonizers' adaptations to pre-settlement Galápagos generally entail some form of *vulnerability* when conditions change. In the case of the Galápagos rail, to which we now return, I hypothesize that its vulnerability stems from the impact of introduced organisms on that most endearing of rail traits: skulking.

Hypothesis 2

As recently as 1979, all seven Galápagos islands over 500 m (1,640 ft) in elevation had Galápagos rails skulking in the dark vegetation of their highland humid zones (Franklin 1979, 203; the tally does not include Pinzón, where rails have only recently arrived). But rails were threatened on the six of those islands where mammals had been introduced —Isabela, Santiago, Santa Cruz, San Cristóbal, Floreana, and Pinta (Fernandina was the exception, with no mammals). What exactly was the threat? The second hypothesis here is that the threat posed by introduced mammals came from habitat change that increased the exposure of rails to predation. The endemic predators didn't change, neither the hawks nor the owls (Figure 4.5)—they had been there for thousands of years. But now, with herbivores eating away the rails' protective cover, whatever balance there may have been earlier shifted in favor of predators' proliferation and rails' demise. Let us now attempt a preliminary test of this proposition on the two islands with the best rail data: Pinta and Santiago.

Consider Pinta first—the small island in the north of Galápagos, with relatively high biodiversity because its central volcanic cone is high enough (650 m, 2,130 ft) to support a substantial humid zone. Pinta thus has the vegetation to allow skulking and, historically, rails "were common in the fern belt of the highlands" (Franklin et al. 1979, 203). But those same moist highlands make Pinta appealing to goats, too, who thrive on humid-zone vegetation. Introduced to the island in 1959, when legal protection for Galápagos was just beginning, goats propagated freely, and their growing numbers had cleared the humid highlands of its vegetative cover by 1970. Consequently, during two separate visits to Pinta

Figure 4.5. A rail's nightmare: endemic aerial predators. *A,* Galápagos hawks, like this adult, became the main predators of rails once invasive species, goats especially, ate away the rails' skulking cover. *B,* Juvenile hawks show almost no fear of humanity: this one allowed me to approach within a few meters. *C,* Barn owls are known elsewhere for almost-silent nocturnal flight, but Galápagos barn owls hunt by day in locations where hawks are rare, such as in the understory of highland forests. *D,* Short-eared owls also hunt by day in areas with few or no hawks, or else they are themselves attacked.

highlands in 1970, "no rails were seen or heard" and may well have been extinct (Franklin et al. 1979, 204, citing a 1970 CDRS report).

In 1971, the GNPS launched a ground-based hunting program that soon eliminated hundreds of goats. Aided by a rainy 1972–1973 El Niño year, the vegetation quickly rebounded. In 1973–1974, ornithologist de Vries visited Pinta and found that rails were again "common" on the central volcano (Franklin et al. 1979, 204). So, in the case of Pinta, we have a "quasi-natural" test of Hypothesis 2: rails abounded before goats; during the free reign of goats, rails became extremely scarce or possibly extinct; and with goat removal, rails rebounded to abundance again. The test would be strengthened by data confirming predation as a leading cause of the decline, but at least we can say that available data for Pinta are consistent with Hypothesis 2.

Interestingly, the Pinta case shows that, even with a sharp decline in numbers, rails rebound quickly. Did some fly in to recolonize, did a few survivors prompt a regrowth, or did both occur? Hard to know for sure, but from even a few adults, rails reproduce rapidly, having generous clutch sizes, 3 to 7 eggs, and rapid maturity (from hatching to adult in 80–85 days). Rails thus offer another example of a more general point about Galápagos flora and fauna: Galápagos organisms generally display impressive resilience, both because they came with general resilience enabling them to survive colonization, and because they have long histories of confronting natural perturbations of the islands, which often endows them with additional "specific" resilience.[22] Moreover, these resiliences can also be described by the roles they play for particular species. We can speak, for example, of *reproductive resilience*, in which survivors of perturbations rebound quickly via a high reproductive rate. Similarly, we can speak of *survival resilience,* conferred by the various features and mechanisms organisms have evolved in the face of survival challenges. Galápagos rails are a terrific case in point because they show *both* reproductive and survival resilience: reproductive resilience in their ability to rebound quickly from low densities, with fairly large clutches and a short generation time, and survival resilience in their low energy needs, high diet diversity, and overall stealth. These features qualify as general resiliences in this case, because they are shared with other, closely related rails (e.g., the eastern black rail; Eddleman et al. 2020). More field studies are needed to see what specific resilience(s) may have evolved in *el pachay.*

Alas, on Pinta, the gains against goats did not last long. Hunters in the 1970s missed a few isolated individuals, and so a major goat rebound occurred by 1990. We mustn't forget, goats too have resilience. Shortly thereafter, Pinta was included in a larger goat eradication campaign, Project Isabela, to which we return later.

Consider now the second island, Santiago, almost 10 times larger than Pinta (577 vs. 59 km^2), with a much larger humid zone. Unsurprisingly, it historically abounded with Galápagos rails, as Darwin noted.[23] He did not mention

---

[22] Introduced in Chapter 3, specific resilience evolves over many generations and thus may be subtle or difficult to detect among species relatively new to the archipelago. The same is true of recent anthropogenic challenges: do they promote the evolution of specific resiliences today among Galápagos organisms? Probably, but most such changes are less than 200 years old, and thus very difficult to detect. There is evidence to suggest that some of the Darwin's finch species are evolving in response to human-induced changes—especially food supply—on inhabited islands (see Hendry et al. 2006; De León et al. 2011, 2018).

[23] Darwin noted that the rail is "the only bird which is exclusively found in the high and damp parts of Charles [Floreana] and James [Santiago] Isld. It frequents *in numbers* the damp beds of [Galápagos sedge] & other plants, uttering loud and peculiar crys [*sic*], . . . is said to lay from 8 to 12 eggs—[and its] iris [is] bright scarlet" (Darwin [1836] 1963, 264). Today's meager data point to clutches of 4 eggs on average (Franklin et al. 1979, 206). The smaller clutch size makes the rail's net reproductive resilience even more noteworthy.

goats on Santiago, though they were introduced there in 1813, but noted that there remained plenty of "damp beds" of sedge, ferns, and more. But on an island with a larger humid zone, one expects the emergence of an even larger goat problem, especially since goats had an early start on Santiago, owing to the visit of the *USS Essex* described previously. An even larger problem indeed. In 2002, I was fortunate to co-lead a "Darwin's Footsteps" expedition to Galápagos, which included some time in the humid zone of Santiago (Figure 4.6*A*).[24] We hoped to see areas forested with giant daisies, much as Darwin had seen, shrouded in the cloudy mist of the highlands, as well as pools of water hosting the Santiago tortoise, the Galápagos subspecies of pintail duck, and colorful water lettuce common to the area.

When we arrived at the humid zone, we were astounded to find the equivalent of a gigantic gravel yard cleared of all vegetation (Figure 4.6*B*). Everything had been stripped away and eaten, leaving only a few sticks and straggling plants inside goat exclosures built by the Park Service. Later, we learned we were not the first to see Santiago in this state (e.g., Adsersen 1989), but we were shocked by the magnitude of the damage. We could see feral goats foraging in the distance, eating and uprooting as they went. The most affecting experience of the expedition, it left us speechless and teary-eyed. Ironically, as soon as we came to a remnant patch of healthy ferns, sedge, and Miconia, we nearly stumbled over a skulking Galápagos rail.

To be sure, the GNPS and Darwin Station were not standing quietly by as goats proliferated on Santiago; they had mounted a valiant effort to control the goats' numbers by hunting. During our 2002 expedition, we took shelter one afternoon from the thick *garua* (misty fog of the highlands) in one of the hunters' huts, marveling at the impressive collection of pig jawbones along the wall. Pigs had been hunted systematically since 1964, and it was becoming clear that the hunt-to-control strategy was not working: pigs and goats are just too prolific when they are the sole large mammalian inhabitants of a major island.[25]

---

[24] The expedition was a collaborative effort by co-leader Frank Sulloway, conservationist Roger Lang, entomologist and photographer Mark Moffett, geologist Ed Vicenzi, and botanist Greg de Nevers. I am grateful to each of them.

[25] I say "large mammalian inhabitants" because Santiago has an endemic rice rat (Harris and MacDonald 2007).

Figure 4.6. Landscapes of Santiago Island. *A,* View toward the highlands from the coast. Visible are a rugged recent lava flow (*dark brown, bottom*), the dry transition zone (*light brown, center*), and the humid zone (*top*), with the misty cloud layer that keeps the area moist and verdant. *B,* A portion of the Santiago humid zone in 2002, where ravenous goats had stripped away a forest of giant daisies (*Scalesia pedunculata*). In the center, a goat-proof exclosure built by the GNPD protects remaining vegetation.

Project Isabela

Another key piece of evidence pertinent to Hypothesis 2 emerged in the 1990s. Somehow, in the previous decade or two, feral goats from Isabela's southern agricultural areas had managed to cross the 12 km (7.5 miles) of desolate, sharp lava of the Perry Isthmus (the narrow strip of land between Alcedo and Sierra Negra Volcanoes). Goats gained new access to the vegetation-covered slopes of Alcedo Volcano, where again they had no natural predators or competitors. They proliferated, spread, and moved on to Darwin and Wolf Volcanoes as well. A turning point came in 1996. Godfrey Merlen, Galápagos Director of WildAid, the international NGO, visited Isabela and reported seeing "two or three goats on the upper flanks of Isabela's Alcedo Volcano in 1992. When he returned 3 years later, he saw hundreds. 'It was total chaos,' Merlen says" (Guo 2006, 1567). The goats had given a virtual buzz cut to the volcano's vegetation. They were everywhere in great numbers, at times surrounding tortoises struggling to find a few blades of grass (Figure 4.7). It reached the point of visible competition with tortoises: "'We saw many more tortoises falling into the volcanic craters,' trying to reach feeding grounds or because of erosion, says [Victor] Carrion, subdirector of GNP" (Guo

Figure 4.7. In the mid-1990s, goats greatly outnumbered tortoises on Alcedo and other volcanoes of Isabela and Santiago, leaving the tortoises only stubble for food and triggering widespread concern over the future of these iconic species. *Source*: Copyright Mark W. Moffett.

2006, 1567). News of the state of Alcedo Volcano fueled a growing campaign to eliminate, not just control, the goats.[26]

An international workshop in 1997 agreed and proposed eradication efforts to clear three islands, Pinta, Santiago, and Isabela. In 1999, initial methods were tested on Pinta, and in 2000, CDRS and GNPS received a 6-year, $18 million Global Environment Facility fund to begin Project Isabela. The goals were, first, to eradicate feral pigs on Santiago, *then* goats, so that goats would keep vegetation down for pig spotting. Eventually the last goats on Santiago were pursued using a technique from New Zealand, with sharp-shooters tracking down radio-collared "Judas goats" from specially equipped helicopters.[27]

Building on lessons learned on Santiago, the same techniques were applied with improved efficiency on Isabela—at the time, the largest eradication effort ever undertaken anywhere. Even with the Judas-goat strategy, the last few goats were the hardest and most expensive to reach. Finally, in 2006, Project Isabela was declared a success: from Pinta, an estimated 50,000 goats had been eliminated; from Santiago, 89,470 (plus 11,000 pigs and donkeys); and from Isabela, 62,820 goats plus an untold number of feral dogs (Lavoie et al. 2007). Project Isabela included other objectives, too, such as improving the inspection and quarantine programs for goods transported to the islands and establishing a monitoring system with a few Judas goats left behind, just in case.[28] The monitoring system proved critical: in 2009, six goats were deliberately reintroduced to Santiago as a form of protest, costing $32,000 for monitoring and removal (Carrion et al. 2011). Observers suspected this might have been an action by disgruntled fishers, upset about fishing restrictions, which we turn to later.

How did this enormous eradication effort affect the Galápagos rail? Does it show that the threat to rails came from habitat change and vegetation removal that exposed the skulking rails to native aerial predators? First, in Figure 4.8, one can see that rail densities were low on Santiago in the years before goat eradication, yet rebounded quickly after Project Isabela—a more than 13-fold increase. As the thicket of vegetation for skulking returned, so did the rails. But did the rebound stem from a change in predation? Was at least one rail predator adversely affected by the return of vegetation?

The answer is Yes, as shown by Jaramillo et al. (2016): the diet of Galápagos hawks changed during the same period. Following goat eradication, hawks

---

[26] For instance, the British periodical *The Guardian* published two related articles in 2006 alone, with apocalyptic titles like "Darwin's Paradise in Peril" and "Devouring Darwin's Islands."

[27] Judas goats were sterilized and tracked via radio collar to help aerial hunters find and eliminate remaining goat populations. They were named after the disciple who betrayed Jesus in the New Testament.

[28] More information on Project Isabela can be found in Cruz et al. (2009) and Lavoie et al. (2007). The latter is a "thematic atlas," with graphic depictions of project sites, strategies, and results. For critical analysis, see Bocci (2017).

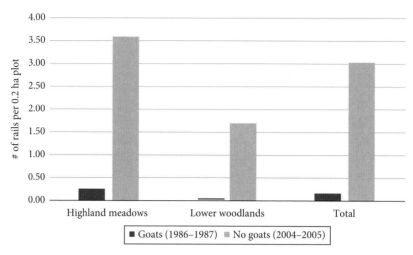

Figure 4.8. The impact of introduced goats on Galápagos rails in different habitats on Santiago. The plot shows average rail population density per 0.2-hectare sampling plot before and after goat eradication. With their protection from predators restored by goat eradication, rails rebounded quickly, increasing more than 13-fold in the highland meadows (compare *red bars* for "Goats" and *blue bars* for "No Goats"). Changes in rail density with goat eradication are statistically significant in both habitats ($p < 0.001$), attesting to rails' survival and reproductive resilience. *Sources*: Data from Rosenberg (1990) for 1986 and 1987; data from Donlan et al. (2007) for 2004 and 2005, plus chi-square tests.

showed a notable a decrease in consumption of land birds like the rail, and a sharp increase in consumption of introduced black rats (*Rattus rattus*). Introduced rats actually "buffered" the impact on hawks of reduced rail visibility.[29] Another analysis (Rivera-Parra et al. 2012) of Santiago hawks after goat eradication found a "drastic decline" in survivorship of nonterritorial juveniles—that is, of young, temporary "floaters" in the collective territorial mating system of the hawks—together with decreased survivorship of nonterritorial adults, because small prey (i.e., rails) were now hard to find. Paradoxically, survivorship actually improved with the number of hawks in territorial breeding-groups, owing to "the benefit of shared defense and offspring provisioning during harsher conditions" when the rails were protected anew (Rivera-Parra et al. 2012, 1197). In short, Hypothesis 2 is supported by the Santiago data, including an impressive shift in hawk diet. But there's more to the story.

---

[29] Interestingly, black rats may have also buffered the impact of hawk hunting on the endemic Santiago rice rat, which hawks clearly ate before the goat problem (De Vries 1975). Black rats were introduced to Santiago by 1835, most likely from the Villamil colony on Floreana (Hoeck 1984). For a comprehensive look at the intricate social life of Galápagos hawks, see De Vries 2015.

The goat eradication on Santiago in 2005 created another problem for the rail, this time via habitat change. Without goats, the invasive Himalayan blackberry (*Rubus niveus*) began to spread rapidly, because it was now free from constant goat "mowing." We saw small patches of blackberry on our 2002 visit to the Santiago highlands, and we surmised then that bird droppings most likely brought blackberries over from San Cristóbal (where blackberries were introduced in the 1970s, presumably for their fruit; Atkinson et al. 2008). By 2010, goats had been eliminated from the island and blackberries had spread into enormous tangles covering 920 hectares (Rentería et al. 2012b), taking over large areas of former rail habitat. As helpful as the goat eradication had been, it clearly had unanticipated costs. What would it now take to eliminate the spreading blackberry and fully restore the skulking area of Santiago?

> With [a] combination of techniques [including surveying on horseback and helicopter] it is estimated that eradication could be achieved within 15 years, at an estimated cost of US$150,000 per year, totaling US$2.25 million. Although this appears to be less costly than eradication of goats [on Santiago] at US$5.5 million (LaVoie et al. 2007), blackberry is only found in the humid zone. Thus, on a per hectare basis, its eradication will cost six times that of goat eradication. The release of the introduced blackberry from herbivory clearly shows that the use of single-species eradications for ecosystem restoration is only a first step in a long-term process. (Atkinson et al. 2008, 124, brackets added)

This example underscores the importance of broadly defined local monitoring after eradication efforts to ensure that equally—or more—difficult and costly problems do not arise.

Red Quinine

The Galápagos rail's vulnerability to introduced goats and their side effects is bad enough, but on Santa Cruz, rails are threatened by a different introduced species: red quinine trees, *Cinchona pubescens* (Figure 4.9). At first glance, they seem an unlikely menace: they are not poisonous to rails or to their insect food, and they provide a dense canopy of vegetation. But here, once again, there are unexpected consequences. Quinine stands bring with them "significant changes in the plant communities, habitat structures, and light, water, and nutrient regimes" (Jäger 2018, 70). These changes adversely affect the rail's shelter and food supply.

The first hints of rail decline on Santa Cruz attributable to quinine came from a study comparing rail densities between the years 1986 and 2000 (Gibbs et al.

Figure 4.9. Invasive red quinine (*Cinchona pubescens*) in the highlands of Santa Cruz. Once valued as a treatment for malaria, red quinine was introduced in 1946 for possible export production just before word reached Ecuador of its artificial synthesis in 1944. Demand soon collapsed, and neglected trees spread via wind-dispersed seeds and sucker sprouts, causing major change in the habitat of the Galápagos rail. In the background, two fields of cattle pasture are visible. Before human settlement, this area was completely covered in *Scalesia* forest; there is one straggling *Scalesia*, dark with epiphytes, left of center. Since this 2002 photo, many of the quinine trees in the area have suffered dieback from a fungal disease; hopefully, rails will rebound with the return of native vegetation.

2003). More conclusive evidence followed when a second study, using identical methods at the same time of year, found a 35% decrease in rail density: numbers declined from 78 rails sampled in 2000 to 51 in 2007 (Shriver et al. 2011). Moreover, the second study found a decrease at all sample sites *except* in "areas where *C. pubescens* has been removed" (Shriver et al. 2011, 224) and at elevations over 700 m, where *C. pubescens* does not grow. Where quinine is thick, there is less cover for skulking and much less incoming energy at ground level to support local food webs. More work is needed on this interaction, but evidence indicates that rail survival and reproduction are both affected by quinine (Jäger et al. 2009; Shriver et al. 2011).

As before, neither the Park Service nor the Darwin Station stood idly by. In late 1970s, an eradication project was initiated on Santa Cruz; by 1981, over 30,000 quinine trees had been destroyed. But quinine continued to spread via wind and sucker sprouts. An improved eradication program began in 2006, but "never really progressed beyond a concept" because of the expense of the required spatial scale (Buddenhagen and Tye 2015, 2905). With the growth of tourism, agriculture was viewed as much less lucrative, leading to less intensive agricultural land use, and sometimes abandonment altogether, allowing further aggressive expansion of quinine and other introduced plant species (e.g., blackberry, guava, alien passion fruit, and Spanish cedar). Consequently, as of 2011, "*C. pubescens* presently covers at least 11,000 ha in the highlands of Santa Cruz Island and is now the dominant vegetation in the Miconia and fern-sedge zones" (Shriver et al. 2011, 222). Just in time, unexpected help may be at hand: in the last few years, dense stands of red quinine on Santa Cruz have, on their own, shown signs of dieback or debilitation due to a fungal disease (Jäger 2018; visible in Figure 3.11*B*).

Where does this tale leave the Galápagos rail? The rail has healthy populations today on five of the seven islands it originally occupied, minus two with especially heavy human impact, Floreana and San Cristóbal. The rail is no longer seen in Galápagos mangrove forests, as was reported years ago. It clearly fits the same general pattern as many Pacific rails, being ground nesting, insectivorous, and a hesitant flyer. Plus, we have now seen its vulnerability to "hungry humans and the dogs, pigs, and rats that accompanied them." Yet it lives on, in contrast to rails on many other Pacific islands. So, returning to the question we started with, what's different?

Certainly, one factor, mentioned before, is that the Galápagos has been settled by humans and their companions for fewer than 200 years, compared to human settlement up to 10 times as many years on other Pacific islands. But the Galápagos rail was clearly on its way out, as happened on other islands, until its recent rebound, which points to two other key factors. The different fate to date of the Galápagos rail can be credited largely to heroic and costly efforts to eliminate three especially impactful invasive species brought by people: goats, blackberry, and red quinine. For three of five islands where rails live today, Project Isabela was especially influential in restoring a viable population. A final factor in the rail's survival is surely its own flying ability; although the rail is reluctant to take to the air, dispersal flights do seem to occur—witness the recent rail appearance on Pinzón—and likely contributed to its post-goat rebounds on Isabela, Pinta, and Santiago. The rail is now classified by the IUCN as Vulnerable, one step above Endangered.

Conservation Implications

The physically small Galápagos rail offers some big lessons about evolution and conservation. In its own way, the rail shows again how specialized adaptations to Galápagos—namely, intermediate flight capability generated by short wings on a large body, and skulking behavior that is normally advantageous in dense understory vegetation—rendered the rail vulnerable when humans arrived with their animal and plant companions (see Figure 4.10). We have distinguished three corresponding vulnerabilities: (1) a new vulnerability to invasive terrestrial predators, including rats, pigs, dogs, and cats; (2) an exacerbated vulnerability to endemic aerial predators when introduced herbivores reduced protective vegetation cover, making skulking difficult; and (3) a new vulnerability to changes in vegetation caused by blackberry, red quinine, or both, that in turn affected rails' food supply and protective vegetation. Note that in the second and third vulnerabilities, the threat was not from the introduced organisms themselves, but from the habitat change they caused, exposing rails to existing threats in new ways. However it happened, the result was the same: humans nearly de-railed the humid zones of six of the seven islands that hosted *el pachay*. And we did de-rail two.

In retrospect, it was not that human activity was primarily predatory, although there was some hunting of rails both by people and by feral pets. As is seen in other examples, human activity more commonly had *indirect* effects by way of habitat change, introduced organisms, or a combination of both. In the

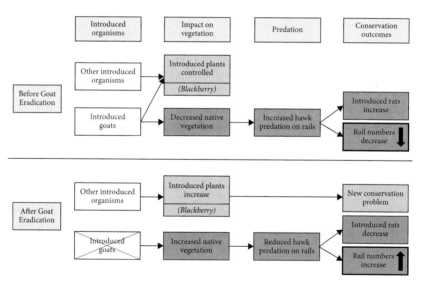

Figure 4.10. Summary framework of major interactions affecting the Galápagos rail. *Top,* Before eradication, goats consumed massive amounts of vegetation, making rails vulnerable to aerial predators, so their numbers declined, while numbers of introduced rats increased. *Bottom,* After goat eradication, native vegetation rebounded, reducing rail vulnerability to aerial predators, who then fed more on introduced rats. However, goat eradication also presented a new problem: introduced blackberry then spread unconstrained, becoming a major conservation issue of its own.

case of rails, moreover, the record to date shows little sign of impact from climate change. It may be that rails are buffered to a degree from changes in rainfall and temperature by both the dense vegetation of their habitat and their protective skulking behavior. Or, more likely, it may simply be difficult to pick up a signal from climate change when so many other forces are complicating rail survival and reproduction. In any case, today's data show that the main threat to rail survival and reproduction is from invasive species.

Let us now conclude with Figure 4.11, an SES summary of the role of invasive species in this example. In their understandable efforts to produce more food and income, resource users (farmers) introduced several new organisms—goats, blackberry, and quinine—to their resource units. Eventually, the introductions escaped from farms or were deliberately released, becoming independent new components of the already complex social–ecological system. On some islands (e.g., Pinta and Santiago), goats were central to the impact on rails, triggering increased predation by native aerial predators. On other islands, one can imagine different green arrows in the figure, representing blackberry and quinine, coming at the rail from farms, with hawks playing a lesser role. In the middle, the rail's evolutionary fitness is like a billiard ball with many green cue sticks. Pushed this way and that by diverse social and ecological forces, how long will the rail stay "in play"? Or will it soon fall into one of the pockets of extinction?

As Figure 4.11 emphasizes, the answer depends on many variables. The rail's fate is another joint social–ecological problem: it certainly depends on the 'force' of all the 'sticks' compared to the rail's own resilience and on any human-initiated conservation interventions. Looking at this SES diagram, with all its interactions and complexity, and thinking of the diminutive rail, one cannot help but feel even more sympathy for their plight. It's fair to say that our introductions have literally blown their cover. In the future, humans must be more careful.

Fortunately, this chapter's lessons also show that the effects of invasives on the rail are largely reversible, as long as a viable rail population survives somewhere, owing to the rail's own evolved resilience and flight capability. However, serious problems remain, with the mice, rats, cats, invertebrates, and invasive plants that are now established in the islands and not going away soon. Many of them remain a negative force on rail fitness. Ironically, some invasive plants benefited from goat eradication and spread widely thereafter, like the blackberry. They make a good point: this chapter serves merely to introduce the problem of alien species in Galápagos by focusing on the rail's story; it does not come close to providing a worthy sample of exotics, their manifold effects, and the options for their control or eradication.[30]

---

[30] A rat eradication effort called "Project Pinzón" has had noteworthy success on Seymour Norte, Rábida, and Pinzón (see Campbell et al. 2015; Rueda et al. 2019). After much experimentation, the

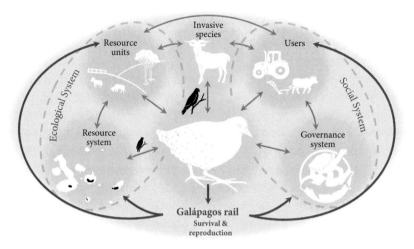

Figure 4.11. Each invasive species (symbolized here by the goat icon) adds a new dimension to its SES, creating a web of new interactions and pressures on the focal species. Resource *users* (diverse farmers, represented by sample technologies) brought in goats as farm animals to their *resource units* (farms, cleared from *Scalesia* forest, two lone daisies still standing) in the greater *resource system* (*shaded highland areas* within the inhabited islands). The goats escaped from farmers' control, despite policies of the *governance system* (*logo of GNPD*), becoming free-ranging *invasive species* with their own interactions within the SES. Lacking competition and predation from other species in their new home, the invaders spread rapidly, eating their way into the skulking habitat of the Galápagos rail. They thereby exposed rails to increased predation from hawks (*green arrows and large silhouette*) plus invasive dogs and cats (not shown), drastically affecting their *survival and reproduction* (*red arrow at bottom, center*). Of course, rail fitness has long been influenced by aerial predation not mediated by goats (*smaller hawk silhouette*). With accumulating pressures on rail fitness, their persistence on five islands is impressive evidence of resilience. *Source*: Adapted from Ostrom (2009).

Even so, the chapter shows that eradications of invasive animals are generally complicated and expensive, and perhaps more so for plants. Many variables contribute to plant persistence: dispersal pathways that are difficult to control (e.g., by native birds, introduced birds, and wind); aggressive growth patterns, including root propagation and suckering; and seed banks in the soil that can

---

invasive cottony cushion scale—a small insect harmful to native vegetation of the islands—is now under "biological control" via the deliberate introduction of another insect, the Australian ladybird beetle, to prey upon the scale (see Causton 2016). See reviews by Schofield (1989) and Phillips et al. (2012) on animals, and for reviews on plants, see Gardener et al. (2010) and Buddenhagen and Tye (2015).

remain dormant for years. In many areas of the archipelago today, exotic plant removal seems to be losing ground.

Ironically, agriculture, which brought many invasive organisms to the islands, helped keep invasive species like quinine at bay before land abandonment began. We'd see definite benefits again today if farms resumed agriculture but did so without further species introductions. A revival of agriculture in sanctioned areas would mean increased food security throughout the islands and fewer supply ships bringing in more exotics. In addition, a "payment for conservation services" scheme, financed by a separate, small "sustainability assessment" for visitors (distinct from Park entrance fees; see Chapter 6), could help farming become more profitable and conservation oriented. With the restoration of some humid-zone habitat on their land, farmers could well add tourism revenues to their agricultural income. Visitors already pay to visit highland farms and see tortoises grazing in the fields; many tell me they would pay more to see the elusive rail in its skulking runways, and gorgeous vermillion flycatchers amid reforested daisies. Such measures would also help relieve visitation pressure on the accessible highland areas of Galápagos National Park. These benefits are certainly worth considering. Most of all, such low-tech, everyday measures would go a long way toward helping with the overall goal advocated here: to put the Galápagos rails back on track.

5

Beautiful on the Inside

[Marine iguanas] assuredly well become the land they inhabit.
Charles Darwin (1835 in Keynes 1988, 353)

As shown in the last chapter, the Galápagos rail provides a poignant example of the ways that human activity, especially agricultural activity following early settlement, impacted the endemic fauna of the archipelago. The rail serves as an indicator species, whose dynamics in the face of introduced plants and animals offer important lessons for efforts to control or eliminate exotic species. This chapter explores an indicator species of a different kind: the famous marine iguanas of the archipelago—*Amblyrhynchus cristatus* ("blunt snout with a big crest"; Figure 5.1)—not simply because they are the world's only ocean-foraging lizard, but also because several of their specialized adaptations illuminate the demanding ecology of the archipelago. Marine iguanas prove an unusually good indicator of the local impact of ENSO, for one. Although Darwin described them as "a hideous-looking creature, of a dirty black color, stupid, and sluggish in its movements" (Darwin 1845, 385), they are actually quite beautiful on the inside.

Black is certainly a useful color for ectothermic ("cold-blooded") reptiles that derive their food from the sea, often submerging in the cold Pacific waters. When they crawl back onto land, black absorbs sunlight and helps them quickly regain a body temperature suitable for digestion and movement, as does heat conduction from the lava on which they lie. But as Figure 5.1 shows, not all marine iguanas are of a "dirty black color" or even "as black as the porous rocks over which they crawl" as Darwin put it (Keynes [1832–36] 1988, 353), noting also that "Somebody calls them 'Imps of darkness.'"[1] In some locations, males turn bright red and turquoise in December and January, during the mating season, giving them the name "Christmas iguanas."

Color in marine iguanas is more complicated than meets the eye, varying by age, sex, size, location, and season. Throughout the archipelago, hatchlings are dark grey to black, as is beneficial for both warmth and cryptic, antipredator camouflage. Since their primary predators on land—hawks and snakes—are

---

[1] That somebody was Lord Byron, Commander of the British ship *HMS Blonde,* that visited Galápagos in late March, 1825, just over 10 years before Darwin (see Olson 2014).

Exuberant Life. William H. Durham, Oxford University Press (2021). © Oxford University Press.
DOI: 10.1093/oso/9780197531518.003.0005

Figure 5.1. The world's only ocean-foraging lizard, the Galápagos marine iguana (*Amblyrhynchus cristatus*). Variation is a salient feature of this species, beginning with color, which differs by island, sex, and season. Colors range from solid black in both sexes among northern islands, to bright mottled red (*A*) among males on Española (shown) and Floreana in the south. During breeding season in the south (December–February), turquoise becomes a prominent secondary color (*B*) among males, prompting the nickname "Christmas iguanas." The iguana in *B* is not the same individual as in *A*, but similarities in their black splotches indicate they are from the same population. Marine iguanas show additional variation in average adult size from island to island, some being only about half the size of those shown here. Given this variation, it's fascinating that marine iguanas remain one single species in the archipelago.

diurnal, little ones are jittery, hiding in cracks or fissures, and they are always on the lookout. In locations where adult males have a small average body size, as on Darwin, Wolf, Pinta, and Genovesa in the north, they, too, are black. In more central islands, where adult males are large and nearly impervious to terrestrial predation, the males' backs and sides feature splotches of muted red mixed with black, sometimes with hints of green along their spiny crests. Sharks are the likely predator of males in these central areas, as predicted by Darwin (1839a, 468) and confirmed by Heller (1903, 90). The males' colors become more pronounced during mating season (see Trillmich 1983), presumably enhancing their attractiveness to females, when they stay on land, well away from sharks, for weeks at a time anyway. In the south, the breeding season produces the brightest colors of all, especially on Floreana and Española, where iguanas are known for gaudy hues (Figure 5.1). Brighter mating colors in the south likely reflect both diet and lower shark densities, judging from the careful census by Acuña-Marrero et al. (2018).

In contrast to the range of color in males, adult females are "more uniform across the archipelago and are remarkably darker, less contrasted and more homogeneous than males" (Miralles et al. 2017, 12). Females also have shorter, less pronounced spines down their backs. Such contrasts in color and ornamentation between males and females are likely the products of sexual selection—a form of natural selection driven by competition for mates. Among marine iguanas, males fight with one another for the best territories (the small land areas that have the qualities females seek for digging burrows). Consequently, some males mate with many females, and some with none, in the same breeding season (Trillmich 1983). Both size and color influence that success.[2]

Some part of adult male coloration appears to be structural, arising from reflectance off proteins in their skin and scales. That component is visible year-round. In contrast, the bright seasonal colors of central and southern males stem from seasonal hormonal changes and from plant-based pigments called carotenoids in their diet. Local naturalists sometimes refer to carotenoids as "the cosmetic coloration of Galápagos" because of their widespread importance in local flora and fauna. In the case of marine iguanas, the relevant compounds are produced by large macrophytic algae (red and green seaweed), marine iguanas' main food (Rassmann et al. 1997, 449). The bright red of Española and Floreana iguanas traces to "a particular seaweed that blooms during the summer months, which also coincides with the iguanas' mating season."[3] Chapter 7 explores how

[2] Eibl-Eibesfeldt (1984, 164) notes that, during nesting and egg-laying on Española, "a remarkable change [occurs] in coloration of the females, which become almost as brightly colored as the males; [this is] linked with . . . increased aggression of the females" in defense of nest sites.
[3] Quotation from Marine Biology Conservation Center website, https://marinebio.org/species/marine-iguanas/amblyrhynchus-cristatus/

evolution sometimes favors seasonally bright colors in males, and sometimes both sexes, as a signal of foraging ability and health status. For now, let me simply raise this possibility for marine iguanas.

In addition to color, differences in average adult body size are also impressive across the archipelago. The average snout-to-vent length, a reliable measure of body size, varies almost twofold between the big iguanas of Santa Cruz (averaging 350.6 mm, 13.8 in.) and the smallest iguanas of Darwin in the north (189.5 mm, 7.5 in.; Miralles et al. 2017, 10). Considering both morphological and genetic differences among populations, Miralles et al. (2017) identified 11 subspecies of *Amblyrhynchus cristatus* and suggested that still more may discovered as additional samples are studied from remote locations. Marine iguanas thus offer yet another example of radiation, like that of the tortoises. But here radiation has not generated species-level differences among the branches, only subspecies differences. Is it because marine iguanas are simply new as a species, and thus have not had enough time for differentiation?

Iguana Phylogeny

To answer this question, again let us begin with phylogeny, the branching tree of evolutionary relationships among related species. The best phylogeny available for marine iguanas (Figure 5.2*A*) shows that the ancestor of *all* Galápagos iguanas split off about 8.25 MYA from an ancestor shared with the genus *Ctenosaurus* of spiny-tailed iguanas of Mexico and Central America.[4] The location back then of the newly separated branch of iguanas and how they got there is unknown, but possibly the ancestor swam or rafted on floating debris out to an island in the Pacific, perhaps along the Cocos ridge or even at the Galápagos hot spot, which would make it now a submerged seamount to the east.

About 4.5 MYA, roughly when the oldest of the contemporary islands emerged from the sea, marine and land iguanas split apart and began evolving as two separate lines.[5] The details are unknown, but we can guess that the *Amblyrhynchus* ancestors wound up in an area with little or no terrestrial vegetation—perhaps

---

[4] MacLeod et al. (2015) used the black spiny-tailed iguana of Central America (*Ctenosaura similis*) to construct this phylogeny, which is a reasonable assumption. But there are more than a dozen other species in the genus in Mexico and Central America, including some with stronger morphological resemblance to Galápagos iguanas. From data on four nuclear genes, Malone et al. (2017, 36) suggested that Galápagos iguanas may be most closely related to Yucatán spiny-tailed iguanas (*C. defensor*), but they also noted that the conclusion is "not unequivocal with our current data."

[5] The 95% credibility interval for this estimate, like a confidence interval around a mean, ranges from 2.76 to 6.67 MYA. The oldest of the current islands of Galápagos—San Cristóbal and Española—emerged from the sea between 2.4 and 4 MYA, and 3.0 to 3.5 MYA, respectively (Geist et al. 2014, 151).

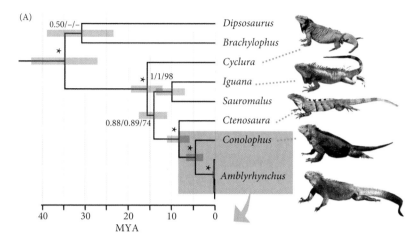

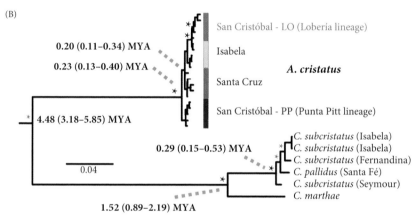

Figure 5.2. The phylogeny of Galápagos iguanas. *A,* The genetically closest relatives of land and marine iguanas belong to the genus *Ctenosaura* ("spiny-tailed iguanas"), which has 15 species today, mostly in Mexico and Central America. The common ancestor to this far-flung group lived until approximately 8.25 MYA (highlighted in the *gray box*). A statistical measure of confidence in the estimate (CI; *purple bars*) indicates 95% certainty that the split occurred between 5.85 and 11.06 MYA. Later, about 4.5 MYA (CI 2.76–6.67), the common ancestor of Galápagos land and marine iguanas split up, forming separate populations that evolved into the genera *Conolophus* and *Amblyrhynchus*. Numbers at the nodes represent statistical support for the branching shown (*asterisks* indicate strong support). *B,* A close-up of the last 4.5 million years, showing the time course of divergence into subspecies of marine iguanas (*top*) and into full species of land iguanas (*bottom*). Today's marine iguanas are all survivors of a single lineage that is more than 4 million years old. Differentiation only within the last 230,000 years produced the various subspecies of *Amblyrhnchus cristatus* shown by island. Today's land iguanas started out in a similar

Figure 5.2. Continued

manner, all from a single evolving line that endured 3 million years until *Conolophus marthae* split off 1.5 MYA. That divergence was followed by a second land iguana differentiation around 290,000 years ago producing two more species, *Conolophus pallidus* and *C. subcristatus*, and several subspecies of the latter. The length of the branches in *B* represents not time per se but the number of replacements in mitochondrial DNA (again, *asterisks* indicate strong statistical support for the pattern shown). *Source*: Republished with permission of S. Steinfartz from Amy MacLeod et al. 2015. "Hybridization Masks Speciation in the Evolutionary History of the Galápagos Marine Iguana." *Proceedings of the Royal Society, Series B: Biological Sciences* 282 (1809). License conveyed through Copyright Clearance Center, Inc.

on a sandbar or a recent, barren lava flow. We infer that, in the absence of adequate terrestrial vegetation, they fed on what was green and available: coastal algae beds at low tide (Figure 5.3*A*). At the start, these terrestrial herbivores from Central America would literally not have had the guts to live like that, not easily anyway, taking in so much seaweed and salt. Worse yet, there was likely little or no fresh water. Mortality was probably high then, but at least a few survived and, with accumulating mutations and the passing of generations, their descendants adapted to that new algae-eating "niche." As the population(s) grew and intertidal algae became scarce, they evolved features enabling adults to reach deeper algae, and eventually to swim out through the surf to dive and feed as much as 20 m (66 ft) under water (Figure 5.3*B*).

Meanwhile, we infer that the other line to split from the ancestral stock about 4.5 MYA settled on one or more islands with dry-zone plants and evolved adaptations to that niche, eating flowers and fruits of such plants as yellow cordia (Figure 5.3*C*) and *Opuntia* (Figure 5.3*D*). It is tempting to think this second line settled on the oldest of the current islands, San Cristóbal and/or Española, where *Opuntias* were also early colonists (Helsen et al. 2009). About 1.5 MYA, this second line split, one lineage evolving into the "pink iguana" (*Conolophus marthae;* Figure 5.4*A*), found today only on Wolf Volcano.[6] The other branch continued intact until it split again about 290,000 years ago—most likely diverging via a dispersal event (the swimming or rafting of a small founding population to a new island). One of the sister species evolved into today's *Conolophus subcristatus,* of yellow and brown coloration (Figure 5.4*B*), now inhabiting

---

[6] The pink land iguana was first noted in 1986 by staff from the GNPD and CDRS, but it was only fully described and genetically analyzed in 2009 (Gentile et al. 2009, see also Gentile 2016, Gentile at al. 2016). The population has been estimated at between 155 and 260 adults. A few juveniles were found during a September 2019 expedition to Wolf Volcano, the first evidence of reproduction in the population, but it remains listed as Critically Endangered by the IUCN (https://www.Galápagos.gob.ec/en/scientists-seek-to-identify-nesting-sites-for-the-pink-iguana/).

Figure 5.3. The feeding habits of two Galápagos iguanas. *A,* A Galápagos marine iguana, *Amblyrhynchus cristatus,* eats green algae growing on a rock at low tide. Here one can readily appreciate the advantages of a blunt nose, flat facial scales, and adjacent sharp teeth that permit extra-close grazing. *B,* Large individuals of the same species are able to feed under water, sometimes diving 10 m or more to feed. Their diet can include red algae, depending on season and island location, from which iguanas on southern islands derive the carotenoid pigments that seasonally enhance reddish areas of their skin and scales. *C,* A Galápagos land iguana, *Conolophus subcristatus,* eats a flower of yellow cordia (*Cordia lutea*). The bright yellow and orange colors of these iguanas, too, stem from carotenoid pigments, such as zeaxanthin, which they obtain from these and other yellow flowers. *D,* A darker land iguana of the same species feeds on tissue inside a fallen *Opuntia* cactus pad, a source of moisture in an otherwise dry environment. Here one can appreciate the advantages of its cone-shaped head in this habitat.

Santa Cruz and its formerly connected satellite islands (North Seymour, Baltra, and South Plaza) plus neighboring—and much younger—islands, Isabela and Fernandina. Formerly, the same species also inhabited Santiago.[7] The other sister species evolved into what we recognize today as the species *Conolophus pallidus,*

---

[7] In Darwin's time, *C. subcristatus* was also found in abundance on Santiago (James) Island: "I cannot give a more forcible proof of their numbers, than by stating, that when we were left at James Island, we could not for some time find a spot free from their burrows, on which to pitch our tent"

Figure 5.4. The three species of Galápagos land iguanas, all in the genus *Conolophus. A, C. marthae,* the pink land iguana—found exclusively on Wolf Volcano, northern Isabela—was described only recently by Gentile and Snell (2009). Its pink color comes from blood showing through its otherwise weakly pigmented scales. *B,* The brightly colored land iguana *C. subcristatus* is found in central and western Galápagos. *C,* In contrast, a pale species of land iguana, *C. pallidus,* is found only on Santa Fé. Worthy of closer study, color differences between the latter two species may stem largely from their diets. *Source: A,* Tui de Roy/Minden Pictures.

of paler, khaki coloration (Figure 5.4C), which is found exclusively on Santa Fé, where we might predict that bright yellow flowers were scarce.

The result today is four Galápagos iguana species in two genera: one genus and species of marine iguanas covering the whole archipelago, and one genus with three sister species of land iguanas confined to the central and western islands. To return to the question whether there was enough time for divergence in marine iguanas, the answer is that, in principle, there was plenty—as fully

(1839a, 469). The population died out in subsequent decades until stock from other islands was reintroduced in 2019 from the Santa Cruz breeding center. Suspected causes of the local extinction include introduced dogs (Heller 1903, 85) and feral pigs (Tzika et al. 2008), but debate continues (Snell et al. 1984, 189). On nearby Rábida, Steadman et al. (1991, 129) identified fossil remains of *C. subcristatus* dated to between 5,700 and 8,500 years BP. It would be fascinating to know where those remains belong in Galápagos iguana phylogeny.

confirmed by the radiation in land iguanas. And yet, the marine iguanas have not diverged—a puzzle worth our consideration.

One obvious possibility relates to their unique aquatic mobility: theoretically, inter-island swimming could disrupt the reproductive isolation of ancestral populations, eliminating a crucial ingredient of diversification. The evolutionary history of marine iguanas could be parallel in this respect to that of the Galapagos rails, the reluctant flyers of the last chapter whose inter-island flights are probably frequent enough to prevent speciation. We know today, from the careful analysis done by MacLeod et al. (2015), that marine iguanas do successfully move between islands (indeed they successfully hybridize in their new homes). But there's a big difference in outcomes to date: genetics and morphology differentiate 11 subspecies of marine iguanas, whereas there is "no indication that the various island populations [of rails] are diverging genetically from one another" (Kricher 2006, 129). Rails remain one species, with no subspecies.

Another idea seems equally plausible: maybe speciation of marine iguanas was well underway when some catastrophic event occurred—perhaps a long-lasting El Niño, an enormous tsunami, or a series of eruptions—killing off all surviving marine iguana lineages except one, from which the current subspecies then recently evolved. Events of this kind and magnitude are known to occur in the archipelago (see below and Chapters 7 and 8). A third scenario is also possible: perhaps it took several million years for the complete ocean-foraging algae-eating ensemble of adaptations to evolve to where they supported a growing and resilient population, which then spread out to different islands and began to differentiate. These and other hypotheses have been proposed to explain the absence of marine iguana speciation, but the data available today seem insufficient to clearly favor one argument over the others (see Online Appendix C).

But what we can say with certainty is this: an impressive "starter kit" of homologous features came with the ancestral iguanas and aided their colonization and spread amid the difficult conditions of the archipelago. At first, one might take for granted such things as protective skin and scales, strong legs and claws, covered ears, the capacity to float at rest, and tails of undulating propulsion. But consider a truly stunning accomplishment for this same starter kit: related iguanas from mainland America made an ocean crossing all the way to Tonga and Fiji in Polynesia, surviving 11,000 to 12,000 km of sun and salt water. As investigated by Gibbons (1981, 255), "The ancestral form of *Brachylophus* [Fiji's 'crested iguana'] probably arrived in the South Pacific from the Americas on rafts of floating vegetation on a course determined by the South Equatorial Current" and southeast trade winds.[8] This most impressive record suggests that iguanas deserve a place

---

[8] There are today two surviving species in Fiji, with evidence of several more species in Fiji prehistory, most likely eaten to extinction by human colonizers (Pregill and Dye 1989; Pregill and Steadman 2004).

among the great "improbable journeys" in the history of life (de Queiroz 2014).[9] Reaching Galápagos, by comparison, must have been a breeze.

Adaptations to a Hot and Cold, Salty World

But even an amazing starter kit was insufficient to soften the marine iguana's "struggle for existence" in Galápagos. Instead, several impressive specializations evolved in relation to its ocean-foraging algae-feeding lifestyle. Several of these adaptations are of the more "usual kind" one finds in the study of vertebrate evolution; for example:

1. The aforementioned dark colors of marine iguanas most of the year facilitate rewarming in the sun—classic among ectotherms in cold places—after foraging in the normally cold waters around the islands (see, for example, Bartholomew 1966).
2. "Behavioral thermoregulation" (White 1973) is used by marine iguanas to reorient their body position either to maximize insolation and warmth early and late in the day or to minimize overheating in the midday sun (Figure 5.5).[10]
3. The marine iguanas' short, blunted snout, lined with sharp 3-pointed teeth, flattened vertically and thus resembling a serrated knife (Figure 5.6), facilitates the close-cropped gleaning of green and red algae from rocky substrates.[11]
4. Marine iguanas have a distinctive flattened tail that promotes energy-efficient swimming, as opposed to the round tails in their close relatives (Tracy and Christian 1985; Figure 5.7). Other iguanas swim in a similar manner, with undulation of the tail, but much less efficiently and frequently.

Several other adaptations also deserve consideration. First, consider the marine iguana's "long and sharp claws with a powerful ability to grip onto the lava,"

---

[9] There are also iguana relatives ("iguanids") in Madagascar. But Okajima and Kumazawa (2009) concluded that their presence traces to the breakup of Gondwanaland, not to an ancient ocean crossing.

[10] Their behavioral thermoregulation includes piling up or "cuddling" at night into what local naturalists sometimes call iguana hotels. There appear to be two benefits of this behavior: it surely promotes a few hours of shared warmth, until the day's heat fully dissipates from the (cold-blooded) ectotherms, but Wikelski (1999) also found that nocturnal infestations of mobile ticks declined by 59% in experimental aggregations of marine iguanas.

[11] Darwin is the original source for this inference of effective grazing on marine algae, as noted by Vitousek et al. (2007, 502). Although posterior teeth are also tricuspid (3-pointed) in related *Ctenosaur* and *Conolophus* iguanas—only in marine iguanas do the anterior (front) teeth have a fully developed tricuspid shape . . . razor sharp at that (Berkovitz and Shellis 2017, 196).

Figure 5.5. Behavioral thermoregulation. Like all ectotherms ("cold-blooded" animals), marine iguanas regulate their body temperature behaviorally, but it takes special maneuvers to work with equatorial sun. Here, iguanas from Española line up facing the sun to reduce the heat of midday insolation. As shown by White (1973) and others, they adjust the angle of their bodies to the sun throughout the day to gain or lose heat.

Figure 5.6. Sharp three-pointed (tricuspid) teeth in a marine iguana victim of El Niño.

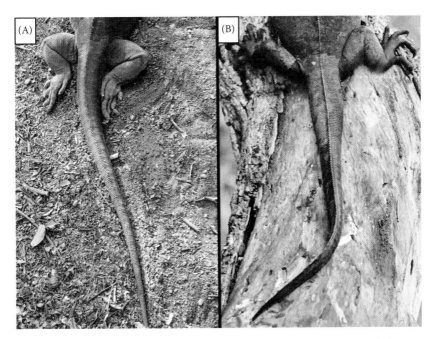

Figure 5.7. The tail of the Galápagos land iguana, *Conolophus subcristatus* (A), is much rounder than that of the marine iguana, *Amblyrhynchus cristatus* (B). Tails of ctenosaur iguanas (not shown), closest relatives of the Galapagos iguanas, are also round. Such evidence confirms that the marine iguana's flattened tail is an evolved adaptation for swimming speed and efficiency.

as noted by Vitousek et al. (2007, 502; citing Carpenter 1966). The strength of the claws is well established (Trillmich and Wikelski 1994, 267), but how do they compare in length, curvature, and other details with the claws of closely related iguanas? In a "reappraisal of the aquatic specializations" of marine iguanas, Dawson et al. (1977, 895) argued—though no specific measures were presented—that the marine iguana's feet "do not differ greatly from those of *Iguana iguana*" (the mainland green iguana). Lacking comparative data, especially with still closer *Ctenosaur* relatives, it is hard to say with certainty, and the topic warrants a fresh look, with careful measurements.

Second, consider the marine iguana's ability to submerge under water for long periods. This ability was described by Darwin (1839, 467), who noted that when a *Beagle* crewman deliberately sank an iguana for an hour with a heavy weight, it survived. This capacity would be a logical, specialized adaptation for ocean foraging. However, studies of common green iguanas on the mainland show that, as part of an ordinary escape response near rivers, they "voluntarily tolerate submergence for up to 4 hr" owing to a combination "of low metabolic rates and high tolerance" of low-oxygen conditions and heat loss (Dawson et al. 1977, 894). Thus, contrary to appearances, toleration of submergence is another characteristic that is *not* a specific adaptation to Galápagos. It's another piece of the starter kit that iguanas brought with them to Galápagos. This example makes a general point that is good to keep in mind: as marine iguanas—or any other organism, for that matter—adapted over time to Galápagos conditions, not every feature that looks like an adaptation to Galápagos is one. Some are shared with close relatives living elsewhere and thus are simply useful homologies.

Third, what about all the macrophytic algae the iguanas ingest? How do they deal with this specialized diet? Like other members of the iguana family, marine iguanas depend on specific fermentative bacteria in the stomach and hindgut for digestion of plant tissue (Mackie et al. 2004). Evidence indicates that these endosymbionts (from the Greek for "living together inside") are themselves specialists at digesting the green and red macrophytic algae that marine iguanas normally ingest, and thus they are different from those of other Galápagos reptiles (Hong et al. 2011). In contrast, they are largely incapable of digesting the brown alga that grows rapidly in marine iguana habitat with warm El Niño waters. How do marine iguanas obtain their specific cold-water endosymbionts? Troyer (1982, note 21) and Vitousek et al. (2007, 498) reported that for the first several months after hatching, juvenile marine iguanas feed on the feces of older iguanas in their colony, taking in the beneficial bacteria of the region. It may not be their most endearing feature, but it appears to be an adaptation serving their unusual diet. However, this behavior, too, is not special to marine iguanas, having been studied extensively by Troyer (1982) in green iguanas of Panama. Other evidence suggests its origin lies deep in iguana phylogeny.[12]

Fourth, what about all the salt they ingest when eating green algae in or near salt water? Again, marine iguanas have an appropriate feature: a nasal salt gland that allows them to get rid of the salt by snorting, which they do with considerable frequency (Figure 5.8*A*). But their land iguana "cousins" also snort (Figure 5.8*B*), though not nearly as often, and they do not eat marine algae at all. Moreover, a long list of other lizards also snort salty wastes from their nostrils (Hazard 2004). So, what's the marine iguana's adaptation? Careful analysis by Hazard showed that evolution has fine-tuned the marine iguana's salt gland in two ways: (1) the gland is enlarged, with most of it up over the eye (in other iguanas, it is part of the nose), thereby allowing (2) the highest rates of sodium and chloride snorting of all reptiles. As two leading researchers put it, the marine iguana has "one of the most efficient salt-excreting glands in the terrestrial realm" (Romero and Wikelski 2016, 107; see also Dunson 1969).

---

[12] Consumption of feces has also been reported among the young of a distant relative, the leopard tree iguanas of the Chilean Andes (https://blog.nationalgeographic.org/2014/01/02/surprise-lizard-feeds-its-babies-feces/).

Figure 5.8. Iguanas are salt-snorters. *A,* A marine iguana snorts highly concentrated salt from its nasal gland, a crucial adaptation for an organism that eats marine algae. The small ground finch flew in just before the snorting, looking for insects or molting skin. *B,* Land iguanas also snort, although neither as often nor with as concentrated salt content.

Adult body size is another feature showing signs of being an evolved marine iguana adaptation, and one that varies between islands, as noted earlier. The full story is complicated, involving a balance between two forms of natural selection (Wikelski 2005). One form, sexual selection, often proceeds via head-to-head combat among adult male iguanas for breeding territories, a process that favors large male body size (Wikelski and Trillmich 1997). The other form is natural selection via thermal inefficiency: large males may hold heat longer during underwater feeding bouts, but they also take longer to warm up between bouts, limiting their return trips for algae, compared to smaller, faster-warming males. Sexual selection thus favors larger sizes, while selection for foraging efficiency favors smaller ones (Wikelski 2005). The optimum will be a trade-off that varies with local conditions, favoring larger body sizes where algae are more abundant, allowing big males, foraging less often, to feed themselves well.

If we thus focus on algal density, we expect a correlation between average body sizes of adult marine iguanas and two variables: (1) proximity to the Cromwell

Current with its upwelling of iron and other micronutrients, which promote a denser algal food supply; and (2) size of the coastal shelf around an island, which correlates roughly with the amount of local algae. Other variables being equal, one expects more shelf around larger islands.

The prediction works quite well, as shown in Figure 5.9. The figure includes two different data sets from the same islands—data collected in 1906 and in 1997—showing very parallel trends by island area (we return below to the gap between them). Small adult body size is found on small islands far from the Cromwell upwelling, such as Plaza Sur, Seymour Norte, and Genovesa, where the iguanas' preferred algae, red and green, are relatively scarce. In contrast, the really big marine iguanas, almost twice the length and up to 10 times the mass of the Genovesa ones, are found on the largest islands near the upwelling—Isabela and Fernandina—where

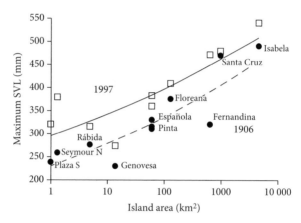

Figure 5.9. Maximum body size of an island's marine iguanas, measured as snout-to-vent length (SVL), is correlated with the island's land area, as shown here for samples from 1906 and 1997. Bigger islands tend to have more coastal shelf with a larger algae supply. Maximum body size is also roughly correlated with distance from the nutrient-rich Cromwell upwelling, which also supports algal growth (e.g., Isabela and Fernandina are on the long end of the plot, Genovesa and Seymour Norte are on the short end). Data for 1906, from the California Academy of Sciences Expedition (*solid dots*), are included together with 1997 data (*open squares*) from Wikelski (2005). The curves for 1906 and 1997 (*dashed* and *solid*, respectively) summarize the general trends (from the best-fitting quadratic regression). Evidence presented later in this chapter suggests that the body size differences between 1906 and 1997 may have evolved in response to changes in El Niño produced by global warming. *Source*: Republished with permission, island names, and corrected data from M. Wikelski based on his 2006 publication, "Evolution of Body Size in Galápagos Marine Iguanas." *Proceedings of the Royal Society, Series B: Biological Sciences* 272: 1576. License conveyed through Copyright Clearance Center, Inc.

macrophytic algae are far more abundant. Recent work confirms an inverse correlation between average adult body size and average sea-surface temperature (SST; Chiari et al. 2016). In areas of very cold ocean water, iguanas tend to be large, highlighting a second advantage to large bodies near the upwelling: heat retention for foraging in cold waters. All considered, the evidence supports body size as another evolved adaptation of marine iguanas.

Unusual Adaptations

For a lizard that evolved such a unique foraging pattern, it's not too surprising that there would be other, truly unusual adaptations. First, a field study on Santa Fé found that marine iguanas "eavesdrop" on the chatter of nearby mockingbirds, and thus run for cover when the birds' warning calls signal a predatory hawk's approach (Vitousek et al. 2007).[13] Hawks are major terrestrial predators on marine iguanas, much as they are on the Galápagos rails. But eavesdropping by a reptile on bird calls? This is believed to be the first example of a species without its own vocal communication listening in on the communications of another species, correctly discriminating its warnings, and reacting accordingly.

A second unusual adaptation has to do with limb bones. Marine iguanas are the first, and maybe only, lizard with extra dense bone deposition in their limbs, particularly front limbs. By all indications, it's an evolved feature that reduces buoyancy and requires less energy for staying under water during foraging (Hugi and Sánchez-Villagra 2012). Buoyancy was surely an advantage among ancestral iguanas, facilitating their colonization of the archipelago in the first place. But as marine iguanas evolved into specialized algae-feeders, natural selection favored reduction in that buoyancy, though not completely, the better to feed at depth. As noted earlier, MacLeod et al. (2015) found convincing evidence that marine iguanas do still float, swim, and even hybridize between islands.

The third unusual adaptation brings us to the major hypothesis of this chapter, concerning the marine iguana's response to the ENSO cycle. How does *Amblyrhynchus cristatus* respond to El Niño's drastic local effects, and how effective is the response?

---

[13] Colony location mattered: in the colony closest to the hawk's nest (300 m away), almost 80% of the marine iguanas exhibited vigilance behavior when the mockingbird alarm call was played, vs. about 45% when a regular song was played. The figures were lower for colonies farther away: roughly 35% showed vigilance behavior to alarm calls, vs. 22% and 28% to regular songs at intermediate (450 m) and greatest distance (600 m), respectively.

Hypothesis

In all of Galápagos, few perturbations rival El Niño events for the magnitude, frequency, and geographic range of their impact—events that go back thousands of years (to 9,200 BP at least; Conroy et al. 2008; see also Zhang et al. 2014). El Niño is a major threat to marine iguanas because their main staples of green and red algae die off in warm water, replaced by indigestible brown alga. Even a mild El Niño today kills some 20% of marine iguanas, and a strong one kills upward of 60%— potentially acting as a strong selection pressure.[14] With such frequency and impact, many changes are possible . . . provided the species survives.

True to form, an unprecedented response to El Niño appears to be evolving among marine iguanas: male and female bodies shrink to as much as 20% shorter—up to an amazing 6.8 cm or 2.7 in. shorter—during El Niño events (Wikelski and Thom 2000, 37). The iguanas' skeletons actually shrink in length, a response previously unknown in vertebrates. According to the authors, shrinkage aids the iguanas' coping by giving them a slightly lower metabolic rate (due to smaller size) and a longer period of survival before starvation. But is shrinking a bona fide adaptation? Might it merely be a byproduct of starvation? The work of Romero and Wikelski (2001, 2010; also Wikelski and Romero 2003) lends support to the hypothesis that *marine iguana shrinkage is a unique hormone-mediated adaptation to El Niño, unknown among related species living elsewhere.* Two lines of evidence support this conclusion.

First, marine iguanas have evolved a special capacity that allows them, when confronted with an El Niño-diminished food supply, to shut off their normal stress response before starvation. Normally, in response to stressors like diminishing food supply, lizards, including marine iguana relatives, generate high levels of the hormone corticosterone (hereafter CORT), which produces arousal and heightened activity. But heightened activity in the face of truly prolonged food scarcity, as with El Niño, does not spell reproductive success. Instead, natural selection has favored shutting down the normal CORT response, causing marine iguanas to become highly inactive when El Niño conditions persist (Romero and Wikelski 2010), allowing them sometimes to ride out the storm.[15] In two island

---

[14] These statistics come from the marine iguana populations on Santa Fé. The severe El Niño of 1983 resulted in a mortality rate there of 60%–70% (Laurie and Brown 1990, 529); in 2002–2003, a more moderate El Niño killed 23% of the marine iguana population (Romero and Wikelski 2010, 3158).

[15] Romero and Wikelski's (2010, 3160) analysis drew upon the separation of responses to starvation into three sequential phases: "phase 1 typified by carbohydrate breakdown, phase 2 by lipid metabolism, and phase 3 by protein metabolism." They suggested that, during phase 2, evolution has favored "damping any increase in corticosterone and thus delaying the transition from phase 2 to phase 3; . . . turning off that response appears to be a key feature for allowing long-term successful coping with prolonged stressors."

populations, Wikelski and Thom (2000, 37) found that "larger individuals of the two island populations shrank more than smaller individuals (even if calculated as a percentage)," that "females shrank more than males of the same size," and that "lizards that shrank more survived longer" during harsh periods.

Moreover, we know the shutdown to be a specialized response because El Niño conditions appear to be just about the only stressor to produce this effect. For example, when the fuel-oil ship *Jessica* tragically ran aground just off San Cristóbal in 2001, the marine iguanas of Santa Fé—downwind and downcurrent from the ensuing oil spill—unfortunately did not respond with inactivity. Instead, Wikelski et al. (2002) found, they went about normal foraging, feeding on oil-contaminated algae, and soon showed elevated CORT levels. When researchers returned to Santa Fé a year later, they found high marine iguana mortality (over 60%) compared to a control site on Genovesa outside the oil spill zone (Romero and Wikelski 2016). The researchers concluded that the marine iguanas' digestive endosymbionts were killed by petroleum toxicity. It was likely a cyclical sequence: the loss of some endosymbionts generated digestive inefficiency, triggering hunger, which prompted high CORT levels that stimulated more foraging, which killed more endosymbionts, triggering more inefficiency and higher CORT until the iguanas died of digestive failure. Once again, this demonstrates the dramatic vulnerability of a Galápagos endemic to a human perturbation. Fortunately, El Niño evokes, in contrast, a serious quiescence in marine iguanas.

The second line of evidence for the adaptation is suggestive rather than definitive, simply because the field of study is new. During low food supply, an additional metabolic pathway may well be activated in marine iguanas, mediated by another hormone, leptin. Leptin has complex effects on vertebrate physiology, including a "co-regulatory role in bone remodeling and energy metabolism. . . . Co-regulation could be an adaptation to control metabolism via changes in body size during long periods of starvation" (Bendik and Gluesenkamp 2013, 5; see also Wei and Ducy 2010). Leptin secretion induced by food shortage is believed to promote bone resorption, allowing marine iguanas to use up some of the energy stored in tissues and simultaneously reducing the body mass to be sustained without food—both advantageous during El Niño events. Best of all, the hypothesized leptin-mediated bone resorption pathway has a reversible structure: with adequate energy supply, leptin also stimulates bone regrowth (Reid et al. 2018). This duality may well be what allows marine iguanas to regrow once the seas are cold, food supply returns, and the shrinkage pathway is shut down.[16] As Figure 5.10 shows, marine iguanas can grow longer, shorter, and longer again, indefinitely, like a concertina playing a long, slow waltz.

[16] Central neuronal control via leptin of both components of bone remodeling in vertebrates—proliferation and resorption—is a very active area of research at present (see Ochi et al. 2019).

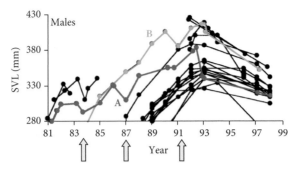

Figure 5.10. Large male marine iguanas shrink with El Niño events and regrow afterwards. Body length of Santa Fé iguanas over 280 mm (on *y* axis) are shown over a time interval that included El Niños in 1982/83, 1987/88, and 1991/92 (*arrows*). Colors have been added to highlight two individuals and their body length changes over time (A in red, and B in blue). Note that all large males shrank during the years 1993 to 1998, which the authors described as a "high-temperature and high population density period." Large iguanas may very well shrink whenever food gets scarce for prolonged periods. *Source*: Republished with permission of M. Wikelski from Wikelski, M., and L. M. Romero. 2003. "Body Size, Performance and Fitness in Galápagos Marine Iguanas." *Integrative and Comparative Biology* 43, no. 3. License conveyed through Copyright Clearance Center, Inc.

This work by Wikelski, Romero, and colleagues suggests that reversible shrinking in marine iguanas is a bona fide adaptation to El Niño events. It adds weight to the other advantages of body size in these animals (i.e., to the advantage for sexual selection among males and for enhanced fertility among females). Moreover, it brings up an interesting and important hypothesis: that the overall lengthening of marine iguanas from 1906 to 1997, as in Figure 5.9, reflects the advantage of body size as El Niños become more frequent and more severe with climate change. In other words, the figure suggests that intensifying El Niño events may have caused natural selection for larger bodies in nearly every island population over that interval.[17] It is early to say for sure, but as I consider leading alternatives,[18] natural selection driven by climate-changing El Niños seems so far the most likely explanation.

[17] The exception of Santa Cruz Island—where the average body sizes of 1906 and 1997 are roughly equal—nicely tests the rule: measurements both years were taken in Academy Bay at Puerto Ayora, where, during the hard El Niño rains, lava fissures bring runoff with "large amounts of nutrients" into the bay, enhancing algal growth and biomass well above other locations during the same events (Romero and Wikelski 2001, 7370).

[18] For example, there may be advantage to increased body size in escaping predation from sharks and marine mammals. But given different densities in predators among the islands (see Acuña-Marrero et al. 2018), would that generate parallel increments? Or perhaps warmer ambient temperatures from climate change help marine iguanas forage more efficiently, and thus simply grow

In short, the marine iguana represents not just the world's only ocean-foraging lizard, but also the first vertebrate documented to show reversible body shrinkage during periods of famine-induced duress.[19] Together, the specialized adaptations of the marine iguana are a convincing testimonial to the harshness of conditions in Galápagos, especially El Niño. Just as Darwin put it (in the opening quote): they do very much "become the land they inhabit." Marine iguanas also demonstrate just how much evolution can favor *survival resilience* in these islands. Although we have seen survival resilience before—in tortoises for example, who carry their own energy and water with them for months, and in rails, who construct a dark world of protected skulking passages—we have not yet seen such a far-reaching example. Here is an organism whose resilience applies to its very own body length.

Amazingly enough, there is more. As marine iguanas rebound from El Niño, they also show a degree of *reproductive resilience*, especially among females. Speaking of marine iguanas generally, Wikelski and Nelson (2004, 193) put it this way:

> [After an El Niño event,] they reproduce more frequently (every year), at a younger age (females), and lay larger clutches (e.g., 3 instead of 2 eggs).... These adaptations allow marine iguanas to increase their population numbers after dramatic population declines, such that mortality rates of 30%–50% after an El Niño event can be compensated within four years.

The enduring stamina of these special lizards illustrates the old adage, "When the going gets tough, the tough get going." In this case, however, the tough wait until the tough going is over, and then they go in a big way.

Worthy of our consideration is one exception from Seymour Norte. It's a most unusual marine iguana behavior that was studied before, during, and after the severe 1997–1998 El Niño. Some of the iguanas there had, for as much as 25 years, supplemented their diet of marine algae with consumption of a less desirable— and less nutritious—coastal *land* plant, *Batis maritima* (commonly, "saltwort"). Again, with careful measurements, Wikelski and Wrege (2000, 113–14) found on this single island that,

---

longer each generation from the same starting point (Wikelski 2005, 1987). But in Plaza Sur and Seymour Norte, would warmer ambient temperatures be enough to add a third or a half to the body length of big iguanas? The cumulative effect of selection over the decades seems more likely.

[19] Following the initial report of duress-induced shrinkage in marine iguanas, there have been reports of similar shrinkage in a Texas salamander (Bendik and Gluesenkamp 2013). And there is also evidence for famine-induced reversible shrinkage in the braincases of vertebrate weasels and shrews (Dechmann et al. 2017). Observations of marine iguanas have opened an important new area of physiological inquiry.

Batis was mainly exploited by the larger animals of the population. Animals that fed on land realized a caloric intake over and above what could be extracted from intertidal and subtidal sources of algae, and attained maximum body sizes well beyond those found elsewhere on the island [where no *Batis* was consumed]. . . . In populations that eat only intertidal algae, large marine iguanas are disproportionately prone to starvation because of their high absolute energy requirements. . . . By contrast, on [Seymour Norte] Island, larger iguanas clearly survived these selective events.

With such clear benefits of supplemental, if less nutritious, land-plant feeding, one might wonder why the practice was not spreading—whether by copying, natural selection, or some of each—across North Seymour or even to other islands?

Wikelski and Wrege (2000, 114) proposed the following answer to the question. They suggested that "conservative foraging-site traditions by individual animals" (returning to the same place to feed each time) is partially responsible. So strong is this conservatism that "individual iguanas have been observed starving to death [from trying to feed in the usual spots] during El Niño periods, in spite of resting with fat individuals that feed only meters away in tide pools." Moreover, they hinted that this feeding-site conservatism is actually beneficial over the long haul. During the 1997–1998 El Niño, the high sea-surface level brought by the warm El Niño waters caused flooding that killed the *Batis* plants in this area of North Seymour. The impact was devastating: none of the large iguanas survived (Wikleski and Wrege 2000, 114), and the *Batis*-eating tradition of North Seymour was lost. In short, the shrink-and-wait strategy described previously outlived this other, more active feeding strategy. For marine iguanas, it *does* seem best to ride out the storm.

Conservation Implications

Despite substantial total numbers of marine iguanas today (estimates reach as high as 280,000 across the archipelago; Wikelski and Nelson 2004), they are listed by the IUCN as Endangered. Much of the IUCN concern stems from human perturbations, whose effects are additive with those of natural perturbations. One serious issue is the elevated mortality associated with oil spills, as discussed earlier, together with other forms of pollution. Another issue is tourism, especially in light of the iguana's sensitive CORT response to stress and the growing number of tourists potentially adding to that stress. This response was studied by French and colleagues during a neutral period in the ENSO cycle (French et al. 2010; Neuman-Lee and French 2017). In a comparison of two sites, one with frequent tourists and one without, at each of two locations, they found:

> stress-induced elevations in plasma corticosterone [CORT] among tourist-exposed populations [in both locations] relative to undisturbed populations. We also found changes in multiple immunological responses associated with stress-related effects of human disturbance, including bacterial killing ability, cutaneous wound healing, and hemolytic complement activity [a measure of overall immunological integrity], and the responses varied according to reproductive state. (French et al. 2010, 792)

In other words, the everyday presence of tourists stimulates the same CORT-mediated stress response as is affected by oil spills and predators. Furthermore, they found that the healing of skin wounds—a common affliction in marine iguanas, who live generally "between a rock and a hard place"— "is significantly suppressed at tourist sites compared to undisturbed sites, in both (a) non-breeding and (b) breeding seasons" (p. 796), as is bactericidal activity during the nonbreeding seasons.

Especially germane to this book is the demonstration that tourism (following GNPD rules, we assume) "alters basic physiological functions likely to affect . . . [reproductive] fitness in many species" and specifically marine iguanas (French et al. 2010, 796).[20] Such a link between human activity and the fitness of the focal organism recalls the SES framework, in a manner akin to that for the waved albatrosses. However, in this case, the resource users (tourists) are observers, not resource harvesters, and are thus not involved in competition and predation with the focal species. Still, human presence has a measurable fitness-threatening impact on the focal species. Making matters worse, marine iguanas have been kept in zoos more than 10 years but have never yet mated in captivity, possibly because a colony may be required "to stimulate mating activities" (Wikelski and Nelson 2004, 196; another contrast with land iguanas, who have been reared for years in breeding centers). Research is needed on this topic, so that breeding and "head start" programs like those for tortoises might be used for marine iguanas.

An even larger concern is the impact of human pets, especially feral pets. One study by Barnett and Rudd (1983) found, in tragic Darwinian irony, that

---

[20] Future studies of this kind would do well to carry out simultaneous assessment of tourist behavior, to verify the assumption that "hundreds of people daily" cause only "low-level human disturbance" (French et al. 2010, 793). It would also help to know what aspects of tourist presence—sound, movement, density, rowdiness, technology, or proximity (including violations of the 2-m separation required by GNPD rules)—most elevate CORT responses of iguanas. Companion analysis of human behavior could lead to better guidelines and healthier marine iguanas.

Figure 5.11. Recently introduced mammals are a major threat to the Galápagos marine iguana, as these dramatic photographs attest. *Source*: *Left*, Tui de Roi/ Minden; *Right*, Copyright John Conn, courtesy of the photographer.

groups of feral dogs in the southern regions of Isabela had evolved, since introduction, specific foraging patterns according to their prey: one group specialized on the individual hunting of marine iguanas (see Figure 5.11) while another group specialized in collective foraging of tortoises and introduced cattle. Some idea of the impact of this predation on iguanas comes from another study in southwestern Isabela (Kruuk and Snell 1981, 200). "The dogs hunted mostly at night, . . . [often] in packs of up to eight animals; the mean pack size . . . was 2.6 (*n* = 29). . . . We estimated the dogs' daily food requirements at 0.75 kg [each, of iguana]. . . . Extrapolating from [a field census, the 20 dogs in the study site would hunt among] a population of about 4,900 iguanas, *taking approximately 27% per year*. Dogs also ate many iguana eggs, which they dug up from the shallow nests in the sand." The researchers' tallies show that risk of predation was greater for large iguanas (80 to 100 cm total length, including tail), especially big territorial males, setting up a selection pressure opposite to sexual selection and El Niño-driven selection. Big males were "the only [iguanas] which could be found exposed on the rocks at night, fast asleep in the same place where they were in daytime. The other iguanas were deep down [in] crevices or under rocks" (p. 202). Sadly, mortality from uncontrolled pets is fully additive to mortality from other sources like ENSO or disease.

Cats have been studied less extensively than dogs, but existing data confirm that cats are also a serious menace for juvenile marine iguanas. Berger et al. (2007, 655, citing an unpublished thesis) reported, "acute predation mainly by feral cats," on an iguana population near human settlements on Santa Cruz, has produced "a population structure that is age-biased toward adult marine iguanas because cats mainly prey on small-bodied hatchlings and juveniles." In another study at two sites over two seasons, Konecny (1987) saw cats approaching adult marine iguanas but not attacking them, presumably to focus on juveniles and on insects, lava lizards, and invasive rats. Dramatic photos leave no doubt that cats, too, prey upon the species (Figure 5.11).

In terms of SES, the iguanas' situation is similar to the situation highlighted in the rail example (Figure 4.10). Species that humans have introduced to Galápagos, in this case household dogs and cats, impinge on the survival and reproduction of the focal organism, but here they do so as direct predatory consumers. So let's add it up: there is our impact on iguanas from tourists in ever greater numbers, to which we add the growing impact of climate change via El Niño, and then we factor in the stark effects of predation from feral pets. It does not take advanced math to realize the net effect: there is little room in the band for the concertina.

Happily, there are sources of hope. In some areas of GNP, such as southeastern Isabela, dogs were eliminated by Project Isabela in 2006. In other areas, like Fernandina and the northern islands, dogs and cats have never been established. On most inhabited islands, on the other hand, dogs and cats remain a menace. To help with this problem and others, Ecuador established a Galápagos Biosecurity Agency (with the acronym ABG in Spanish), built on the experience of an earlier inspection and quarantine agency for the archipelago.[21] ABG works with NGOs like Animal Balance and Darwin Animal Doctors, plus the GNPD, in an ongoing program to spay and neuter cats and dogs, as well as to provide a range of veterinary services and to offer community education programs that promote responsible pet ownership.

Another important contribution has come from a recent "conservation priority assessment" for marine iguanas based upon their genetic differences within and between islands (MacLeod and Steinfartz 2016). Setting aside several small populations (on Darwin, Wolf, and Pinzón), and using what they term "a wealth of molecular data" from more than 1,200 individuals, the authors identified 10 population clusters to serve as "management units" for marine iguana conservation. The effort exemplifies an evolutionary approach to conservation, providing priorities for action that factor in both the severity of anthropogenic threats and effective population size as a measure of evolutionary potential. Surprisingly, though, the list of threats by island lacks one very important item: El Niño, which deserves a prominent place for each island, especially those with low effective population sizes, for two reasons: (1) its historical impact on the species, unrivaled by any other threat to date; and (2) the mounting evidence that, under the

---

[21] Established in 2000, the earlier System for Inspection and Quarantine in Galápagos, SICGAL, tried valiantly to stem the tide of exotics, but it was beset with problems. A study by Zapata (2007) found that, from 2001 to 2006, "the number of inspectors was cut by 20%, compared to a 100% increase in the number of inspection units," such as passengers, suitcases, and cargo containers. Today, there is a stronger economic and institutional commitment to the new ABG system, with over 180 employees and greatly improved technology (see Cruz et al. 2017b).

influence of human-induced climate change, El Niños are now more frequent, longer lasting, and more severe in their temperature changes than ever before (Chapter 2). As impressive as its endocrine toolkit may be for coping with food scarcity, the concertina cannot help with an El Niño event lasting beyond a few months. Shrinking is an evolutionary gamble: it can only do so much.

Figure 5.12 summarizes key arguments of this chapter, illustrating the conceptual links between challenges to marine iguanas of Galápagos past and present.

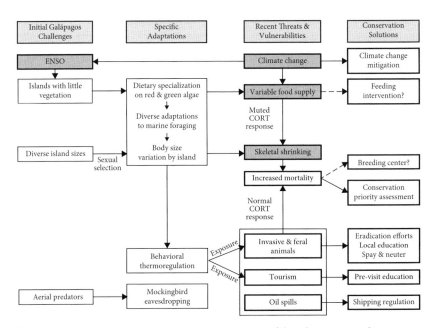

Figure 5.12. Schematic summary of key arguments of this chapter regarding adaptations, vulnerabilities, and conservation solutions for marine iguanas. From a terrestrial herbivorous ancestor, iguanas evolved a whole package of adaptations enabling them to live as foragers of marine algae, including changes in their teeth, snout, tail structure, and salt glands. Body size is another key adaptation, representing a trade-off on each island between advantages of large size for reproduction (sexual selection in males) versus disadvantages of large size for foraging efficiency. Hypothesis 1 (*shaded*) views the shrinking of body size as an evolved adaptation, mediated by a muted CORT response, to the hardships of El Niño, allowing larger iguanas to survive longer periods of famine. *Solid arrows* are links with mounting empirical support; boxes caused or affected by human activity are *bolded*. Key implications for conservation action, actual or potential, are shown in the right-hand column. Educational efforts are helping both locals and visitors to appreciate the many ways human activity impacts the viability of marine iguanas and other Galápagos endemics. The evolutionary fitness of these organisms today is a function of both natural conditions and human activities in the archipelago.

It also links their adaptations today to the initial challenges of the archipelago and to their radical algae-eating lifestyle. A prominent example is body size adaptation to local food supply, related in turn to island size and distance from the Cromwell upwelling (Figure 5.7). Body size in males is also a product, as noted earlier, of sexual selection mediated by the aggressive defense of territories where females come to mate and lay eggs. Furthermore, body size is a key component of the marine iguanas' evolving adaptation to El Niño, mediated by the evolved modification of stress response (the "muted CORT response"). Finally, marine iguana body size is also among the variables making big iguanas more susceptible to human impact via feral pets (risk of predation grows with size), oil spills, and probably even tourism, given the popularity of viewing larger iguanas. Body size thus serves to underscore the key takeaway message that survival and reproduction in marine iguanas are now as much a function of human activity in the islands as they are of natural conditions.

Figure 5.12 also makes clear that extinction remains a very real possibility for the world's only ocean-foraging lizard. The column of threats is particularly staggering, although equally noteworthy efforts are underway to manage many of the dangers, like oil spills and feral pets. Tourism is one of those threats that deserves particular attention, since marine iguanas are often among the first endemics that visitors encounter. There is a need to educate visitors *before* they arrive in Galápagos about the stress effects that their visit can cause, perhaps via an addendum to existing GNP rules and, even better, an educational video with such messages at the start of visits to the islands—perhaps during the outbound flight. Figure 5.12 also shows that the threat from climate change is especially direct, with few local options for treatment. It's not even clear that a breeding center would work to help repopulate endangered management units.

An important conclusion follows: it is crucial to take action where we can, to curb and reverse the rising concentrations of greenhouse gas before more severe El Niño events come along—events that could well put an end to the fabled "imps of darkness." This threat is not vague and distant: it could happen any year, and with each of our carbon emissions, we're integral to the problem. Once again, the fate of a Galápagos endemic, seemingly esoteric and remote, traces right back to each of us. This is not someone else's problem. This is something we can do. This time we, too, must shrink, in our habits of atmospheric carbon production.

6

"An Inexplicable Confusion"

Amongst the species of this family [of finches] there reigns (to me) an inexplicable confusion. Of each kind, some are jet black, & from this, by intermediate shades, to brown. . . . Moreover a gradation in form of the bill, appears to me to exist. . . . There is no possibility of distinguishing the species by their habits, as they are all similar, & they feed together (also with doves) in large irregular flocks.

Charles Darwin (*Ornithological Notes*)[1]

Two of the most famous examples of evolution in Galápagos are the adaptive radiations of giant tortoises and Darwin's finches, each with many species sprung from a common source. The tortoise radiation produced an amazing outcome: a different tortoise species on every major volcano in the archipelago, and two on the largest one. The finch radiation produced an even greater number of species—currently, 18 are distinguished—reinforcing the view that an isolated cluster of small islands out at sea is a good evolutionary incubator for species diversity.[2] But the differences between the finches are subtle, even when clearly presented in artistic reconstruction (Figure 6.1). In Galápagos, visitors are often told, "Only God and Peter Grant can tell all the species apart."[3] No wonder Darwin, first to describe their subtle variation, found them to be a source of "inexplicable confusion."[4] How ironic that their radiation has become a textbook classic around the world.

---

[1] Quote from *Darwin's Ornithological Notes* (Barlow, 1963, 261). This passage, and most of Darwin's ornithological notes about Galápagos, were carefully dated by Frank Sulloway (1982a) to 9 months after the *Beagle* left the archipelago to return to England.

[2] Because the tally of 18 species is based on recent, detailed genetic analysis, most older depictions feature a lower number of species, typically 15 (with 14 species in Galápagos—see Figure 6.1 and caption—plus the Cocos finch on a separate island belonging to Costa Rica). Using genetic data, Figure 6.2 shows 16 species, but the two species boxed in color—large cactus finch (blue box) and sharp-beaked finch (red box)—each represent two genetically distinct species.

[3] Peter Grant is the leading contemporary authority on Darwin's finches. Together with his spouse, Rosemary Grant, Peter is the author of many acclaimed publications about the ecology and evolution of the finches.

[4] No doubt he would have found them even more confusing had he recognized all of the finches in the sample of songbirds he collected. Darwin mistook the large cactus finch for a blackbird, and the warbler finch for a wren, easy mistakes to make. On Darwin's difficulties identifying finches, see Sulloway (1982b). Mistaking these birds for something else is easy to do: only since 1990 has it been

Exuberant Life. William H. Durham, Oxford University Press (2021). © Oxford University Press.
DOI: 10.1093/oso/9780197531518.003.0006

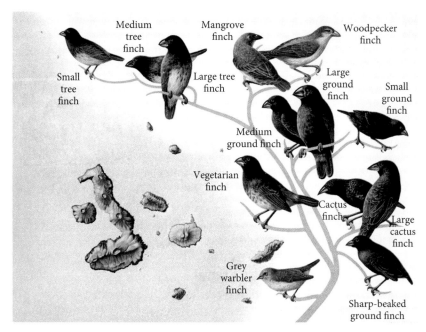

Figure 6.1. Artfully rendered evolutionary tree of Darwin's finches, with common names and a reference map of Galápagos. Each species is represented by a single adult male for 13 of the 14 most widely recognized morphospecies in Galápagos today (the olive warbler finch is not shown, nor is the Cocos finch, a close relative on Cocos Island, Costa Rica). Trees like this one, based on trait comparisons, illustrate the overall similarity among these closely related species—a similarity that prompted Darwin's "inexplicable confusion." The five species at the top are the "tree finch group," featured in this chapter; the six darker birds below them and to the right are the "ground finch group." *Source*: Adapted by permission from NGS Art Department/National Geographic Creative.

The finches are one of the most carefully studied of all groups of organisms in the archipelago, thanks to the heroic efforts of field researchers like David Lack in the 1930s, whose book *Darwin's Finches* (1947) popularized that name, Robert Bowman in the 1950s and 1960s, who focused on the finches' diverse songs, and, most recently, Peter and Rosemary Grant, who have worked with finches for more than 40 years. The Grants endured the hardships of life on the tiny desert islet called Daphne Major (nearby is the even smaller Daphne Minor) to understand evolutionary dynamics of the ground finches—those "avian evangelists

evident that Darwin's finches are actually tanagers, members of the family Thraupidae, as opposed to Frangillidae, true finches (see Burns et al. 2002).

for the power of evolution by natural selection" as Nicholls put it (2014, xiii). The Grants' project, born of meticulous quantitative fieldwork, stands out as one of the crowning achievements of modern evolutionary research (see Grant and Grant 2014). A couple of the Grants' key findings are especially germane to our efforts here.

First, through painstaking beak measurements of resident medium ground finches (*Geospiza fortis*) before, during, and after a drought in 1977, the Grants showed that there had been a small but heritable change in beak depth from a single year's bottleneck. When comparing average adult beak depths before and after the drought, they found that the survivors had an average beak depth about 9% larger. The Grants inferred that this seemingly small difference had given a survival advantage to birds during the drought: their larger-than-average beaks enabled them to process larger- and harder-than-average seeds after the easier seeds were gone. Birds with smaller beaks, meanwhile, died of deprivation, and the population plummeted from something like 1,200 birds to 200. Because beak depth is heritable in finches, the differential survival of adults with slightly larger beaks then appeared in their 1978 offspring as slightly larger beaks, compared to the average beak size of 1976 offspring, before the drought. The difference is small—about 0.9 mm—but heritable, and was crucial to surviving the drought. It represents the evolutionary change caused by 1 year of natural selection. Prior to the Grants' discovery, many researchers regarded evolutionary change as a slow and lengthy process, producing visible results only over many years by the accumulation of "insensibly fine gradations" (Darwin 1859, 171). Now we know that, under certain circumstances, natural selection acts fast.[5]

Second, through careful genetic analysis, the Grants and collaborators generated a new and informative phylogeny of the finch clade (Figure 6.2). It reconfirms several main branches of the finch radiation and adds several surprises. The branches correspond to clusters of species that have been recognized for years as morphologically distinct species, or what we can call morphospecies.[6] For example, one can pick out a warbler finch group of two morphospecies, a tree finch group of five morphospecies (matching Figure 6.1), and a ground finch group of six morphospecies, including two cactus finch species (also called "cactus ground finches"). In addition, there are other finch species that don't fit easily into these groupings: the Cocos finch

---

[5] For a lively account of the first several decades of the Grants' work, and of other recent studies in evolution, I recommend Weiner's *Beak of the Finch: A Story of Evolution in Our Time,* winner of the 1995 Pulitzer Prize.

[6] Using morphology to distinguish finch species is nothing new: for example, it was the basis for John Gould's 1837 identification of 13 distinct, but closely related, species in Darwin's collection (see Sulloway 1982b).

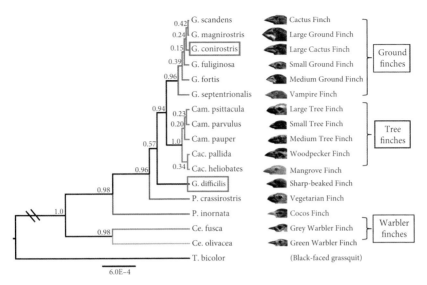

Figure 6.2. A simplified phylogeny of Darwin's finches based upon comparative genetic analysis, with conventional morphospecies at right. Branch lengths are proportional to time and genetic distance between species (except for the bottom branch to black-faced grassquits). Numbers on branches represent statistical support for branching points: support is strong for values over 0.75. Thus, tree finches and ground finches are genetically distinct as groups, but their component morphospecies are not, although this was widely assumed. The large cactus finch (*upper box*) and sharp-beaked finch (*lower box*) are each considered today to be multiple species, shaped by convergent evolution from different ancestors (see Lamichhaney et al. 2016). The vegetarian finch is shown with its own genus, *Platyspiza* (abbreviated P.), as is the Cocos finch, *Pinaroloxias* (also P.). Other genera are *Geospiza* (G.), *Camarhynchus* (Cam.), *Cactospiza* (Cac.), and *Certhidae* (Ce.). *Source*: Used with permission of K. Petren from Farrington et al. 2014. "The Evolutionary History of Darwin's Finches: Speciation, Gene Flow and Introgression in a Fragmented Landscape." *Evolution* 68, no. 10, with added morphospecies names and groupings. Copyright 2014, The Society for the Study of Evolution.

from Cocos Island, Costa Rica, 630 km away, the large cactus finch (boxed in blue), which is actually two similar-looking species, and the sharp-beaked morphospecies labeled *G. difficilis* in the figure (boxed in red). The latter looks morphologically like one species of finch, but genetic analysis shows that it, too, represents two species (for a phylogeny with full complexity and all the branches, see Lamichhaney et al. 2015, Figure 2). The two finches evolved similar morphology starting from different ancestors—another example of convergent evolution, as seen in Pacific island rails.

In all, the new finch phylogeny adds three species to the classically recognized 15. To that total one might add the first bird species to form while researchers watched: the newly documented "Big Bird" lineage of Daphne Major, a population of reproductively isolated hybrids descended from the mating of two morphospecies.[7] Although a species name has been proposed (see Grant and Grant 2014), which would put the finch total at 19 species, I agree with Lamichhaney et al. (2018, 5) and prefer to call it a "lineage" for now, because it currently involves fewer than 50 birds showing reproductive isolation for just a few generations. Either way, at 18 or 19 species, that's a lot of finches for this little cluster of islands!

Paradoxically, another important result of the phylogeny is that genetic analysis does not support all the species-level distinctions that naturalists and visitors try to identify in the islands (the small images in Figure 6.2, at right). A key reason is that speciation is far from complete among the recognized morphospecies. They are "incipient" or "incomplete" species, continuing to hybridize with one another, mixing their genes and diluting their accumulating differences (see Grant and Grant 2008). Most observers still distinguish the traditional taxa, or try, and there is little harm in that, so long as we remember their not very distinctive genetics: in a very real sense, these are mutt-like finches! No wonder they were confusing to Darwin in 1835—the "inexplicable confusion" is inherent.[8]

One further inference from Figure 6.2 is helpful here: the time course of diversification in the "tree," represented by branch length from any split. According to Lamichhaney et al. (2015), the first finch diversification began about 900,000 years ago, with the splitting off of an ancestral warbler finch (labeled here with the statistic 1.0). The vegetarian finch—the specialized leaf-eater in the group—then diverged about 546,000 years ago. Finally, the tree finches and ground finches began diverging about 412,000 years ago (labeled with the statistic 0.94), which is roughly 82,400 generations. Compared with the diversification of giant tortoises, for example, in the last 1.74 MY (first dispersal after

---

[7] The new lineage formed in 1981 when a male large cactus finch (*Geospiza conirostris*) flew in from Española and mated with a resident female medium ground finch (*G. fortis*). Their large hybrid offspring learned the distinctive song of the father and that, plus large beaks, have kept them reproductively isolated from other finches on Daphne Major. In just three generations, they became a reproductively isolated lineage of "Big Birds" with their own distinct genome (Lamichhaney et al. 2018).

[8] But Darwin did discern an important pattern within four of the ground finch morphospecies assigned to the genus *Geospiza* by John Gould. In 1845, in a hint to readers that he had more to say on the topic, Darwin wrote: "Seeing this gradation and diversity of structure in one small, intimately related group of birds, one might really fancy that . . . one species had been taken and modified for different ends" (p. 380). Evidence today shows that Darwin was right and links the modifications to differences in food supply.

colonization at 3.2 MY, see Figure 3.3), these data imply that the finches are quite "newly minted" in the archipelago, and incompletely "minted" at that. No wonder they are so difficult to distinguish!

We also understand today more about the dietary dimension of Darwin's confusion, at least with respect to ground finch species, which "feed together (also with doves) in large irregular flocks" (opening quote for this chapter). Careful analysis of ground finch diets at four different locations on Santa Cruz brought De León et al. (2014, 1093) to the conclusion that "the ground finches are 'imperfect generalists' that [feed on] overlapping resources under benign conditions (in space or time), but then retreat to resources for which they are best adapted during periods of food limitation." In other words, they resort to seed sizes that match their beak sizes only during times of scarcity. Their diets completely overlap during times of relative abundance, feeding quite literally into Darwin's confusion.

To make matters worse, there are new signs that the ground finch overlap may even be increasing. The problem stems from ongoing urbanization in Galápagos, and the increasing—if mostly unintentional—sharing of human food with birds. Sharing takes the edge off seasonal scarcities and is already starting "to erode ecological differences that promote and maintain adaptive radiation in Darwin's finches" (as reported by De León et al. 2018, 1; also McNew et al. 2017). It's another example of softening of the archipelago's harsh conditions described in Chapter 4, undermining the food limitations that would one day complete the finch speciation process. So tightly linked are the social–ecological systems of Galápagos that humans are already changing the course of finch evolution.

But note that these arguments and data, going back to Darwin, pertain to the most common group of birds in the islands, the ground finches. Relatively less is said, and known, about tree finches, apart from the phylogenetic inference above (Figure 6.2) that they are indeed distinct relatives of the ground finches, that their main habitat is trees and forests, and that their diets generally include a good number of insects (see Figure 6.3). One goal of this chapter is to shed further light on the *arboreal* branches of the finch tree.

Let me also pick up on a point made in Chapter 3, where it is noted that Galápagos provides an almost ideal setting for adaptive radiations, including many beyond the finch and tortoise cases. These radiations include smaller, less well-known organisms like the tiny endemic land snails (genus *Bulimulus*; see Parent and Coppois 2016), Galápagos snakes (Zaher et al. 2018), wolf spiders (De Busschere et al. 2012), and lava lizards (in two separate radiations; Benavides at al. 2009). A second goal of this chapter is to explore one of the most fascinating of these additional Galápagos radiations, and a personal favorite: a tree of trees.

Figure 6.3. A small ground finch female (*Geospiza fuliginosa*) compared to a small tree finch female (*Camarhynchus parvulus*). *A,* The small ground finch is a granivorous finch, specializing in small plant seeds, like grass and morning glory seeds, which it breaks open with its beak. *B* and *C,* The tree finch, in contrast, eats both insects—as in *C,* eating a larva—and seeds, which it also disperses readily, as do other tree finch morphospecies.

The *Scalesias*

The *Scalesias* (*lechosos* in Spanish, from *leche,* Spanish for "milk," owing to their sticky sap) are a group of daisies, most of them shrubs, related to similar Andean daisies of the greater family Asteraceae (from Latin for "star," the usual shape of flower heads in this plant family, composed of many small flowers).[9] These are among the most unusually beautiful of all Galápagos organisms in my view. The leaves of the 15 different Galápagos species are stunning, as different as salad leaves of romaine and frisée, and much easier to distinguish than finch beaks. Yet, like finches, all have radiated from one common ancestor related to mainland genera (*Heiseria, Pappobolus,* and *Syncretocarpus)* by way of a single colonization (Fernández-Mazuecos 2020). And 4 of the 15 have evolved to be tree species (Figure 6.4*A*), reaching as high as 20 m (66 ft), with woody trunks to 30 cm (12 in.) in diameter, and endearingly little composite flower heads up in the canopy (Figure 6.4*B*). In the humid zones of six islands, these splendid giants form dense, single-species forests, which in pre-settlement times covered thousands of hectares and formed rich, diverse ecosystems (Figure 6.4*C*). The 11 species that are not trees are, when fully developed, small shrubs in the dry and

---

[9] Asteraceae are sometimes also called Compositae, emphasizing the fact that their inflorescences are composite assemblages of many small flowers. The first *Scalesia* was collected by British "collector prince" Hugh Cuming in 1829. His botanist friend George Arnott named the plant. Arnott intended to christen it after another Scottish botanist, William A. Stables, but got the last name wrong, filing with "Scales"—hence the name *Scalesia* (see Noltie 2012). Happily, that helps us pronounce it.

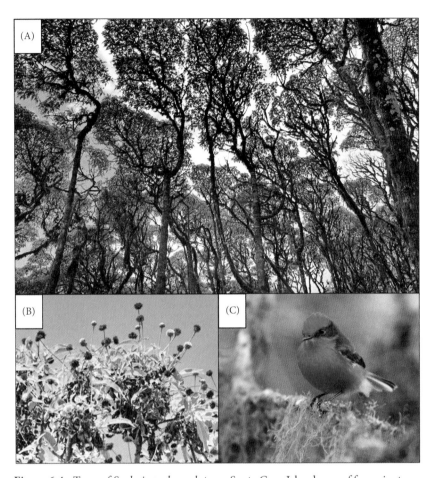

Figure 6.4. Trees of *Scalesia pedunculata* on Santa Cruz Island, one of four giant daisy species in the archipelago. *A,* Note the impressive height of woody trunks in the *Scalesia* forest. Note also the big gaps in the canopy between adjacent trees, so-called crown shyness. *B,* A closer look at canopy vegetation shows daisy flower heads atop the long peduncles (stems) that give the species its name, and the characteristic dead leaves just below the new ones. *C,* Insectivorous vermilion flycatcher (male) on a *Scalesia* branch next to a small sample of the epiphytes (mosses, lichens, ferns, and liverworts) that abound in this forest habitat. Sightings of this attractive bird in the highlands of Santa Cruz, quite common 20 years ago when this picture was taken, have sadly become increasingly rare with the decline in *Scalesia* forest. *Source*: *A,* Courtesy of Kirk Zufelt, Ontario, Canada.

transition zones of most other islands. All but five of the archipelago's main islands have resident *Scalesia* species, and several have more than one; Santa Cruz
has six, including the most impressive of the tree species, *Scalesia pedunculata*.[10]
This tree's flower heads grow above the canopy, projecting from the leaf layer by
extra-long stems or peduncles, hence its species name.

How did four *Scalesia* species become trees from what we infer were bushy
little ancestors? As is so often the case, Darwin had a theory for that:

> Oceanic islands are sometimes deficient in certain classes [of plants & animals],
> and their places are apparently occupied by the other inhabitants. . . . Islands
> often possess trees or bushes, belonging to orders which elsewhere include only
> herbaceous species; . . . an herbaceous plant, though it would have no chance
> of successfully competing in stature with a fully developed tree, when estab-
> lished on an island [without trees] and having to compete with herbaceous
> plants alone might readily gain an advantage by growing taller and taller and
> overtopping the other plants. If so, natural selection would often tend to add to
> the stature of herbaceous plants . . . and thus convert them first into bushes and
> ultimately into trees. (Darwin 1859, 391–92)

Other theories have been proposed (see Carlquist 1974, Ch. 10), but today's evidence largely supports Darwin's scheme—except we know today that the starting
colonist for *Scalesia* was woody, not herbaceous, which can greatly speed up the
evolution of height. The first evidence for Darwin's theory is phylogenetic: the
ancestor of all *Scalesias* diverged from a clade of its closest shrubby relatives
along the coast of mainland South America about 3 MYA (Fernández-Mazuecos
et al. 2020).[11] Second, a shrubby origin is also suggested by the preponderance of
herbaceous *Scalesias* in the drier zones of the archipelago, in contrast to but four
tree species of the humid highlands.

---

[10] Debate continues over what makes a tree (see Ehrenberg 2018), tracing back to Joe Hooker's
overly cautious labeling of Darwin's tree sample from Santiago as "frutescens," meaning "shrubby"
(Thisleton-Dyer 1905, plate 2717). Following the lead of botanists Wiggins and Porter (1971),
Hamann (1993), and McMullen (1999), I consider four *Scalesias* to be bona fide trees: the sibling
species *S. pedunculata* and *S. baurii*, plus the sibling species *S. cordata* and *S. microcephala* (species
pairs from Fernández-Mazuecos et al. 2020). Before humans introduced so many herbivores to the
islands, it is possible that still other *Scalesia* species had an "evident trunk" and indefinite growth (criteria that seem to me the most reliable). *S. affinis* may be one example: it is a shrub today, and Wiggins
and Porter (1971, 354) mention a plant 9 m tall. Moreover, in recent genetic analysis, *S. affinis* forms
a clade with tree species *S. cordata* and *S. microcephala* (Fernández-Mazuecos et al. 2020). Certainly,
Scalesia morphology warrants further research.

[11] A comparative study of terrestrial bird species in Galápagos (Valente et al. 2015, using finch
data from Farrington et al. 2014) gives the parallel date for the separation of the ancestor to Darwin's
finches from related mainland species as 3.028 MY. There are generous confidence limits around
these so-called stem or colonization estimates for both daisies and finches, such that their close similarity may be purely coincidental. Still it seems fair to raise the possibility that the two colonizations
might have happened together, as would be the case if finch ancestors somehow carried ancestral
Scalesia seeds with them at colonization.

Figure 6.5. Flower heads of selected *Scalesia* species. *A*, A flower head of *S. pedunculata* on Santa Cruz, one of four tree daisy species in Galápagos, all lacking in ray florets ("petals"). *B*, A small herbaceous relative, *S. affinis*, from Isabela, with more ray florets than most *Scalesias*. *C*, One of the principal pollinators of Galápagos, a female carpenter bee, visits another *S. affinis* flower head.

Third, consider an unusual feature of the tree species *S. pedunculata:* its forests exhibit impressive crown shyness, inter-crown gaps between adjacent trees, such that trees rarely touch one another, let alone overlap (Figure 6.4*A*). As Nicholls (2014, 59) put it, such a forest seems a "paragon of mutual respect." Among explanations for crown shyness, studied elsewhere in other species, is kin selection—a form of natural selection that could well favor this means of reducing competition for sunlight among close genetically related individuals. The suggestion is that adjacent *S. pendunculata* trees are close relatives that have evolved crown shyness through the inclusive fitness benefits of having healthy neighboring kin. The mutual survival and reproduction benefits—including perhaps mutual wind protection—would have contributed to their rapid and successful adaptation to the humid highlands from shrubby ancestors, consistent with Darwin's theory.[12]

There is evidence for an additional evolutionary change as the daisies adapted to Galápagos, a topic explored experimentally by Nielsen et al. (2002). Back when the ancestral daisy colonist arrived in the islands, there were very few pollinating insect species, and among them a single bee species, the endemic carpenter bee (*Xylocopa darwini*, Figure 6.5*C*), which is visually attracted to yellow and white flower heads (in turn explaining why these colors are common among Galápagos flowers). Judging from the white flower heads of most *Scalesias*, the colonizing ancestor probably had white inflorescences, too. Back far enough, they were most likely ringed with ray florets that look like petals, giving the daisies their starlike appearance. Over the years, in habitats where carpenter bees

---

[12] Further study is clearly warranted, including fine-grained genetic analysis of *Scalesia* stands. But pre-settlement distribution of "canopy shy" tree *Scalesias* in nearly homogeneous, one-species forests is consistent with this interpretation, as is their method of reseeding (described in this chapter), which places sibling seedlings in close proximity.

were scarce, one would expect big, showy, flower heads to confer a fitness advantage via their ability to attract bee services. The *Scalesia affinis* flower head (Figure 6.5*B*) is such an example, averaging six to ten ray florets per inflorescense. Meanwhile, in habitats where carpenter bees were plentiful, one would expect little or no fitness advantage to the showy florets; daisies that, by mutational change, had no florets to sustain would then have had more resources to put into their seeds, to evolutionary advantage. In experimental tests, Nielsen et al. (2002, 139) handily confirmed these expectations: "In *S. affinis,* rayed [flower heads] received more pollinators and more pollen [than experimental flower heads with florets removed], which resulted in a significantly higher embryo production. In *S. pedunculata,* no effect [of experimentally added florets] on embryo production was found." Big showy flower heads are thus beneficial to *Scalesias* in areas of low pollinator density. Today, in areas of high pollinator density, *Scalesia* flower heads have no florets, only the central disk of tightly packed small flowers (Figure 6.5*A*).[13]

The story of giant daisies is further complicated by the human history of the archipelago, specifically by the two immigration 'waves.' As mentioned previously, the first wave, roughly 1830 to 1890, consisted of Ecuadorian settlement and colonization efforts, even penal colonies, many of them short-lived. The second wave, about 1930 to 1970, included Europeans—especially Norwegians and Germans—as well as Ecuadorians, who in many cases came searching for a tropical Eden. Alas, Galápagos was far from most notions of tropical paradise, but it did offer both waves a couple of productive livelihoods: one in fishing, and one in cultivation and cattle. The latter required clearing, settlement, and home construction in the humid zones of the higher islands—especially San Cristóbal, Floreana, and Santa Cruz, as well as on the slopes of Sierra Negra Volcano. Agricultural livelihoods inevitably brought the colonists into the habitat of the tree *Scalesias*, which they easily cleared for crops and pasture, and sometimes split in half for construction wood. As daisies, after all, their wood is soft and has a central pith.

By 1905–1906, botanist Alban Stewart of the California Academy Expedition to Galápagos lamented that, in southern Isabela, a "considerable amount of

---

[13] Another candidate adaptation to Galápagos: all 15 *Scalesia* species are marcescent, meaning the plants cling tenaciously to their own dead leaves of previous seasons, draped along their branches (as in Figure 6.4*B*). Hypothetically, such a prominent feature, not shared with *Scalesias*' closest mainland relatives, may have offered a number of advantages, singly or in combination: moisture collection (old leaves provide surface area for moisture condensation, guiding drips toward the plant's own roots), herbivore deterrence, color contrast in attracting pollinators, efficient recycling of scarce nutrients (they eventually fall at the base of the plant), or reduced air flow and desiccation around new leaves (a particular benefit in lowlands). In the highlands, protecting tree *Scalesia* flower heads from encroaching epiphytes might be an additional advantage of dead leaves, but this would be an exaptation, since dead leaves also occur in lowlands where there are few epiphytes. The whole topic of *Scalesia* marcescence is ripe for further study.

S. cordata forest had already been cleared away on Sierra Negra [volcano] and estimated an original range of over 17,300 hectares" or 42,700 acres (in Mauchamp and Atkinson, 2010, 109). Years later, Mauchamp and Atkinson (2010, 110) estimated that "there has been a loss of 99.9% of *S. cordata* forest" in the area.[14] Sadly, their assessment produced parallel findings on other islands with tree species. There remains, they said, "a maximum of 100 ha of *Scalesia* forest on Santa Cruz. This tiny area represents 1.1% of the original forest [of 9,600 hectares, see Figure 6.6]. . . . In San Cristóbal, there is no *Scalesia* forest left (0% of the original distribution), while in Santiago the remaining area is restricted to within the five fenced areas that were constructed between 1974 and 1999 and covers a total of 1.1 ha (less than 0.1% of the original distribution). There may be a little more remaining on Floreana, but data collection there is not yet complete" (p. 109). Later studies found that the *Scalesia* forest of Floreana, now just 2.3 km² or about 11.3.8% of its original extent (data from Dvorak et al. 2017, 134), is actually the largest remaining stand in all Galápagos (Peters and Kleindorfer 2019, 321).

Clearing for home sites and agriculture thus had a major impact on the tree *Scalesias*—a loss we regard today as most unfortunate. But was it the work of disrespectful, outlaw farmers, insensitive to the needs of conservation? Not really. We must remember that immigrants to the islands were given land by the government of Ecuador specifically for farming. There was no significant conservation attention to Galápagos until the centenary of the publication of *On the Origin of Species* in 1959, well after the bulk of early immigration. Indeed, the government of Ecuador had a promotional land grants program designed specifically to encourage immigrants to use land for agriculture—which meant development in the humid zones. Yes, *Scalesia* conservation is a serious matter today, with efforts underway on several islands to restore tree *Scalesias* and other species,[15] but we should think twice before condemning the early settlers of Galápagos, who were awarded land for their heroic efforts at colonization in a difficult ecosystem, long before conservation there began.

One day, thinking about the tragic decline in tree *Scalesias*, I read an equally disturbing paper about the decline in several species of tree finches, plus other birds, in the *Scalesia* habitat of Santa Cruz and other islands (Dvorak et al. 2012). Field

---

[14] A study of the tree daisy *S. cordata* in 2004 (Philipp and Nielsen 2010, 497) found that "One of the largest natural stands of *S. cordata* . . . is now in an enclosure." Their study revealed that "a small population size, the low production of viable seeds and the low frequency of young trees, combined with the possibility of low genetic variation due to genetic drift, make *S. cordata* extremely threatened" (p. 502).

[15] One example is the GNPD's work on Santa Cruz using herbicide to combat invasive blackberries, *Rubus niveus*, in favor of *Scalesia* seedlings (Jäger 2015). Another is the Ministry of the Environment's nursery for *S. pedunculata*, which has grown over 15,000 giant daisy seedlings for transplanting.

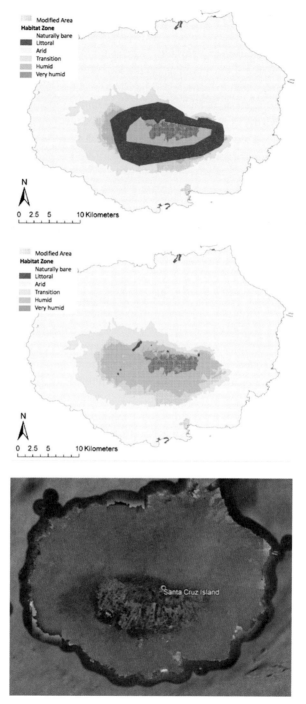

Figure 6.6. Approximate loss of *Scalesia* forest on Santa Cruz Island over a 100-year period (roughly 1905 to 2005). Green shading in top two panels shows vegetation

Figure 6.6. Continued
zones (transition, humid, and very humid) with keys at left. *A,* In 1905, *Scalesia* forest was extensive (*shown in red*) according to a survey by botanist Alban Stewart of the 1905–1906 California Academy of Sciences Expedition. It was estimated at 9,600 hectares by Mauchamp and Atkinson (2010). *B,* One hundred years later, only 100 hectares (1%) of that same forest remains (*also in red*), as per Mauchamp and Atkinson (2010). *C,* The island as viewed in Google Earth 2020, showing agricultural clearings in the humid and very humid habitat areas (scarred tracks in the dark green area). White spots are clouds. *Sources: A* and *B* courtesy of R. Atkinson from Mauchamp and Atkinson (2010, 111), in Galápagos Report 2009–2010. Puerto Ayora, Galápagos, Ecuador; *C,* Google Earth Pro (2020): Image Landsat/Copernicus.

surveys and a growing literature pointed to declining populations of large tree finches, medium tree finches, woodpecker finches, warbler finches, and vermilion flycatchers—all regarded as insectivorous birds. Moreover, they were declining inside *Scalesia* forests as well as in adjacent agricultural habitat with scattered *Scalesias*. I wondered if the two trends—*Scalesia* declines and tree finch declines—might be directly related. I knew from the work of multiple authors, including the Grants, that cactus finches have a close coevolutionary relationship with *Opuntia* cacti in Galápagos.[16] I wondered, were parallel declines in tree *Scalesias* and tree finches hinting at a similar coevolutionary relationship between the species?

Before long, I found another suggestive hint: despite the tree finches' reputation for being mostly insectivorous, a year-long study of feeding behavior among finches on Santa Cruz (Guerrero and Tye 2009) showed that most tree finches are actually omnivores, with substantial numbers of fruits and seeds in their diets. Furthermore, the study revealed that their fruit- and seed-eating behavior is fairly inefficient, an inadvertent benefit to the plants: the small tree finch and woodpecker finch, for example, discarded seeds or swallowed them whole for about 25% of their total feeding behavior.[17] This finding means that roughly a quarter of the total feeding effort of these supposed insectivores could readily disperse plant seeds. Two other finch species, the large tree finch and the olive

---

[16] The slow-growing, long-lived *Opuntia* cacti are classified as six endemic species in Galápagos, all descended from a single colonization event (Helsen et al. 2009). The cactus finches that use their flowers, fruits, and seeds for nourishment have evolved specialized beaks, extra-long and sharply pointed, via evolutionary modification of the calmodulin pathway (Abzhanov et al. 2006).

[17] When all seed treatments, including "crushed," are included in the tally, 61% of small tree finch food was seeds. Tebbich et al. (2004) reported similar values for tree finches, but with variation between wet and dry seasons. The small tree finch went from 31% fruit and seeds in its diet in the wet season to 53% in the dry season. The large tree finch changed from 9% fruits and seeds in the wet season to 25% in the dry season, and the woodpecker finch went from 0% in the wet season to 38% in the dry. This is an impressive amount of seed handling for birds widely known as insectivores. In contrast, the olive warbler finch's food was only 2% fruit and seeds in the wet season and 6% in the dry time of year.

warbler finch, did almost as well—20% and 15% of their feeding, respectively, resulted in discarded and swallowed seeds[18]—and a warbler finch was observed directly feeding from the seeds of the *Scalesia pedunculata* tree. The only outlier was not officially a tree finch: the leaf-eating vegetarian finch was found to have no seeds in its diet. Of all Darwin's finches in the study, including the seed-specializing ground finches, tree finches proved to be *the best* seed dispersers.

This hint seemed important, not only for what it says about the overlooked seed dispersal capacity of tree finches, long regarded as simple insectivores, but also to correct a common misconception about *Scalesias*. The dispersal of the giant daisies has sometimes been improperly equated with that of dandelions, because they are in the same plant family, Asteraceae, and have small round flower and seed heads. Some sources have gone so far as to depict *Scalesias* as wind-dispersed plants.[19] But *Scalesia* seeds look like small versions of sunflower seeds, many times the mass of dandelion seeds, and lack altogether the parasol-like projection, or pappus, that allows dandelions to ride the wind. Botanist Ole Hamann (2011, 410; see also Philipp and Nielsen 2010, 501) put it well: *Scalesia* seeds, he said, without a major pappus, have "no morphological adaptations for long-distance dispersals." Instead, they seem much better candidates for dispersal by birds, as suggested by Eliasson (1974, 10).[20]

From these various hints, it seemed possible that tree finches and their ancestors could have played a major role in pre-settlement Galápagos, transporting tree *Scalesia* seeds around the archipelago. A reciprocal relationship also seemed possible: in pre-settlement times, there were few tree species on the islands, as noted in Chapter 3, and very few in the highlands. What species composed the primary habitat for tree finches and their seed- and insect-foraging activities if not the tree daisies? Could these various services between finches and *Scalesias* be part of a set of mutual benefits that fueled their coevolution? If so, this would mean that tree finch decline today could contribute to tree *Scalesia* decline through lack of seed dispersal, and similarly tree *Scalesia* decline could be part of tree finch decline due to habitat and food loss.

---

[18] Castro and Phillips (1996, 130) had reported, even before the study described here, that the insectivorous large tree finch also "feeds upon fruits of native plants," although they did not mention swallowed or discarded seeds.

[19] In a beautiful video, *Galápagos 3D* (Colossus Productions, 2013), British naturalist David Attenborough is explicit on the matter. During a scene in "Episode 2, Adaptation" that shows both *Scalesia* forests and dandelion seeds floating across the screen, he says "This very special kind of dandelion is called *Scalesia*. It's unique to Galápagos."

[20] Eliasson (1984, 105) offered a second pathway for tree finch dispersal of *Scalesia* seeds. As he noted, small insect larvae often "bore tunnels in the dense flower aggregates of the heads," whereupon "the injured heads secrete a sticky, gummy exudation" that makes seeds sticky. "Sticky [seeds] might adhere to the bill or plumage of Darwin's finches that regularly perch in the *Scalesia* canopy and search for insects," and thereby be dispersed.

It is important to mention that some causes of finch population decline are only remotely related to *Scalesia* forest decline, if at all, including rats, mosquitoes, and the dreaded parasitic fly mentioned earlier, *Philornis downsi* (Figure 4.4). But to the extent that "habitat modification" is a problem for the finches—often listed as the #1 threat (as in Mauchamp and Atkinson 2010)—a coevolutionary relationship between *Scalesias* and tree finches could change the way we think about their natural histories and conservation.

Hypothesis

The remainder of this chapter offers preliminary tests of the hypothesis that the four tree *Scalesia* species, native to the humid-zone habitat of Galápagos islands, coevolved with four species of tree finch: small, medium, and large tree finches and the woodpecker finch.[21] I call the hypothesis "loose coevolution" to avoid implying a tight species-by-species association (one *Scalesia* to one finch). Rather, I hypothesize that the daisies and birds incrementally, interactively, opened new niches for one another. Today, tree finches are insectivorous omnivores who collect their insect prey from dense epiphytes and decaying wood in the *Scalesia* forests and who are also seed eaters and dispersers to an important extent. Before the evolution of tree daisies, there would not have been forests in the humid zones, dense with epiphytes, insect inhabitants, or seeds. And before tree finches, few *Scalesia* seeds would have reached the humid zone of any island, much less colonized similar zones throughout the archipelago.[22] Here I explore the possibility that these species emerged together as a "chicken and egg" coevolutionary product.

Testing the Coevolution Hypothesis

I came up with three different preliminary tests of the hypothesis that tree *Scalesias* coevolved with tree finches. First, it seemed important to test whether

---

[21] I omitted one member of the tree finch group, the mangrove finch (*Cactospiza heliobates)* because it specializes in coastal mangrove forest habitat, with a completely different insect fauna and, tragically, only a few dozen of these birds remain in the archipelago (see Lawson et al. 2017).

[22] Early on, I considered giant tortoises as candidate dispersers of tree *Scalesias* within and between islands. Domed tortoise migration could potentially explain, for example, how tree *Scalesias* moved between islands and reached the humid zones of diverse islands. But tortoises have access to tree *Scalesia* seed heads only when mature trees topple during winds or El Niño rains. Further doubt was cast by the thorough analysis of tortoise dung by Blake et al. 2012 (see Chapter 3), who found no *Scalesia* seeds amid seeds of 45 other plant species. Evidence is good, however, that shrubby *Scalesia affinis* is dispersed by land iguanas (*C. subcristatus*; see Traveset et al. 2016).

the distributions of tree *Scalesias* and tree finches show close correspondence in space and time. In space, to begin with, one can compare distributions by island of tree finches and tree *Scalesias,* asking if they do indeed co-occur today. Drawing on reliable sources, including shrub *Scalesias* and small tree species, and noting simply presence or absence, I arrived at the match shown in Figure 6.7. One finds tree finches today *only* on islands with *Scalesias,* with one exception (as shown, Marchena has large tree finches and no *Scalesias*).[23] Moreover, every island supporting four species of tree finches also has full tree-size *Scalesias.* There are only three islands, of 17 in my tally, with at least one *Scalesia* species and no tree finches, but one of them—Wolf—has potential seed-dispersing services of the grey warbler finch.[24] And the other two islands, Baltra and Seymour Norte, were part of Santa Cruz as recently as the Last Glacial Maximum, 20–30,000 years ago (Geist et al. 2014), and then had dispersal services of tree finches, most likely in the dry season. Thus, evidence points to a reasonably solid concordance in space.

With respect to time, a new phylogeny of the *Scalesias* (Fernández-Mazuecos et al. 2020) shows striking concordance with the temporal diversification of tree finches. As noted in the discussion of Figure 6.2, the line leading to tree finches split from that of ground finches at roughly 412,000 YA, and began further diversification soon thereafter. The new daisy phylogeny is a splendid match to that figure, with the first split from a shrub to a tree lineage at 430,000 YA, well within confidence limits of the tree finch date.[25] That tree lineage went on to differentiate at roughly 380,000 YA, hypothetically by tree finch dispersal to different islands, into what became *S. pedunculata* and *S. baurii,* the tallest and shortest of today's arboreal daisies. Then about 350,000 YA, another *Scalesia* lineage diversified from its shrubby past into a second tree-line ancestor, which then differentiated—again hypothetically, via tree finch dispersal to separate volcanoes—into tree species *S. cordata* and *S. microcephala.* Concordance in

[23] Castro and Phillips (1996, 130; also Grant 1999, 62) noted that, on most islands, the large tree finch "prefers the highlands, although it can occasionally be found feeding in the transitional zone and the coast. In Marchena [alone, it] breeds in the lowlands." Further study of the population on Marchena is certainly warranted.

[24] As noted in Figure 6.2, the two species of warbler finches, olive (*Certhidea olivacea*) and grey (*Certhidea fusca*), were the first species to diverge and are thus the oldest finch species. They occur throughout the archipelago, one species per island. The olive species occurs on central islands where it is "confined to moist upland [*Scalesia*] forest," whereas the grey species occurs on peripheral islands that are generally low and dry (Tonnis et al. 2005: 824). Both species are primarily insectivores, but the olive species includes a small number of fruits and seeds in its dry-season diet (Tebbich et al. 2004; Guerrero and Tye 2009), and the grey species probably does as well.

[25] Mario Fernández-Mazuecos kindly pointed out to me (email of November 27, 2020) that a tree *Scalesia* line may well have evolved a bit before 500,000 YA, by the time of the split of ancestors to *S. cordata* and *S. microcephala* from ancestors of *S. affinis,* but even that "would still be a good match to the split of the tree finches."

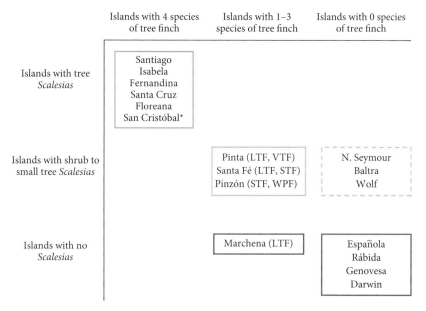

	Islands with 4 species of tree finch	Islands with 1–3 species of tree finch	Islands with 0 species of tree finch
Islands with tree *Scalesias*	Santiago Isabela Fernandina Santa Cruz Floreana San Cristóbal*		
Islands with shrub to small tree *Scalesias*		Pinta (LTF, VTF) Santa Fé (LTF, STF) Pinzón (STF, WPF)	N. Seymour Baltra Wolf
Islands with no *Scalesias*		Marchena (LTF)	Española Rábida Genovesa Darwin

Figure 6.7. Concordant spatial distribution of tree finches and *Scalesias* in Galápagos today. Tree finches are found only on islands with *Scalesias* (*orange and green boxes*), with the exception of Marchena (*small red box*). *Scalesias* are found only on islands with tree finches, except for three low islands, North Seymour, Baltra, and Wolf (*dashed orange box*). Wolf has the seed dispersal services of the grey warbler finch; the others were connected to Santa Cruz as recently as 20,000 YA and likely had dry-season dispersal services of tree finches. (LTF = large tree finch; VTF = vegetarian finch; STF = small tree finch; WPF = woodpecker finch.) *Sources*: On *Scalesia*, Wiggins and Porter (1971), Eliasson (1974), and Itow (1995); on finches, Harris (1974) and Castro and Phillips (1996). *Note*: Vargas (1996, 23) says of the fourth species—LTF—on San Cristóbal, "common in the past . . . [was] not observed at all" in 1996.

time means that the evolving tree finches could have promoted these tree *Scalesia* dispersals and diversifications, as is consistent with their impressive record today at interisland flight.[26] The match in time also supports a hypothetical reciprocal influence: the differentiation of tree *Scalesias* into distinct species—along with the co-resident plant and animal species that evolved with them—could well have shaped the diversification of tree finches. The best available evidence thus points to substantial concordance in both space and time between the tree daisy and tree finch species in question.

---

[26] A recent study of interisland gene flow (Lawson et al. 2019) found tree finches to show substantial migration rates, and the large tree finch to have the highest migration rate of all Darwin's finches.

My second test examined important ecological relationships between tree finches and tree *Scalesias* today, drawing upon the literature for both taxa. Ideally, one would like to show that the tree and finch species have interdependent fitnesses—i.e., that their abilities to survive and reproduce directly depend on one another. Interdependent fitnesses would be the case, for example, if tree finches dispersed tree *Scalesia* seeds within and between islands at the same time as tree *Scalesias* provided tree finches with key food resources and shelter. Showing that they do so today (i.e., in recent decades) is a necessary test; by the same token, it is insufficient, for one also needs to establish that their fitnesses were interdependent historically, as the species differentiated to their current forms. The latter is easier said than done: one often has no choice but to *assume* that at least some of the key interactions observable today extended indefinitely into the past, and here I, too, make that assumption.

As a test for interdependent fitnesses, I followed parallel examples in the literature (e.g., see Heleno et al. 2011; Traveset et al. 2015) and carried out an interaction analysis between tree finch species and tree *Scalesia* species, eventually focusing on *Scalesia pedunculata* because it offered ample data. I included three finches outside the tree group for comparison: the vegetarian finch as a "control species" (which rarely, if ever, eats seeds) and the two mostly insectivorous warbler finches because Guerrero and Tye (2009) observed one of them eating *Scalesia* seeds. With the able help of research assistants, I thoroughly surveyed the literature for all reported ways that the finches use the main tree species of the humid zone, and I did the same for a set of control birds not hypothesized to have coevolved with *Scalesias*. The set of humid-zone tree species included *S. pedunculata* and three control tree species not hypothesized to have coevolved with tree finches.

Our survey of the literature focused on published evidence of the following seven fitness-related interactions: fruit eating, pollination and/or flower feeding, seed predation, seed dispersal, nesting, insect foraging, and "other" feeding, such as leaf eating.[27] We went to great lengths to survey all available literature and to eliminate double counting (as when authors reported the same observation in multiple publications). Later, a different assistant and I repeated the literature review as an accuracy check and resolved any discrepancies. To prepare the final figures, we tallied the numbers of pairwise "bird × tree" interactions

---

[27] I have assumed that each of these interactions has equal impact on the fitnesses of the host plants and on the fitnesses of the birds. Admittedly a big simplification, this assumption allows me to simply add up the number of different interactions for a measure of their total fitness impact. Someday, with a larger data set, one could break out categories of interactions with greater vs. lesser impact for each of the partners in the interaction.

and represented the totals for each pair by proportional shading from white (0 interactions) to black (highest tally in the data). (For methodological details and published sources, see Online Appendix D.)

To provide a model of what close interdependence looks like with these data and shading, I used the same procedure to portray the long-recognized interaction between cactus finches, plus other ground finches, and various species of *Opuntia* cactus. It's an especially good model for comparison because earlier analysis confirmed, on the basis of anatomy, that cactus finches have had a close evolutionary relationship with *Opuntia*. Studies by Abzhanov et al. (2006; further elaborated in Grant and Grant 2014) showed that the cactus finch evolved an extra-long and pointed beak, via an identified genetic mechanism, with advantages for feeding from the flowers and fruits of the *Opuntia*. So, in this case, too, we tallied—and checked—the interactions of cactus finches and their other ground finch relatives with the various *Opuntia* species of different islands. To that matrix of interactions, we added a set of three control birds, not hypothesized to have coevolved with *Opuntias,* and two control tree species, not hypothesized to have coevolved with cactus finches or other ground finches. We used the same interaction coding scheme and the same subscripting convention, keyed to the publications we found in the literature (data and details in Online Appendix D).

Consider, first, the arid-zone interactions summarized in Figure 6.8 by proportional shading. The area in the highlighted green box at center shows the "thicket" of interactions between test finch species and the *Opuntias.* These data are convincing that ground finches show many and diverse fitness-related interactions today with *Opuntia,* especially cactus ground finches, whose interactions with the cacti form a darkly shaded row.[28] Surrounding that central thicket is a much lower density of interactions among the control birds and the *Opuntias,* and among the control plants and the full suite of finches.[29]

---

[28] Although cactus finches are part of the ground finch group (see Figure 6.2), I was initially surprised that *other* ground finches also show high densities of interaction with *Opuntia,* particularly since earlier researchers inferred a special relationship between *Opuntias* and cactus finches (see text). Detailed interactions outlined in Online Appendix D1 confirm that there are differences in the *kinds* of interaction hidden within the sums presented in Figure 6.8. Online Figure D1 shows that cactus finches are specialists in pollination and flower feeding, interactions coded P in that figure. Nearly half of all P interactions (44%) reported for ground finch species were from cactus finches, which is more than twice the proportion (20%) expected by chance among five ground finch species.

[29] These results make it look like an "intermediate" density of interaction takes place between various species of *Acacias* and ground finches, and between *Opuntias* and mockingbirds. However, that density is an artifact of the categories *Acacia* species, which aggregates three or more species of *Acacia,* and "mockingbirds," which aggregates four species into one count, because the source literature was not consistent about which species were included.

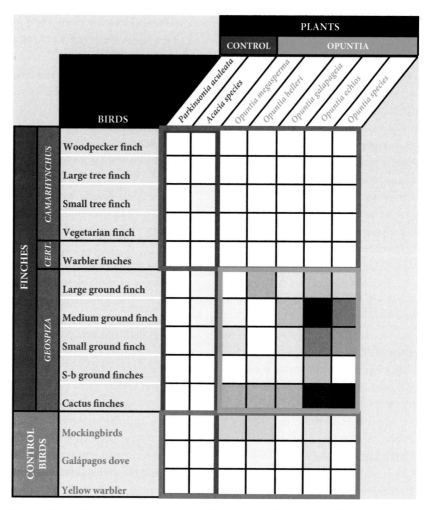

Figure 6.8. Arid-zone interaction matrix. Summarized here are the diverse interactions described in the literature between the bird species (*on the left*) and the plant species (*across the top*). Both birds and plants include test species hypothesized to interact heavily with each other, plus control species with no special hypothesized interactions. Shading is proportional to the total number of interactions reported. All forms of interaction in the analysis are simply summed together here (for a breakdown of specific interactions, see Online Appendix D). Here, data on cactus finches and large cactus finches are combined as "cactus ground finches." Shading confirms that the group of ground finches (genus *Geospiza*) have a "thicket" of diverse interactions (*outlined in green*) with the four *Opuntia* test species, plus unspecified *Opuntia* spp. (*at top*). Cactus ground finches show particularly dark shading across the row of *Opuntias*, reflecting their special relationship. Interactions are less dense between ground finches and control trees, *Parkinsonia aculeata* and *Acacia* spp.

Figure 6.8. Continued

They are also less dense between control birds and *Opuntias* (*outlined in orange*), and between warbler finches (Cert., for *Certhidea*), the tree finch group (*Camarhynchus*), and *Opuntias* (*outlined in blue*). For simplicity, all tree-dwelling finches are shown here and in the following two figures under the heading "*Camarhynchus*" (usually, vegetarian finches and woodpecker finches are assigned to other genera, as in Figure 6.2). This measure was taken to make these figures easier to read and interpret. The several sharp-beaked ground finch species are also here combined and abbreviated as S-b ground finches. *Source*: See Online Appendix D.

Now consider the humid-zone interactions of Figure 6.9, represented by the same shading system. The figure confirms, with another, smaller thicket of fitness-related interactions, that a special connection exists between *S. pedunculata* and the group of tree finches—all *except* for the leaf- and bud-eating vegetarian finch (serving as an intra-*Camarhynchus* control). There are also high-density interactions between tree *Scalesias* and warbler finches. By the same token, the humid-zone control birds generally do not interact with tree daisies, and the tree finches generally do not interact with the control trees. Thus, a key finding of this analysis is that the density and kinds of interactions between tree finches and *Scalesias* are on par with the density and kinds of interactions between cactus finches and *Opuntias*, interactions strong enough in the past for evolution to reshape cactus finch beaks.

This conclusion is especially clear in Figure 6.10, which omits the control plant species from both arid and humid zones and focuses on the test *Opuntia* and *Scalesia* species, along with all birds used in this analysis, including controls. One can see that the interaction densities are similarly high for the test species in all three green boxes (*Opuntias* and ground finches, *Scalesias* and tree finches, plus *Scalesias* and warbler finches) and that they are much lower in the control birds' orange boxes. Note that bird species not in the test group for a given zone can be considered control species for that zone. Also note that the interactions of control birds in the humid zone (labeled C1) show even greater contrast with the green test boxes for that zone—that is, they show very few *Scalesia* interactions—than the same comparison of control birds with test plants in the arid zone (marked C2). One reason, I suspect, for this latter difference is the general scarcity of food resources in the arid zone, which has likely selected for a more generalist plant-use strategy among birds there.[30] In any case, the findings of interaction analysis are clear: tree finches and *Scalesia pedunculata* show a density of interactions on

---

[30] Birds on islands often interact with a wider than normal range of plants, as is termed "interaction release" by Traveset et al. (2015). The suggestion here is that interaction release is greater in arid zones, known for their acute food shortages, than in normally lush humid zones.

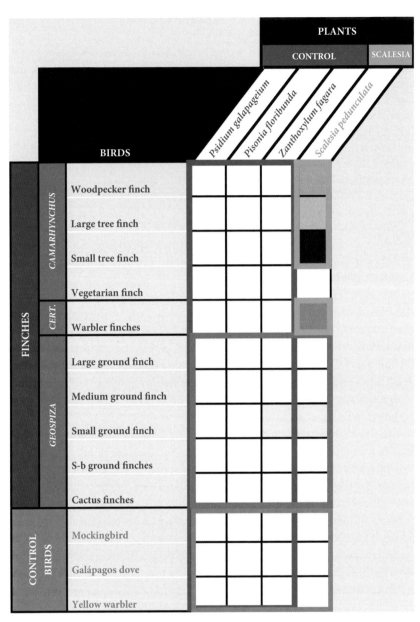

Figure 6.9. Humid-zone interaction matrix. The array of species follows the form of the previous figure: bird species at left, plant species across the top. However, here the test birds are the tree finch group (labeled "*Camarhynchus*") and the warbler finches ("Cert.," for *Certhidea*). The test plant here is the most widespread humid-zone tree, *Scalesia pedunculata*. Shading is proportional to the total number of interactions reported. Here there is a thicket of interactions (*green boxes, upper right*) between warbler finches and *Scalesias* (*lower green box*), and the tree finch group and *Scalesias*, except the vegetarian finch, as expected. *Orange, blue*, and *purple boxes* represent interactions with control species for comparison. *Source*: See Online Appendix D.

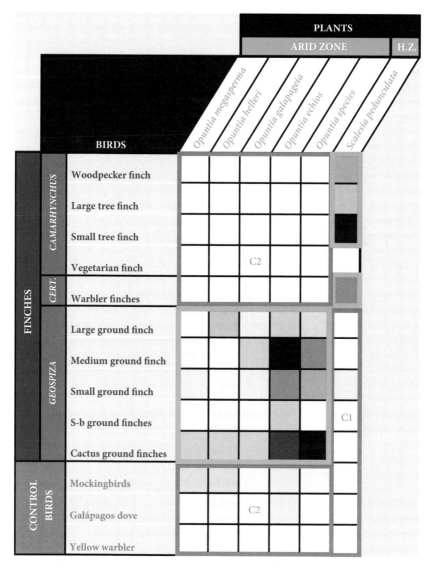

Figure 6.10. Summary interaction matrix for arid and humid zones combined, omitting the control plants for simplicity. As in previous figures, bird species are shown at left, plant species across the top (where H.Z. stands for humid zone). Shading is proportional to the density of interactions reported. *Green boxes* represent dense thickets of interaction. Note especially that the shading in the row of cactus ground finches across the *Opuntias* is similar to the shading of the column of *S. pedunculata* across the tree and warbler finches, minus the vegetarian finch (a control). Other control species' interactions are shown in *orange boxes*, labeled C1 and C2, as explained in the text. *Source:* See Online Appendix D.

par with those of cactus finches and *Opuntias*, which have an established coevolutionary history.

The third test of the proposed hypothesis asks: Are the population dynamics (increases and decreases) of tree *Scalesias* and tree finches correlated? When one of the proposed partners declines in number or density in an area, for example, does the other also decline in that area? Given the postulated mutual benefit of their coevolution, one expects at least a rough correlation.

In fact, such a correlation was one of the first clues to the possibility of coevolution. As noted above, the declining area of *Scalesia* forest on Santa Cruz is associated with the decline documented by Dvorak et al. (2012) in the numbers of woodpecker finches of the tree finch group (see Figure 6.11). The association holds both inside *Scalesia* forest habitat as well as in adjacent agricultural habitat (where *Scalesia* density is, of course, lower), and it holds for warbler finches and the insectivorous yellow warbler as well—all shown in the figure. The authors also report "significant decreases," not shown in the figure, "in at least one vegetation zone, of the large tree and vegetarian finches, but sample sizes were low" (Dvorak et al. 2012, 83).[31] In contrast, the "small ground finch, a foraging generalist [which thus serves as a control species for this purpose], *increased* significantly from 1997/1998 to 2008 in both the agricultural and *Scalesia* zones" (p. 82). A more recent study (Fessl et al. 2017, 155) added the small tree finch to the tally of populations declining in the *Scalesia* zone. The same can be said for correlated decreases of *Scalesia* trees and insectivorous vermilion flycatchers (Merlen 2013); on San Cristóbal, where human activity has all but eliminated *Scalesia* forest, both the large tree finch (Vargas 1996) and the vermilion flycatcher (Carmi et al. 2016) are now listed as extinct.

Thus, all three preliminary tests turned up support for the hypothesis: tree finches and *Scalesia pedunculata* fit the pattern expected of coevolved species. Of course, the findings warrant further testing, including field experiments in Galápagos, with both tree and non-tree *Scalesias*—confirming their tree vs. shrub mature form, for example, in the absence of invasive herbivores—and with tree and non-tree finches, plus other bird species, closely monitoring their interactions with *Scalesias*. Moreover, the new phylogeny of daisies, plus techniques for inferring gene flow among populations of tree finches (Lawson et al. 2019), may soon make possible detailed biogeographic models, enabling

---

[31] The yellow warbler's decline is noteworthy. In a study of seed dispersal by birds on Santa Cruz (Guerrero and Tye 2011), the mostly insectivorous yellow warblers did swallow a few fruits and seeds (as shown in 6% of 179 fecal samples). Yet none of the seeds were from tree *Scalesias,* which were present in three of the study's four sample sites. So, the yellow warbler's decline in the *Scalesia* forest and agricultural zones of Figure 6.10 offers another kind of control for the arguments here. Given the warbler has no recorded direct interactions with *Scalesia* trees (Figure 6.9), the warbler's decline most likely reflects the loss of understory habitat for their preferred insect prey.

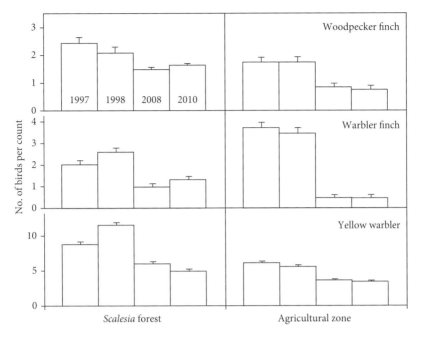

Figure 6.11. Population trends in woodpecker and warbler finches, plus yellow warblers, between 1997 and 2010 on Santa Cruz Island. Bars represent mean number of birds per point count (plus or minus standard error). Note that y-axis scales are different by species, and x-axis scales are not continuous. As *Scalesia* forest diminishes, woodpecker and warbler finches also decrease, as do insectivorous yellow warblers. For some species, the decrease is steeper in agricultural zones where *Scalesias* are few and scattered. *Source*: Used with permission of B. Fessl from M. Dvorak et al. 2012. "Distribution and Abundance of Darwin's Finches and Other Land Birds on Santa Cruz Island, Galápagos: Evidence for Declining Populations." *Oryx* 46, no. 1: 78–86. Reproduced with permission of Cambridge University Press through PLSclear.

one to ask just how concordant the two group's histories of interisland dispersal and evolution were.[32] For now, the best we can say is that the adaptive radiations of tree finches and tree *Scalesias* appear to have been interdependent in broad outline. As the tree finches diversified, the tree-sized *Scalesias* did, too, each contributing to the survival and reproduction of the other.

[32] I am grateful to the production team at Oxford University Press for allowing late modifications of this chapter, to incorporate implications from the *Scalesia* phylogeny of Fernández-Mazuecos et al. (2020). Prior to that publication, the only study of all *Scalesias* I had to work from was the classic morphological cladogram (i.e., a phylogeny without dates) of Blaschke and Sanders (2009), using 63 morphological characters—such as leaf margin, petiole, and ray morphology—across all *Scalesia* species and several daisies of the mainland. A noble effort, it is also misleading on such topics as the number of tree species. Similarly, the best chronology I had to work with came from Schilling et al.

We might go on to speculate, though I emphasize it's just a scenario for now, that sometime after the vegetarian finch differentiated (546,000 years ago; Farrington et al. 2014), the common ancestor of later finches might have found suitable food in the seeds of shrub *Scalesias* of the lowland and dispersed them far and wide in the archipelago. When a population of those finches established themselves in the humid zone of an island (perhaps Santa Cruz, as a guess), with their lowland *Scalesia* tag-alongs, Darwin's model for the evolution of taller and taller plants could well have come into play, facilitated by the evolving specialization of tree finches. As a forest of ever-taller *Scalesias* took shape, there would be more branches, shade, and surface area for epiphytes. An increase in epiphytes would, in time, increase the diversity of insect habitats of the highlands, potentially contributing to differentiation among the evolving tree finches. The mutualism of diversifying *Scalesias* and tree finches would then offer habitats to a growing number of other humidzone denizens, eventually giving rise to a full new ecosystem on high islands. Further advances in phylogenetics and biogeography will one day fill in many of the blanks, and it's a certainty that there will be surprises. For now, we have a plausible evolutionary pathway behind the thicket of interactions between tree finches and tree *Scalesias* and the emergence of the larger daisy-forest ecosystem, home to many other Galápagos endemics.[33]

Here let me note another hypothesis concerning the *Scalesias*. Like the marine iguanas (see Chapter 5), tree *Scalesias* may also have evolved their own adaptation to the ENSO cycle. Careful longitudinal study by botanists has established a pattern of "self-cyclic" stand mortality of mature tree *Scalesias* (for example, see Hamann 2001; Itow 2003; and Mueller-Dombois and Fosberg 1998) followed by mass seed-setting when conditions are right. One way this happens is for the mature daisy trees in an area to topple over during the soaking of an El Niño event, and for the new generation to begin at the next La Niña, because "*Scalesia pedunculata* will not seed properly except in dry

---

(1994, 252) who estimated "approximately 1.9 to 6.3 million years . . . for the total divergence time" between mainland relatives and Galápagos *Scalesias*. That study was based on dating chloroplast DNA divergences, a kind of work known for large error terms. In earlier drafts of this chapter, I felt I was putting myself out on a limb to suggest that all four tree daisies evolved from shrubby ancestors within the last 400,000 to 500,000 years, on a par with the tree finches. I am grateful for the chance to confirm here that the limb proved to be woody.

[33] Given the unique species of this case, one could fairly ask, is this mutualism a unique outcome of the special "creative force" of Galápagos, or are there equivalent bird–tree mutualisms elsewhere? In *Made for Each Other,* Lanner (1996) explored the equally fascinating relationship between pine trees in the North American West and a group of jays and nutcrackers (of the crow family, Corvidae). The pines are unable to reproduce without the services of the birds, who in turn depend upon nutritious pine seeds to rear their young. In performing services for one another, the mutualists also create a full ecosystem with resources and opportunities for many other organisms.

weather" (Kastdalen 1982, 8).[34] Another way for this to happen is for the die-off (of the earlier generation) to be triggered early in the course of a prolonged drought, such as those prompted by La Niña, that then also prompts the mass seeding of the next generation.[35]

No matter how it happens, this unusual pattern of disturbance-induced stand mortality and drought-induced self-replacement is a common feature of the tree *Scalesias*, *S. cordata* and *S. pedunculata,* and likely of *S. microcephala* as well (Lawesson 1988). It's an almost ideal setup for kin selection to promote crown shyness, as discussed earlier, because many closely related seedlings will grow up near one another, almost like a clone. Furthermore, the role of both ENSO stages (El Niño and La Niña) as triggers suggests that the life-cycle sequence of *Scalesias* represents another adaptation to ENSO. Such a hypothesis certainly warrants further study, including careful attention to the physiological mechanisms underlying synchronous stand mortality and massive seed set. Could tree *Scalesias* be as well tuned to ENSO's climatic swings as the concertina-like oscillations of marine iguana body length? Given the deep history and dramatic impact of ENSO cycling in Galápagos, we should not be too surprised to find multiple corresponding ENSO adaptations among organisms there. If the hypothesis holds, there is another intriguing implication: the daisy-forest-dependent tree finches may also have evolved some "rebound" resilience to *Scalesia* diebacks, perhaps as simple as a behavior change to feeding on seeds and insects on the ground, among fallen *Scalesias,* rather than in the canopy. If so, prompt action today to restore anthropogenically damaged tracts of *Scalesia* forest may also help tree finch populations recover, as if there had simply been a rather prolonged La Niña.

Alas, the proposed ENSO adaptation of tree *Scalesias* highlights an additional vulnerability. Shade-tolerant invasive trees—like red quinine (*Cinchona pubescens*), the guava tree (*Psidium guajava*), citrus trees, and Spanish cedar (*Cedrela odorata*)—do not wait for a stand dieback or dry spells to germinate; their seedlings get the jump on *Scalesia* seedlings, which are seriously shade intolerant. As a result, *Scalesias'* evolved wait for stand dieback and subsequent drought—if such it be—inadvertently gives advantage to the invaders. Adding to their woes, the invasive blackberry, *Rubus niveus*, is rapidly advancing into the last remaining piece of *Scalesia* forest on Santa Cruz (the large red spot in Figure 6.6*B*). One assessment projected a dark picture:

[34] As noted in Chapter 2, the Walker cell connects the ENSO events of Galápagos with those in Indonesia, producing opposite effects at the two ends. With that in mind, the El Niño-induced dieback of *Scalesia* can be seen as the Galápagos counterpart of El Niño-induced mast-fruiting of Dipterocarp trees in Indonesia (see Janzen 1974).
[35] Yet a third way for stand die-off and regeneration to occur is following a fire, whether natural (e.g., from a volcanic eruption) or human-induced (Lawesson 1988). But fire is much more likely to cause substantial stand dieback during the drought conditions of a La Niña event, and so even fire-based dieback is probably tied to ENSO.

While *R. niveus* has only reached high densities in the *Scalesia* forest in the last 5 years, . . . we can already see that these high densities are correlated to low light levels in the ground stratum, causing changes in the micro-climate normally present in the forest. An abundance of canes and a dense foliar layer produced by *R. niveus* creates a dark and wet habitat that is unlikely to be suitable for the recruitment of shade-intolerant native species [including *Scalesia pedunculata*]. . . . In addition, *R. niveus* is a scrambling species that may smother native plants, leading to a dense monotypic thicket with little other vegetation present. (Rentería et al. 2012b, 8)

These words proved prophetic: a follow-up study found, 4 years later, no natural regeneration of the *Scalesias* in the area invaded by blackberry, whereas there was "spectacular recovery" in an adjacent area where the GNPD carried out control of the invader (CDF 2019, 37).

The impact of invasive species here parallels the case of Galápagos rails. Nonnative species brought in for use by highland farmers have escaped into growing feral populations with substantial impact on the fitness of the focal organism. In this case, they work their way into highland forests via shade tolerance and aggressive growth patterns (see Shimizu 1997; Hamann 2016), turning former *Scalesia* adaptations into stark vulnerabilities. Employing the SES framework, we see that human activity has rapidly complicated habitats where *Scalesias* once were dominant, especially via the fitness impacts of the introduced species. Can giant daisies handle the dynamic new SES that humans' actions have created for them? With the *Scalesias'* prominent role in humid-zone ecology, tree *Scalesia* conservation warrants high priority.

Conservation Implications

As with all hypotheses early in testing, we can reach only tentative conclusions about this chapter's candidate example of interdependent evolutionary radiations. But until we know better, even these preliminary results suggest we'd do well to consider the conservation of tree finches and tree *Scalesias* to be closely related. We can reasonably expect that there will be advantages to working on joint solutions, rather than fully separate interventions, especially considering scarce resources of funding and person-power. Certainly, there are species-specific issues on both sides, such as parasites and disease, where joint solution efforts may be too slow and indirect to help in a timely manner. But with respect to population restoration efforts, our findings here suggest advantages to working together on the conservation of tree finches and tree *Scalesias*.

Consider reseeding efforts, for example, to restore stands of *Scalesia pedunculata* in the highlands of Santa Cruz and San Cristóbal (and, ideally, soon Floreana). Human intervention for planting, weeding, and even watering may be inescapable at the start, along the lines of several ongoing efforts (see also Shimizu 1997). But if such efforts gave priority to areas with substantial—even if declining—tree finch populations nearby, finch activity could soon obviate the need for human-assisted planting and epiphyte introduction. Greater areas and densities of mature *Scalesias* with epiphytes, in turn, will improve food supply and nesting locations for the finches and the availability of suitable territories for vermilion flycatchers. In a few years, the full effects of positive feedback may be realized, as the coevolved mutualists work their various benefits to one another. That is part of the beauty of mutualisms, after all: the benefits on both sides are self-reinforcing. With less than 2% of the original *Scalesia* forest remaining in Galápagos, and less than 1% on Santa Cruz (CDF 2019, 37), there is urgent need for tree *Scalesia* reforestation, which, to all indications, will also serve as an important measure for tree finch conservation.

There is an important corollary argument to be made concerning human use of the humid zone on several islands. We must be forward-looking in efforts to help tree *Scalesias* and tree finches, searching for effective solutions, rather than focusing over our shoulders on agriculture and agriculturalists as causes of the problem. We must find conservation pathways going forward in active agricultural landscapes, collaborating *with* agriculturalists and highland settlers and recruiting their useful knowledge of the humid zone into a broader conservation effort. Today's endeavors are bound to go further if they build local people into planning and execution.

Several recent efforts along these lines should surely be encouraged and expanded. For example, the program Galápagos Verde 2050 (Jaramillo et al. 2015) uses novel water-conservation and sapling-protection technologies to raise tree *Scalesias* and others of the genus, recruiting the participation of highland farmer's associations. There would be great value in expanding this program, especially to properties sharing—or very near—a boundary with the *Scalesia* forests of the National Park. Contiguous planting would encourage epiphyte colonization from nearby *Scalesia* stands and eventual range expansion by tree finches coming out of Park forests, which would ideally reinforce the further spread of tree *Scalesias*. In some areas, this effort would have to go hand in hand with a campaign to cut back invasive cedar, guava, quinine, and blackberry and to eliminate their seed banks.

Another, complementary, method is suggested by the efforts of the NGO FUNDAR (Foundation for Responsible Alternative Development in Galápagos). This organization has experiments underway with coffee cultivation side by side with, and under, *S. pedunculata* on highland farms, which so far appear successful. Although certainly an introduced species, coffee is not an aggressive invader and is thus easily controlled. "Conservation coffee" is also produced in the shade of *Scalesias* on private coffee farms, such as Lava Java, MonteMar, Semilla Verde, and others. Imagine the special appeal of a marketing campaign for dedicated international consumers: "sustainable coffee, shade grown in Galápagos, in restored forests of rare, endemic giant daisies." If such efforts are successful, their revenues could help fuel their expansion. On San Cristóbal, where *S. pedunculata* forest has been all but eliminated, coffee might well be enough to bring back a *Scalesia* canopy in some areas.

Finally, there is promise in trying a "payment for conservation services" (PCS) program that would use funds from tourists to pay farmers an annual service fee for supervised planting and care of tree *Scalesias*. I suggest that PCS funding could come from a small "sustainability assessment"—say $10 or $20 per person—required from all foreign visitors, separate from Galápagos National Park entrance fees, since the Park rightfully has its own priorities, and this work must be done outside the Park.[36] The goal would be to start the PCS program in farm areas contiguous with existing *Scalesia* forest patches, to provide seed sources, epiphytes, and finches. The farmers would be paid a supplement not for cultivating and grazing in a portion of their land, but for planting it with *Scalesias* from a nursery and caring for them into maturity. With planning and organization, farmers could also make supplemental income from ecotourism and educational programs as visitors come to see tree finches and tree *Scalesias* in their only home on Earth. An analogous program, but without tourism incentives, was introduced with the 1985 Farm Bill in the United States; in the Conservation Reserve Program (CRP), the U.S. government pays farmers to remove marginal agricultural land from production to reduce soil erosion and to promote conservation practices instead.[37] Lessons from the U.S. program could well have value for a PCS program in Galápagos, and thus help speed restoration of daisy forests.

In summary, Figure 6.12 pulls together the main arguments of this chapter, highlighting the conservation implications of recent threats and vulnerabilities for this group of mutualists. Beginning at the upper left, islands with highland areas offered a wide-open tree niche to the initial *Scalesia* ancestor. It seems likely that omnivorous finches provided dispersal services back then, eating the

---

[36] This sustainability assessment could fund a variety of projects over time and also allow, for the first time, direct support from tourism for the work of the Charles Darwin Research Station. This idea is discussed further in Chapter 9.

[37] The CRP is still in place in the United States today. The USDA allots roughly $2 billion annually for contracts that last for 10 to 15 years, reducing soil erosion by over 200 million tons per year (USDA 2016b). Recently, the CRP announced their intention to also include wildlife protection, like thinning the canopy of forest tracts to allow sunlight to reach the forest floor, in their incentive program (USDA 2016a).

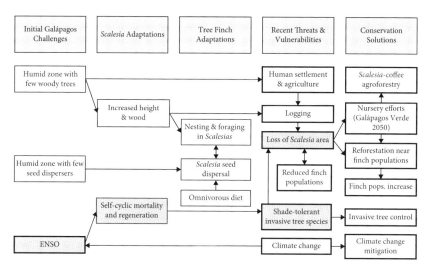

Figure 6.12. Schematic summary of main arguments in the proposed coevolution of tree daisies and tree finches in Galápagos. Highlighted here are the interactive adaptations of *Scalesias* and finches to highland areas, plus the impact of human settlement and agriculture in the same areas. The figure also suggests that *Scalesia's* stand die-off and cyclic recruitment are adaptations to the ENSO cycle. The linkages between the ENSO cycle, spread of woody invasive species, and loss of *Scalesia* forest are shaded. Boxes related to human activity are *bolded*.

seeds and sometimes flying with them to diverse humid zones of the islands. Over time, finches and *Scalesias* then facilitated each other's adaptations to those zones, opening up resources and opportunities for one another as byproducts of their own evolution. Tree *Scalesias* evolved their arborescent shapes, enhancing their capacity to collect moisture from cloud interception, and thus their suita-bility for epiphytes and insects. Those same shapes provided finch nesting sites with good foraging close at hand, thus promoting the evolution of tree finches, and ensuring a ready supply of seed dispersers for the trees. Hypothetically, in response to the ENSO cycle, the *Scalesias* then evolved their synchronous stand diebacks, allowing the next generation to germinate and grow unobstructed in full sun. The mutualism of tree finches and tree *Scalesias* would thus have pro-duced a new forest ecosystem in the archipelago, eventually becoming home to dozens of species. Much later, the area of this ecosystem changed drastically with human settlement, when vast areas of the giant daisies were logged for farmland and construction, as a result of government inducements to settlers. Along the way, farmers introduced shade-tolerant hardwood trees with aggressive growth patterns and no stand die-offs. These hardwoods began replacing the *Scalesias*, further reducing habitat and food supply for tree finches.

Figure 6.13. Another member of the tree finch group, the woodpecker finch (*Camarhynchus pallidus*), feeds on insects, grubs, and seeds, but without a woodpecker's head and neck muscles. Instead, it uses sticks as probes to extract prey. The finch shown here uses a twig of cat's claw (*Zanthoxylum fagara*, one of the control trees of this chapter), with spines oriented for easy insertion and efficient capture, to probe the knot behind it in the tree trunk.

But all is not lost. Some small stands of *Scalesias* remain, and many of them still have tree finch populations, although the finches are declining. With concerted reforestation efforts and a proposed PCS system for farmers, the giant daisies may one day regain large contiguous stands and the full services of the tree finches. It's truly not a matter of somehow making the humid zones "pristine" again, but it is a chance for humanity to share Galápagos with some delightful coevolved mutualists and their associates (Figure 6.13).

In these ways and others like them, it is time for conservation efforts to be scaled up to match the height of the splendid giant daisies. Now that the patterns of relationship have been illuminated, among Darwin's finches and among *Scalesias*, thanks to molecular techniques; now that we have a much better sense of the role of finches in the broader ecology of the archipelago—as key dispersers of certain plants in both arid and humid zones; and now that the vulnerability of daisies to invasion by hardwood species and our agriculture has been well documented, it is clear that daisy conservation is crucial to the archipelago, and equally clear that humans must be the ones to step in. There is no "inexplicable confusion" about it. It is up to us.

7

Caught in a Booby Trap

One of the greatest attractions of the family [Sulidae (gannets and boo-bies)] is its blend of phylogenetic compactness with considerable adap-tive radiation. The sulids are all closely related but yet have diverged enough to occupy widely different habitats and to exhibit substan-tial differences in ecology and behavior. From them, we can learn a great deal.

J. Bryan Nelson (*The Sulidae*)[1]

Galápagos is probably the best place in the world to experience the boobies, a group of closely related seabirds that form their own clade within the family Sulidae. The reason is that three especially colorful species are native to the ar-chipelago and are present in good numbers: the red-footed booby, the Nazca booby, and the blue-footed booby (Figure 7.1). Their feet alone are a visual knockout, with brightly contrasting colors. But the intensity of colors, and of color differences among them, continue over their bodies, heads, and beaks. Researchers don't yet understand the full significance of the boobies' bright, al-most gaudy colors, but we'll make some headway on the topic in this chapter. Let's begin with the blue-foots.

The Booby Parade

Visitors to Galápagos will not soon forget their first encounter with the mating dance of the blue-footed booby. With gorgeous blue-green feet and great energy, an adult male "parades" around an approaching female for minutes at a time, first lifting one foot up to be admired, then the other, back and forth (Figure 7.2). The display is all about the extravagantly bright, webbed feet he is showing her, or an-yone else who will look. It's clearly meant to impress the female, and sometimes it does. She'll honk approval, he'll whistle, and the spectacle will continue, the male

---

[1] The quote is from Nelson's 1978 monograph, *The Sulidae: Gannets and Boobies* (p. 1), a thorough account of each species in this widespread family. See also Nelson's 1968 guidebook, *Galápagos: Island of Birds*.

Exuberant Life. William H. Durham, Oxford University Press (2021). © Oxford University Press.
DOI: 10.1093/oso/9780197531518.003.0007

Figure 7.1. The three native booby species of Galápagos: the red-footed booby (*Sula sula, left*), the Nazca booby (*Sula granti, middle*), and the blue-footed booby (*Sula nebouxii, right*). One would be hard pressed to find three more colorful seabird species anywhere else on Earth.

shifting between uplifted feet. Sometimes, she'll get into the act and lift her feet as well, prompting the male's energetic reciprocity. To us it appears comical, vivacious, even goofy. To them, it's a perfectly serious matter. As Nelson put it (1978, 557), "Even solitary males will parade around their territory with a ridiculously solemn air. In fact, early in breeding, male blue-foots seem hardly able to walk 'normally' so prevalent is Parading in this active species. The act of collecting even a tiny scrap of nest material produces intense Parading, which often continues on the spot as the bird 'shows' this material [to any organism that cares to look]."

When males and females parade together, they are checking each other out, assessing one another as mates, right there for the whole world to see. They could never get away with such flamboyant terrestrial display on the mainland because predators would be there in a flash. There is an innocence about the whole show, a carefree energy right out in the open, that epitomizes the special freedom of island-nesting organisms in Galápagos. Their nonchalance is such that people say the name "booby" is derived from the Spanish word *bobo* (clown or fool). Early sailors thought the birds must be *bobo*: with little or no fear of humans, the birds were readily hand-selected by sailors for the daily soup pot.

Parading and foot-lifting often lead to sky-pointing (Figure 7.3A) by the male or both members of the pair, to ritualized presentations of nest material, a form of symbolic nest construction,[2] and eventually to mating. Or it can end

---

[2] The bits and pieces used in displays of the blue-footed booby—like sticks, grass, and old feathers—are mainly discarded on the floor of their nests, which are little more than guano rings with a small, central depression. In contrast, the red-footed booby makes presentations of the big sticks with which its arboreal nests are constructed.

Figure 7.2. A male blue-footed booby parades in front of a potential mate, foot-lifting his bright turquoise feet, one after the other, to attract her attention. It's not working very well: the female looks elsewhere. (For a sense of the motion, focus back and forth between left and right panels.)

as innocently as it began, and each tries again with another candidate. Should a male temporarily leave the parade ground of his nesting territory—an area cleared of debris by so much foot-lifting—his return landing will entail an exaggerated "salute" with the same turquoise feet, while still mid-air (Figure 7.4). These birds are obsessed with their feet.

The blue-footed booby is not the only Galápagos booby with fancy footwork. A less elaborate version of the mating dance is also carried out by the Nazca booby, largest of the three, who also sky-points modestly (Figure 7.3*B*), and by the tree-nesting red-footed booby, underscoring the homologous nature of these displays within the trio. Sky-pointing by the red-footed booby, up in a tree, is no mean feat but still they do it blithely (Figure 7.3*C*); alas, parading on branches is far more difficult, even with their uniquely prehensile red feet. The blue-foot definitely has the fanciest footwork, on the ground or in the air.

Though native to Galápagos and nesting there, the three species enjoy fairly large ranges. The red-footed booby is found across the Pacific and into the Indian Ocean, while the Nazca and blue-footed boobies occur in the Western Hemisphere alone, from Galápagos to coastal Ecuador and north to Mexico.[3] Yet, the three species have become veritable icons of the archipelago, probably

---

[3] Booby distributions are complicated by small populations on far-flung islands. Nazca boobies, for example, are accurately described as "nesting mainly on the islands of Galápagos and Malpelo [Colombia], with much smaller numbers breeding from Isla La Plata, Ecuador, to San Benedicto Island, Mexico [similar latitude to Mexico City]" (Pitman and Jehl 1998, 156). See Patterson et al. (2011, 182) for a worthy effort to simplify the overall pattern.

Figure 7.3. Sky-pointing in the Galápagos boobies. *A*, A male blue-footed booby sky-points while a female looks on. Her pigmented iris looks larger than his (though they are the same size)—a characteristic feature of the species. *B*, The sky-point of Nazca boobies is tidy and short. *C*, Tidiest of all, the sky point of a red-footed booby is called "four-pointing" with beak, tail, and both wing tips pointing skyward, helping with balance up in the trees.

because, with Galápagos having the world's largest populations of all three, they are easy to see.

Globally, the three Galápagos species are joined by four other booby species—the Peruvian booby, masked booby, brown booby, and, the rarest of the group, Abbott's booby of the Indian Ocean—plus three related gannet

Figure 7.4. A male blue-footed booby "salutes" as he lands, showing off his turquoise feet against a bright white stomach. *Source*: Copyrighted by Matthew Almon Roth, used under Creative Commons License BY-NC 2.0 (from https://www. flickr.com/photos/matthewalmonroth/40379716982).

species at higher latitudes; together, they form the Sulidae family of 10 species (Figure 7.5). The three sulids of Galápagos are thus not an endemic radiation (found only in Galápagos) like, for example, the finches and *Scalesias*. Instead, they are three branches of a broader radiation that find suitable habitat in the archipelago.

Similarities by Descent

True to form, the three species of Galápagos boobies exhibit the full suite of homologous features common to boobies and gannets. These features include:

1. Fish capture by plunge-diving into the ocean from heights of up to 30 m, entering the water at breakneck speeds (Figure 7.6).
2. Aerodynamic head and body, especially with wings folded back in plunge configuration, whereupon they resemble so many darts striking the water.
3. Extrarespiratory air sacs all along the neck and stomach of each bird that cushion the blow of high-speed water entry (see Daoust et al. 2008).
4. Binocular vision that facilitates accuracy in fish spotting and capture, and no external nostrils to fill up with water (nostrils open on the inside of the beak).

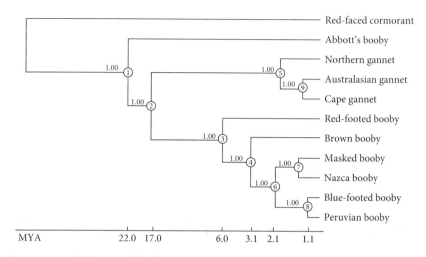

Figure 7.5. Phylogeny of the family Sulidae on a scale of relative genetic differences, with corresponding time line. The family includes three species of gannets and seven species of boobies. The red-faced cormorant, not in Sulidae, is an "outgroup" for comparison. Abbott's booby was the first species to diverge, approximately 22 MYA. The gannet line branched off at 17 MYA, and the remaining boobies, known as the "Sula clade," began diversifying with the separation of red-footed boobies at 6 MYA. Next, brown boobies split off 3.1 MYA. Finally, a division at 2.1 MYA gave rise to two sibling-species pairs, masked and Nazca boobies plus blue-footed and Peruvian boobies, each separating about 1.1 MY according to this analysis. *Source*: Figure 3b, Bayesian estimated species tree (BEST), used by permission of J. Morris-Pocock from S. A. Patterson et al. 2011. "A Multilocus Phylogeny of the Sulidae (Aves: Pelecaniformes)." *Molecular Phylogenetics and Evolution* 58: 181–91. License conveyed through Copyright Clearance Center, Inc. Timeline added with approximate dates for branch points.

5. An active uropygian (preen) gland near the tail that secretes a waxy substance that doubles as waterproofing and insect repellent.
6. Webbing between all four toes, giving them "totipalmate" feet, providing a broad surface of tissue for the colorful foot displays.
7. Relatively long life spans (15–20 years), with a tendency toward philopatry (adults nest in the same general location where they hatched).
8. Highly territorial in nesting and yet gregarious, building colonies often numbering hundreds of nests, but with a buffer of spacing between neighboring nests.
9. Small clutch sizes (one to four eggs), no brood patches (but wide feet can cradle eggs), helpless chicks at hatching, plus prolonged care and feeding by both parents.

Figure 7.6. A feeding frenzy of blue-footed boobies. Various stages of high-speed plunge diving are visible, including birds entering the water adjacent to sharp volcanic rocks. Attacks of the kind are thought to confuse schools of sardines, the boobies preferred species, and other small bait fish.

10. Frequent aggressive behavior, both in defense of nesting territories and in attacks on chicks of other, same-species individuals in the colony.

This is an impressive list of features, showing just how well-adapted boobies are to their flamboyant, plunge diving lifestyle. But the last characteristic, aggression, is surprising, all the more so for a group of organisms that human observers often see as comical, zany, and a little goofy. In reality, booby adults are frequently aggressive, sometimes even without ostensible provocation. In large colonies, for example, they will commonly attack without goading and jab at unrelated chicks until they bleed. Sometimes, their attacks on unguarded, helpless chicks—sitting quietly on their own nests—prove fatal (Nelson 1978, 741). An unsuspecting tourist can experience a few of the same painful jabs if, along a visitors' trail, they step too close to a booby's nesting site. It is fair to say that a booby will ignore you up to a certain, surprisingly close distance. But cross that invisible line and boobies do not take nest intrusion lightly.[4] Sulids, in

---

[4] Evolutionist Steven Jay Gould wrote an insightful article, "The Guano Ring" (1983, 50), about this boundary. As Gould noted, "Any intrusion into a guano ring would alter the behavior of adults from blissful ignorance to directed aggression. A single toe across the ring elicited an immediate barrage of squawking, posturing, and pecking."

short, have a menacing side, one of their most basic shared properties, even as we sometimes chuckle at their other behaviors.

Is such aggression adaptive? As several authors have suggested (e.g., Müller et al. 2008; Nelson 1978), this is a deeper question than it may seem. On the one hand, blatant aggression is widespread among sulids. On the other hand, the aggression seems spiteful, dysgenic, and at times maladaptive. The case of the northern gannet (*Morus bassanus*) is exemplary:

> Whilst the extreme nature of adult hostility to the young may seem dysgenic enough in the colony, it seems a thousand times more so on the sea. Nevertheless, newly fledged young, unable to rise from the water, are often attacked on the sea. . . . The attack may be prolonged and concerted, involving up to 20 or 30 attacking adults for an hour or more. They peck and hold tenaciously to the juvenile, forcing it below the surface and hanging on as it flaps and threshes in its efforts to win clear. As one attacker tires of the activity, another flies in. . . . Almost certainly a few young birds die as a result, but the number is totally negligible [compared to the total local gannet population]. Senseless, in adaptive terms, though these attacks may seem . . . [they] are part and parcel of the gannet's well-developed intra-specific aggression. (Nelson 1978, 208–9)

Nelson's theory—that "adult hostility is an effective and presumably advantageous method of discouraging chicks from wandering" away from their nests (Nelson 1978, 208–9)—assumes that it is "not sufficiently harmful . . . for natural selection to have produced a mechanism by which adults differentiated between young-off-the-nest in the colony (to be attacked as a trespasser) and young-off-the-nest at sea (to be left alone). The risk of diluting intra-specific aggression . . . would seem too great to justify adding refinements to this essential nature" (Nelson 1978, 208–9).

Persuaded by Nelson's eloquent description of sulid aggression, but not by his explanation for it, I take the evolution of booby aggression as a focus for this chapter. In contrast to the hypotheses for marine iguana's physiological and morphological adaptations (see Chapter 5), the goal here is to provide a persuasive hypothesis for the evolution of this important booby behavior. Let us begin by exploring the subtle and not-so-subtle differences between the three booby species found in Galápagos. Then, examination in more detail of one specific difference, foot coloration, and its relationship to sexual selection, takes us back to breeding and population fluctuations in the archipelago, especially in the blue-foot population. This will leave us better situated, I believe, to understand booby aggression.

Booby Differences

Let's begin with the aspects that make Galápagos boobies especially interesting: their differences. Consider, first, the red-footed booby, the smallest of the three: it sports a wingspan of generally only 1.5 m (5 ft) or less, it has by far the shortest legs of the lot, and it surely represents the advantage of a low center of gravity for nesting in trees. The red-foot's coloring stands out in the daylight: glorious prehensile red feet contrast with body plumage that is either starkly white (in the case of the "white morph") or a light brown (the "brown morph"). We still have no convincing explanation for the persistence at good frequency of both morphs in the Galápagos population of red-foots.[5]

Another distinctive feature of red-footed boobies is that, in the trees, they make real nests of twigs procured from the surrounding vegetation (Figure 7.7). The nests are uncomplicated, certainly, but they are genuine stick nests, secure enough for adults to take turns incubating. And incubate they do . . . their *one* egg. Parents make a big to-do over a single egg, which they normally lay at intervals of 15 months, sometimes longer. Why only one egg and so long between clutches? It comes down to a "food squeeze" on the boobies, says Nelson (1968, 53), which can even lead red-foots at times to abandon breeding after nest-building, after egg-laying, or even after brooding a chick. Such evidence compels the view that red-footed boobies are especially sensitive to changes in food supply. With ample food, they feed their single chick past the fledgling stage; without ample food, they abandon the egg, or even the hatchling.

As further evidence of their food squeeze, red-footed boobies are known to fly up to 150 km (90 miles) in pursuit of prey, more than either of the other Galápagos boobies, and they are specialists in catching flying fish and squid, including some very impressive flying squid.[6] In addition, adult red-foots are known for being nocturnal, at least in the sense of returning to the nest at night, which may help both in catching squid (who come close to the ocean's surface at night) and in avoiding kleptoparasitic frigatebirds that attempt to steal their food. The red-foots' smaller size makes them easier to steal from, and the squid

[5] Plumage color differences have a genetic basis (Baião et al. 2007). An estimated 95% of the world's red-footed boobies are the white morph. In Galápagos, white is nowhere more than 30% of the total, as on Darwin Island in the north, and is closer to 8% everywhere else. Founder effect (see Appendix 1) may have a lot to do with it, but so might differential advantage of camouflage when hunting for prey (see Baião and Parker 2008) or when flying back to nests at night, undetected by frigatebirds (who are especially prone to stealing food from red-footed boobies). Or, it may be something else: we still do not know. This is another topic colorfully ripe for further investigation.

[6] Using jets of water quickly expelled from their mantles, squid of several different families can propel themselves into the air very effectively, to heights over 5 m and for distances over 50 m (Macia et al. 2004; Jabr 2010). When chased by tuna or other predators, hundreds of squid may take to the air at the same time, much like flying fish.

Figure 7.7. Red-footed boobies are the only Galápagos booby to nest in trees. For them, presentations of twigs and branches between mates are not simply symbolic, but often fresh contributions of nest material (visible here in the lower nest). These two nests happen to be close together, but that is unusual; nests are generally scattered among neighboring trees, making hatchlings easy prey for aerial predators—one reason that red-footed boobies do not colonize islands that also have Galápagos hawks.

they pursue is among the frigatebird's favorite foods. In fact, squid—whether stolen or honestly harvested—are a prime source of the pigment that makes male frigatebirds' throat pouches turn bright red in their mating season (Juola et al. 2008). The same squid-derived pigment also enhances the color of the red-footed booby's feet during its own mating.

A First Booby Trap

Kleptoparasitism is bad enough, but there is also a predator—the Galápagos hawk—that influences the location of red-footed boobies' colonies. Tree-nesting by these birds, assisted by their prehensile feet, is surely advantageous in other parts of their global range, where they often face terrestrial predators. The arrangement works well since parents forage over long distances and are absent much of the time, leaving chicks unprotected. In Galápagos, however, tree nests backfire, exposing chicks to hungry Galápagos hawks while the parents are away. Even worse, roosting in the trees is a three-dimensional matter, as opposed to nesting on the ground in a mostly two-dimensional layout like the other two Galápagos booby species. Three dimensions reduces the chances of mutually beneficial deterrence by close booby neighbors ("mobbing" behavior), as happens in the ground-nesting species. The red-footed chicks are thus booby-trapped in their arboreal nests. As a result, natural selection seems to have favored heightened sensitivity to hawks among red-foots, who do not roost on hawk-inhabited islands. Today, there are red-foots' rookeries on Wolf, Darwin, and Genovesa islands, where no hawks have been recorded, and on San Cristóbal and Floreana, in the years since hawks were eliminated there.

A bird of a different feather, so to speak, is the Nazca booby—the largest of the three native species (wingspan up to 1.8 m, 70 in.). The Nazca boobies therefore roost near cliff edges and strong updrafts, making it easier for them to accomplish take-off and landing with long wings (reminiscent of the even larger waved albatross). Flying or standing, Nazca boobies are quite beautiful to behold, with a bright orange beak against pure white body plumage with an intensely black fringe (Figure 7.1). Dark black trim at the base of the beak gives the birds the appearance of wearing a mask, and for this reason, Nazca boobies were lumped together until 2002 with masked boobies (*Sula dactylatra*—meaning "black fingers," in reference to black wingtips). That year, genetic analysis (Friesen et al. 2002), building on earlier morphological comparison (Pitman and Jehl 1998), confirmed roughly a half-million years of differentiation between the two.[7]

In contrast to its red-footed relative, the Nazca booby feeds in open ocean closer to its colony, with a maximum daily foraging range of about 100 km (62 miles; Zavalaga et al. 2012). The Nazca booby eats flying fish (like its red-footed counterpart), but it normally depends more upon higher quality, energy-rich sardines, which generally abound in the cool ocean waters around the

---

[7] The Galápagos species was named after the Nazca Plate (see Figure 3.3), which roughly matches its breeding area. The main visible difference is in beak color: masked boobies have a yellow beak, as opposed to the rich orange of Nazca beaks, a contrast suggestive of differences in diet and the carotenoids therein.

Figure 7.8. Nests of the Nazca booby, fittingly called "scrapes," have little in the way of structure or ornamentation beyond a surrounding guano ring. Their simplicity does not prevent adult Nazcas from making symbolic presentations of twigs and plant parts during mating, a few of which can be seen in the center of this nest.

archipelago. Its diet shifts to low-energy sources, however, when the cool ocean waters turn warm during El Niño, prompting concern over the species' long-term welfare amid warming seas.[8] (Later discussion returns to some implications of this shift.)

Nazca boobies nest on the ground in big colonies numbering at times hundreds of nests. But there's normally not much to a nest—just open ground, a guano ring, a few twigs sometimes, and scattered volcanic rocks—accurately called a "scrape" by naturalists (Figure 7.8). In such nests, the Nazca boobies lay either one or two eggs, depending on the food supply (Clifford and Anderson 2001). When two eggs are laid, yet another food-supply dynamic kicks in, and so does booby aggressiveness. "Aggression between nest-mates starts early," noted Nelson (1978, 411), and virtually always results in the expulsion of the smaller or weaker hatchling from the scrape, whose death from dehydration or predation is inevitable. Booby aggression here takes the form of siblicide (the killing of one sibling by another) in nearly all cases.

---

[8] See Tompkins et al. (2017), who show that Nazca breeding success in Galápagos has dropped below replacement since 1997 due to warmer waters and associated poor diet.

A Cromwell Specialist?

Let me conclude this examination of booby differences back where it began, with the famous blue-footed booby. The key difference in this case is the blue-footed booby's association, throughout its range, with cold water (Nelson 1978, 514). The blue-foots thoroughly depend on cold-water upwellings and the prey that come with them—especially cold-loving, energy-rich Pacific sardines. But blue-foots are also inshore feeding specialists, generally plunging within a few hundred meters of shore, at what sometimes seem to be near-suicidal velocities (Figure 7.6). A division of labor in foraging,[9] and a tendency to carry out synchronized plunges in flocks (whereby multiple coordinated dives confuse schools of sardines), allow the blue-footed booby normally to find food "relatively easily—at least compared with the warm-water pan tropical [boobies, like the red-footed and brown boobies]" (Nelson 1978, 521, brackets added).

When conditions are favorable and sardines are abundant, this foraging ease translates into relatively equivalent success in reproduction. Normally, blue-footed females lay two to three eggs in a small depression, roughly like a soup bowl, inside a guano ring that is otherwise much like that of Nazca boobies. Females and males take turns incubating the eggs, often with a bit of ceremony at the changing of the guard. They dedicate roughly 45 days to incubation, followed by a 4-month rearing period, when both parents are involved in feeding the hatchlings. If the sardine supply is good, they can raise two to three chicks, a number unique among the Galápagos booby species. But if the sardine supply falters, one or more hatchlings will be lost to aggressive siblicide—as with the Nazca booby, when food is scarce, the older chick drives the younger chick(s) from the nest, whereupon they die, even as the parents watch. If the sardine supply is desperate, as in a strong El Niño event, they abandon the nest completely, and many fly to the continental coast and north, seeking respite in the Gulf of Panama or farther north.

Judging from these multiple close-fitting adaptations, it seems reasonable to hypothesize that the blue-footed species evolved in Galápagos and dispersed from there to its current range. Its special characteristics—dependence on cold upwellings, inshore foraging specialization, exclusive reliance on sardines, and its flight response to El Niño—all look like features of an adaptation to the Cromwell

---

[9] Males are smaller than females, averaging only 71% of their size, but they have tails nearly as big. The proportionally larger tail allows them to swerve sharply on entering shallow water, which makes them adept at shallow dives and at catching smaller, faster prey. These features enable males to make shorter foraging trips and to be especially efficient at feeding young chicks, while female blue-foots specialize in somewhat larger prey and are more efficient providers for older chicks.

upwelling, including its blockage during warm phases of ENSO.[10] Genetic analysis by Patterson et al. (2011) suggests that blue-foots branched off within the last 1.1 MY from an ancestor shared with the neighboring, and similar-looking, grey-footed Peruvian booby of coastal South America.[11] The diet of the latter swings between sardines and anchovies with the climate shifts of an interdecadal water-temperature system—called the Pacific Decadal Oscillation (PDO)—precluding the specialized sardine diet of its blue-footed counterpart.

By one scenario, this specialization would have evolved when an unknown number of common ancestors (to Peruvian and blue-footed boobies) on the Peruvian coast flew west with strong trade winds and had the good fortune to land in Galápagos. Sardines abounded in the cold water of their new home but coastal anchovies were rare to nonexistent.[12] By this scenario, the common ancestor had greyish feet like the Peruvian booby and the chicks of both species today. But having generally abundant sardines and no anchovies, the Galápagos population evolved a dietary specialization, perpetually rich with energy and with carotenoids, the latter promoting a blue-green color to the skin at high concentrations. Particularly successful booby divers would have accumulated visually detectable levels of carotenoid, which could very possibly have been seen first and foremost as tinting of their grey feet. As an indicator of fishing success, tinted feet would promote the evolution of color-sensitive mate choice by females (the topic discussed next in this chapter). By this scenario, sardine specialization would have evolved in Galápagos, under conditions provided by the Cromwell Current. Eventually, the boobies would have expanded beyond Galápagos to the east and north, probably flying to escape El Niño as they do today, taking them to new locations that also lie near cold upwellings rich with sardines (nicely documented by Taylor et al. 2011). Yes, many other scenarios are possible: blue-foots could conceivably have originated in one or more of these other places, moving thence to Galápagos, but the scenario outlined here better explains their package of specializations and also accounts for their geographical distribution, both west and north of their closest relatives of Peru.[13]

---

[10]   Other Cromwell specialists of Galápagos include the flightless cormorant, Galápagos penguin, the Galápagos fur sea lion, and arguably the marine iguana. Considering the Cromwell's rich and varied food web, this menagerie of specialists is not surprising.

[11]   Taylor et al. (2011, 889) used an even more recent date of 0.2–0.45 MY for this split, but it seems to have been based upon an earlier and less reliable dating technique. Where their ranges overlap today, the two species do sometimes interbreed to form hybrid offspring (Taylor et al. 2010).

[12]   Grove and Lavenberg (1997, 192) noted that, although adults of the coastal anchovy "sporadically arrive in the archipelago, [the species] may not be able to colonize Galápagos waters successfully because their eggs and larvae would be transported away from the islands by strong currents."

[13]   My argument is heterodox in two ways, but it has logical advantages. First, it's just the reverse of Nelson's (1978, 813) suggestion that "the Peruvian [booby is] merely a specialized form of the blue-foot which has become adapted to the particular conditions of the Humboldt area." My reasoning is guided in part by similarities between the brown and Peruvian boobies that are not shared by the blue-footed booby, whose features are arguably more derived. For example, both the brown booby

Bird	Nest site	Feeding & primary prey	Eggs per clutch	Chicks
Red-footed Booby	Trees	Pelagic Flying fish & squid	1	1
Nazca Booby	Ground	Intermediate Sardines & flying fish	1–2	1
Blue-footed Booby	Ground	Inshore Sardines	2–3	1–3

Figure 7.9. Summary of key features among booby species of Galápagos. Note differences in primary prey and number of eggs per clutch. *Sources*: Data from Nelson (1978), plus Anderson et al. citations in the text.

In summary, the full set of Galápagos boobies are the avian equivalent of a musical "theme and variations." Even to the casual observer, the three species have many features in common—a fact that nicely highlights their differences, including subtle ones (as in Figure 7.9).

In due course, this chapter examines patterns among the differences summarized in the figure. But first, let us explore the matter of foot color. Why do two of the three booby species have such brightly colored feet?

Why Turquoise?

Let's start with the blue-footed booby. If you focus on feet in a colony of these birds, you'll see an impressive level of variation: some are bright sky blue, some are a deeper royal blue, and yet others are a blue-green like a piece of turquoise. Looking closely, you may discern a pattern: chicks have grey feet, young adults' feet are often bright sky blue, and adults in the mating parade have blue-green turquoise feet. As you keep looking, you'll see that the birds that are already nesting have royal blues, and elderly blue-foots have deep blue feet. What's this variation all about?

---

and the Peruvian booby often build substantial nests, and both nest on cliffs when space is short, a behavior nothing like the blue-foots'. Second, my reasoning posits an east-to-west colonization, fitting the normal patterns of strong trade winds and the progression rule of east-to-west colonization (as in Chapter 3). Third, it produces the same result as a previous argument for parapatric (side-by-side) speciation: Peruvian and blue-footed boobies "are young sister species that experience limited hybridization where their ranges overlap . . . and no physical barriers to dispersal exist between their ranges" (Taylor et al. 2013a, 2). This distribution may simply result from blue-foots' expanding out of Galápagos during El Niño events. Fourth, according to this proposal, the unique properties of blue-foots are not an uncanny coincidental match: they evolved as adaptations to the archipelago.

Sexual Selection

In nature, bright or gaudy colors are often a hint that the evolutionary force of sexual selection is acting within a population. Darwin (1859, 88) realized this, noting that sexual selection "depends, not on a struggle for existence, but on a struggle between the males for [access to] the females; the result is [usually] not death to the unsuccessful competitor, but few or no offspring." In other words, males evolve traits not with a survival function per se, but instead with a reproductive function via success in accessing mates. Chapter 5 briefly considers an example of *intrasexual* selection among male marine iguanas, wherein head-to-head combat for breeding territories favored large male body sizes. Here we are concerned with *intersexual* selection via competition to attract mates, which favors traits that are attractive in color, shape, ornamentation, and/or behavior to the choosing sex.

In Galápagos, one sees many examples of intersexual selection, especially in birds, ranging from big kleptoparasitic frigatebirds, where males sport bright red gular pouches during mating, to tiny vermilion flycatchers (Figure 6.4C), where only the male is vermilion; the nest-tending female is drab and more cryptic. In many such cases, the colorful attributes of males have been shown to be a reliable signal of a male's diet and health status: the brighter the color, generally speaking, the better the diet and health of the individual. He sports the variations of flight and feather that make him successful at foraging and at provisioning offspring. Such variations, when heritable, will produce the same advantage in his mate's offspring, and thus she—and he—will accrue reproductive advantage. Such variations will increase in frequency in the population across generations (with, of course, no proviso that females are consciously aware of their choices). The result is sexual selection for colorful ornamentation of males.

In blue-footed boobies, if brightness of feet in a male accurately reflects diet and health status, then natural selection will favor females who acquire mating partners with bright feet. The result will feed back on the males, producing sexual selection for the brightest male feet. But we've already seen that, in boobies, both sexes have this colorful feature, not just the males. How does *that* work?

Let's start with the male's ornamentation. As male boobies mature, the proteins in the skin of their feet, especially the protein collagen, align in a way that absorbs many colors of light, but reflects a deep blue. That deep blue is called a "structural" color, resulting from the very structure of the skin and its reflectance. During the blue-foots' mating season, however, a breeding male's feet brighten up to a lovely blue-green or turquoise. Studies have traced the change to the presence in the skin of the aforementioned carotenoid from the booby's diet—this time yellow, as opposed to the frigatebird's red, but no less influential in mating. According to Torres and Velando (2010, 158), further sleuthing revealed that "foot color in blue-footed boobies changes rapidly and varies seasonally, and rapid changes in foot color are influenced by food and carotenoid availability (Velando et al. 2006). Male foot color varies from a bright blue-green turquoise (reflectance peak at 540 nm) to dull blue (reflectance peak at 440 nm), and females prefer males with bright turquoise feet" (Torres and Velando, 2003).

The key carotenoid here is zeaxanthin, which has two noteworthy aspects. First, the blue foot's favorite food, Pacific sardines, are rich in zeaxanthin and related meso-zeaxanthin (Nolan et al. 2014);[14] second, carotenoids are not merely pigments—they are also antioxidants and immunostimulants that facilitate boobies' protective response to infection or disease. Blue-footed boobies with zeaxanthin in their feet are healthy enough that their metabolism diverts concentrations of it from their immune system to ornamentation. So, bright feet mean a well-provisioned, healthy bird, signaling not only a male's ability to provide for healthy offspring, but also any heritable advantages in hunting success. As a result, natural selection favors females who (unconsciously) prefer males with bright feet, featured in those lovely displays of parading and saluting, which, in turn, drive selection of males with pigment-bright feet.

One last piece of this argument is especially convincing. It turns out that the female booby's visual system is extra sensitive to reflectance in the range of 460 to 620 nm, sometimes called "green chroma," which corresponds to the wavelengths of light reflected by turquoise feet that are rich in zeaxanthin (Velando et al. 2006). In fact, zeaxanthin has its highest reflectance at 550 nm—smack in the middle of green chroma. This implies that there have been mutual evolutionary adjustments working on this perception: evolution finely tuned the visual system of the booby to give the females highly sensitive feedback on the dietary adequacy of males, who, in turn, evolved a focus on fishing the richest local source—sardines—of the healthful telltale carotenoid. This match provides a rather colorful example of interactive coevolution. The fine-tuning of the visual system is another of the features suggesting a Galápagos origin for the blue-footed booby: conditions are especially apt for them in the normally sardine-rich Cromwell upwelling.

But, wherever it all started, the blue-foot's color signaling works with surprising efficiency, as confirmed in a clever experiment by Velando et al. (2006), undertaken on Isla Isabel off the coast of Mexico (northern reaches of today's blue-foot distribution). After blue-footed boobies had paired up and laid a first egg, researchers took advantage of the average 3.5-day spacing between sequential blue-foot eggs, and painted all the male's feet in one group a dull blue, leaving a control group's feet unpainted. Painted males behaved normally for the next

---

[14] The sardines obtain these pigments by eating zooplankton in the sea, who get them in turn from eating phytoplankton, including algae, who internally biosynthesize them.

few days, parading around and showing off their now-duller feet. But females *did* change in response, and rapidly. With the painted males, they almost immediately cut back on sky-pointing, presentation-making, and, more to the point, copulation. Most telling of all, the second eggs—laid while males had dull feet—were of much lower volume and smaller yolk than those laid in the nonpainted control group. Thus, females proved amazingly responsive, withholding reproductive commitment in a matter of days when the male's feet changed color—a striking example of sexual selection at work.[15]

But what about adult female boobies' foot color: does it matter in the same way? It turns out that males, whose foraging is a crucial part of parental care of the young, have themselves evolved to be choosy about feet (see Figure 7.10). When Torres and Velando (2005) tested a reciprocal paint job, changing females' feet to a duller blue, the females received a lower percentage of total courting opportunities plus a lower percentage of extra-pair courtship compared to nonpainted controls. The implication is that bright turquoise feet are a meaningful and attractive signal in both directions—males signal their healthy condition to females in their feet, and the females signal it right back. Literally and figuratively, the males and females are on the same wavelength. Bright feet are undoubtedly beautiful: they promise offspring with beautiful feet to come, and they are thus a reliable, 'honest' indicator of evolutionary fitness.[16]

Now, one might reasonably ask, what about the feet of the similarly colorful red-footed booby? Is there a parallel role for pigment-enhanced coloration? To date, red-footed booby color signaling has not been studied in as much detail as the blue-foot case. Booby specialists believe that "the mechanism of color production in the feet of other Sulidae species is probably like that of the blue-footed booby, as [suggested] by similar seasonal and geographic variation" (Nelson, 1978), and by the fact that "the tomato red feet of red-footed boobies (*Sula sula*) [are] more intense during courtship" (Torres and Velando 2010, 162). There is no doubt that the pigments are different from those in the blue-foots, as befits the difference in diet between the species. The red-foots' supplemental pigment is most likely the carotenoid astaxanthin, obtained from dietary squid—the same pigment in the gular pouch of frigatebirds, who frequently obtain it, in turn, by stealing squid from red-foots.

---

[15] The tendency for a blue-footed booby's feet to get darker with time is partially reversible. A study by Velando et al. (2009) found that mating-age boobies who take a 1-year "sabbatical" from the demands of mating gain a more youthful foot color. Indeed, they reported that one 14-year-old male in their study looked 6 years younger after 1 year of sabbatical. I have faculty colleagues who swear that sabbatical does the same for them.

[16] Such sensitivity to color will also reduce the chances of species recognition mix-ups during courtship, making foot color an effective reproductive isolating mechanism.

Figure 7.10 Foot color is important to both partners in a blue-footed booby mating. Here a female (at right, with an apparently larger iris) vigorously parades her lovely turquoise feet while her potential mate (with beady eyes) looks on. At this part of the dance, her display conveys more to him than his display conveys to her; she appears to be oblivious to his feet, even as he also parades.

Finally, one might also ask about the color of the Nazca booby's feet: how do we make sense of their feet, which seem awfully plain by comparison (as in Figure 7.1)?[17] Since the foot-lifting display is subdued in these boobies compared to their blue-footed relatives, one suspects that the role of feet in sexual selection is simply toned down, as it were. The bright orange of their beaks, however, surely enhanced from the consumption of many sardines, suggests that beaks play the bigger role in mate attraction among these beautifully white boobies with drab feet. The logical experiment remains to be undertaken.

Plunging Population

For many years, the blue-footed boobies of Galápagos seemed to be thriving, with only temporary population losses from El Niño events—losses that were

[17] In Galápagos, the Nazca's feet are "olive or khaki in males [and] lead or olive in females" (Pitman and Jehl 1998, 161). They seem likely to be the product of a blue-grey structural color modified by the yellow dietary carotenoid zeazanthin, derived from sardines in their diet, as in the blue-foots. Writing of the related masked booby's feet, Nelson (1978, 321), reported "drab olive or grey seem the basic shades more or less brightened with yellow pigment"—the yellow most likely coming from sardines in the diet.

generally compensated by eventual gains from La Niñas. Sometime around the year 2000, however, people noticed a new plunge-dive in the species—in the numbers of adult boobies in the islands' breeding areas. Galápagos seabird expert David Anderson, of Wake Forest University, and his team were called in to assess the change. In a survey in 2011, they found only about 6,400 adult blue-foots in the archipelago, down from an estimated 20,000 in the 1960s. In 2011 and 2012, they also counted nests in prominent nesting sites around the islands, only to find drastic declines across the board. They concluded "Galápagos blue-footed boobies have attempted to breed very little between August 2011 and June 2013. . . . During this period, no more than 10.9% of the adult population had an active nest at any one time, and only 134 fledglings were noted during this time" (Anchundia et al. 2014, 9). Before that, they found "only two juveniles across the archipelago between May and August 2011, [meaning that] no successful breeding occurred in the previous 2 years." The decline was all the more surprising given that all three Galápagos booby species had long been listed by the IUCN as Least Concern for conservation. The change was sudden and unexpected: What happened?

A Second Booby Trap

A clue comes from an earlier field study by Anderson and colleagues in the days of the 1986–1987 El Niño. In that year, the sardine population of Galápagos plunged, sardines being a cold-water species caught in warm El Niño water. Anderson's comparison of prey items taken that year by Nazca and blue-footed boobies was especially revealing. The Nazca boobies weathered the storm by prey-switching, capturing a higher percentage of flying fish than sardines that year. But the sardine-focused blue-footed booby was caught in the trap of its own specialization. Flying fish were neither part of its foraging strategy nor its diet, and so the population suffered. All nestlings died at five major island colonies, breeding was suspended for months, and most adult blue-foots flew off to unknown sites, deserting established colonies.[18]

The symptoms of the new blue-foot decline in 2011 were similar, but lasting. There was no speedy rebound like that after the 1986–1987 event. Anderson's team suggested that the blue-foot decline had been triggered by another El Niño

---

[18] Mosquitoes become a big problem on many islands during El Niño events (see Chapter 2). As Anderson (1989a, 211) noted, blue-footed booby nestlings of 1986–1987 *all* died of starvation. But on Española, it was not "solely from food limitation. . . . Heavy rainfall provided breeding sites for mosquitoes *Aedes taeniorhynchus*, hundreds of which swarmed around the few blue-booted booby chicks alive in February, and which appeared to make adults reluctant to enter [for feeding them] or remain in the inland nesting areas."

event in 1997–1998, but that this time the warm water lasted (Anchundia et al. 2014). They suggested that the El Niño event signaled a lasting "regime change" or "phase shift" in the ocean, replacing cold-water fish species—including sardines—with warm-water fish and spelling trouble for the sardine-seeking blue-foots. Such a shift is believed to be caused not by El Niño (although it may be triggered by one) but by another, much slower climatic cycle, the aforementioned PDO. As Anchundia and colleagues noted, "Sardines cycle between high and low abundance with a period of 25 years in the Pacific, linked to the Pacific Decadal Oscillation (Chavez et al. 2003)" (Anchundia et al. 2014, 11). Studies of the Peruvian upwelling showed that the PDO was associated with a 25-year shift, back and forth, between sardine abundance and anchovy abundance. The sardine population declines or "crashes" at the higher ocean temperatures that encourage anchovy flourishing, and vice versa.

The problem in Galápagos is that anchovies are scarce even in the best of times. When the 1997–1998 El Niño hit, warming the water, the sardine density plummeted, but there was no anchovy boom to take its place in blue-foots' diet. Some boobies remained in Galápagos and struggled to reproduce, because food was genuinely scarce. Many adult boobies fled to other areas; thus, the population really did take a dive.

It's another case demonstrating the vulnerability of an evolved Galápagos specialization. A plunge-diving sulid ancestor evolved a specialization on abundant sardines in normally cold waters wherever it lived at that time. Eventually, or perhaps from the beginning, it came to Galápagos, and formed dispersed colonies with success most years at surviving and reproducing, fueled by upwellings bearing near-shore sardines. But the very dietary specialization that made blue-footed boobies successful in Galápagos—and with such bright feet—backfired with every El Niño event and PDO, trapping the birds out in the Pacific Ocean without good feeding alternatives.

Fortunately, evolution often cuts both ways. If ENSOs and PDOs have caused hardship and selection among boobies for tens of thousands of years, one might expect some measure of evolved resilience, especially in the blue-foots, whose dietary focus makes them hardest hit. We have certainly seen resilience in other Galápagos species, as in the marine iguana's ability to shrink and regrow, for instance. Such resilience could take many forms, from behavior to morphology and physiology. What is the evidence for resilience in blue-footed boobies?

Brood-Size Reduction

Research in Galápagos has revealed impressive resilience among the blue-footed boobies—indeed among all three species of boobies of Galápagos—in the ways they adjust the size of their broods. Referring to Figure 7.9, one can see that red-footed, Nazca, and blue-footed boobies generally lay one egg, one or two eggs, and two or three eggs, respectively. Because food supply varies month to month in Galápagos and elsewhere in their ranges, especially with sea-surface temperature, all three species have evolved ways to match brood size in a given breeding season to their concurrent foraging productivity. When times are good and food is plentiful, boobies of all three species tend a full brood, enhancing their fitness and incidentally promoting population recovery. When times are bad, they often lay eggs normally but then reduce broods to a smaller number of hatchlings or even to zero. Most interesting of all: the evolved brood reduction mechanism in Nazca and blue-footed boobies yields insights into sulid aggressiveness.

Hypotheses

We come now to the two main hypotheses of this chapter. *Hypothesis 1* is not particularly controversial. It says, first, that variation over time in the local supply of small fish—including sardines, flying fish, herring, etc.—has been a major selective force in the evolution of sulids in Galápagos and elsewhere. It says, second, that such variation, acting across long periods of time, has selected for effective mechanisms of brood-size regulation that adjust the demand for food to the immediate supply. Such mechanisms differ between Galápagos booby species in ways that reflect their feeding ecologies, allowing them to match their reproductive output to their seasonal foraging productivity. It says, third, that siblicide is the main mechanism for brood adjustments in Nazca and blue-footed boobies.

Hypothesis 2 is more controversial, and brings us back to the bizarre, seemingly wanton aggression by adult sulids toward nonrelated, nontrespassing chicks. The hypothesis proposes that adult aggression toward unrelated chicks, to the point of killing them sometimes, is itself an evolved response to food scarcity. But not just seasonal or occasional scarcity: it suggests adult sulid aggression toward chicks is an evolved response to food scarcity extremes, such that it appears to be anomalous, dysfunctional, and bizarre much of the time, but it saves the lives of aggressive adult boobies in truly desperate times.

Hypothesis 1: Booby Siblicide

Let us begin with Hypothesis 1, that abandonment and siblicide are adaptive mechanisms of brood-size reduction, adjusting the number of hatchlings to food supply in a way that, despite appearances (of reducing reproduction), enhances fitness. The case of red-footed boobies is straightforward and easy. The normal brood size is one, except, as noted earlier, in times of food scarcity, when nests, eggs, or even hatchlings are abandoned. Abandonment is neither intricate nor sophisticated, but it is nonetheless a mechanism for brood reduction that all three Galápagos booby species use in times of serious food supply squeeze. It allows the adults to trim wasted effort and mate again in another season when food supply improves.

When the food squeeze is less severe, brood reduction in blue-footed and Nazca boobies is achieved by another mechanism noted earlier: siblicide, the aggressive ouster from the nest of one or more smaller chicks by the chick that has attained the largest size, usually the first hatched. In the case of the blue-foots, who lay two to three eggs in their nests, siblicide is said to be facultative and thus responsive to local conditions in any given year. In years of good sardine supply, parental blue-foots can raise and fledge two of two, or even three of three, healthy chicks.[19] But in years when sardines are scarce, only the largest chick survives, if that. Siblicide does not take place right away, because initially, the largest chick regulates its behavior and shows tolerance toward the smaller chicks. During short-term food shortages, it has been shown that the largest chick displays a "kin-selected tolerance" for siblings, moderating its food intake to allow them to eat (Anderson and Ricklefs 1995, 163). This behavior suggests that sardine shortages are sometimes short-lived, in which case evolution favors allowing siblings to survive because of the indirect reproductive contribution to fitness they represent as close genetic relatives.[20] But, when food shortage is chronic, hungry dominant chicks prevent smaller siblings from eating and soon evict them from the nest. Facultative siblicide in blue-footed boobies appears thus to be finely tuned to seasonal food supply, providing an effective if unusual mechanism of resilience. With the return of sardines, parents rebound to clutches of two or three, thus enhancing their fitness, which thereby allows a plunging population to turn around and regrow.[21]

---

[19] A blue-footed booby nest with three surviving chicks is relatively uncommon in Galápagos. In his 1963 survey, Nelson (1978, 525) noted that only 5.6% of clutches on Daphne Major had three hatchlings. Using data from the 1920's for sites in Mexico, Nelson reported that three-chick nests were 5 to 6 times the Galápagos figure; however, he was unable to control for year-specific differences in conditions at the various sites.

[20] Since siblings share, on average, 50% of their genes by descent from their parents, evolution will favor small sacrificial behaviors by larger (generally first-born) chicks that enhance the representation of their genes in the next generation by way of the survival, and eventual reproduction, of smaller siblings with many of the same genes. The resulting tolerance by the bigger chick is believed to be an example of kin selection. It, too, is facultative, responding to local, seasonal food supply: a well-fed older chick allows more siblings to be fed.

[21] Let's compare siblicidal behavior between the blue-footed boobies, with brood size 2–3, and their sibling species, the Peruvian booby, brood size 3–4, who differentiated roughly 1.1 MYA. Feeding in the Peruvian upwelling, the latter alternates its feeding between anchovies and sardines, following swings in regimes or phases owing to decade-scale changes. The more abundant and stable food supply thus achieved obviates the advantage of siblicidal brood reduction. Even in El Niño years, coastal Peru provides small "upwelling cells" of food supply not found in Galápagos (a food

Among Nazca boobies, the situation is slightly more complicated. First, for reasons not fully understood, Nazca boobies hatch only about 60% of their eggs, compared to more than 90% for other birds, including other boobies (Anderson 2016).[22] Thus, if a pair of Nazcas had only a single-egg clutch, the probability of having no hatchlings is 40%; for a two-egg clutch, the probability falls to 16%, a substantial reduction. The second egg is thus called "the insurance egg," providing a backup against the failure of the first egg or first hatchling to thrive. By experimental manipulation of the Nazcas' food supply, Clifford and Anderson (2001) were able to show that one-egg clutches result from food limitation. So when times are good, the Nazcas are able to lay an insurance egg. But that's where the surprise comes in: when times are good and both eggs hatch, siblicide is the inevitable result. Without fail, "Whichever chick is larger pushes the smaller from the nest" (Anderson 2016, 141), to be taken by a predator or to die from exposure or starvation (see Figure 7.11). In a study of three breeding seasons on Española, Anderson (1989b) found brood reduction from two chicks to one in every case, and 98% of the time within just 10 days of the second chick's hatching (in 94 of 96 sample nests). When Anderson's team observed it as it happened, the larger chick aggressively attacked the smaller chick 82% of the time. As a result, two surviving chicks from a Nazca nest is more than improbable: siblicide is an obligate form of brood reduction in the Nazca booby.

A Third Booby Trap

But what's the advantage in *that*? Why would evolution favor obligate siblicide—a hatchling without a chance in every clutch—among Nazca boobies? Again, Anderson (1990b) tested the hypothesis that siblicide was itself an adaptive response to food limitation. He experimentally suppressed siblicide in Nazca booby nests and found that parents then brought more food to their hatchlings and wound up with higher fledging success rates. At first it looked as if suppressing siblicide could lead to higher parental fitness, but as the researchers watched for a longer period, suppression of siblicide resulted in fewer "grandchicks" (Anderson 2016, 140), probably because the first-generation chicks shared food

source also popular with waved albatrosses, see Chapter 2). Their more reliable food supply is a key reason that, unlike the blue-foots, Peruvian boobies do not practice siblicide (Nelson 1978, 618).

[22] Inbreeding may well contribute to the low hatching rate. Huyvaert and Anderson (2004) showed that Nazca booby dispersal is limited, owing to philopatry, but enough so that the closest kin are normally separated by 10 to 40 intervening conspecific breeding sites. Even so, with strong philopatry and small colony sizes, a generalized inbreeding is inevitable, which elevates the frequency of genetic homozygosity, rendering some fraction of all eggs unviable. This may well be why Anderson (1990a, 340) found that "the proportion of eggs with no visible embryo for the entire population ranged from 19.0% for two-egg clutches in 1985 to 63.8% for one-egg clutches in 1985." Inbreeding is also suspected in the case of hatchling failure among Galápagos cormorants (see Chapter 8).

Figure 7.11. Obligate siblicide by a larger, generally older, Nazca chick against its smaller, younger, sibling (*enlargement at right*). The larger chick aggressively pushes its sibling from the center of the nest as the parent watches but does not intervene. *Source*: iStock.com/"Soopy Sue."

with a sibling and subsequently had lowered survival, reproduction, or both. But note that the fitness advantage of obligate siblicide is a third booby trap. On the one hand, there's a benefit to Nazca parents in having a second "insurance egg" as a hedge against nest failure, but not if it results in more than one surviving chick. The key variable, again, is food supply—primarily flying fish and sardines—which is generally not as productive as that of blue-foots when foraging distance is factored in. As a result, Nazcas are booby trapped into obligate siblicide, generation after generation: lay two, rear one.

The net result, paradoxically, is that the two forms of siblicide in Nazca and blue-footed boobies are bona fide adaptations. In both cases, siblicide has a definite function: in Nazca boobies, if just one chick survives, the parents are more successful as grandparents, and in blue-footed boobies, the reproductive advantage goes to a variable number of chicks, adjusted to the annual food supply. Siblicide has a history of natural selection for that function, to the point of there being species-specific differences in chick hormone levels. Analysis by Müller et al. (2008) found that Nazca hatchlings had three or more times the androgen levels of blue-foot hatchlings across all possible nest compositions (one chick, two chicks, one egg fails to hatch, etc.).[23] Truly, Nazcas hatch into the world prepared for aggression.

Figure 7.12 summarizes the main arguments of Hypothesis 1, relating the mechanism of brood reduction for each of the three native Galápagos boobies to its foraging specialty and variations in food supply. The other three non-Galápagos species—brown booby, masked booby, and Peruvian booby—are included to show how they fit, hypothetically, into the logic of the arguments, given their different ecological circumstances. As shown in Figure 7.5, these six species form the Sula clade and share a common ancestor within the last 6 MY or so.

Hypothesis 2: NAV Behavior

We come now to the most unusual aspect of booby behavior: that of nonparental adult visitors (NAV), especially prevalent among Nazca boobies, but also found at reduced rates among blue-foots. NAVs are nonbreeding adult birds, male or female, who search a given breeding colony for unguarded nestlings. They approach the hapless chicks and carry out one of the following actions:

1. *Parental* behavior, such as preening.
2. *Courtship* behavior, with the usual symbolic nesting presentations of twigs, feathers, etc.
3. *Sexual* behavior, attempting to mate with the chicks.
4. *Aggressive* behavior (Figure 7.13), such as biting, shaking, or plucking out down, while the chicks often adopt a submissive posture.

---

[23] Androgen is a steroid hormone found in both males and females, known to play a key role in avifaunal aggression, especially in breeding adults (see Wingfield et al. 1990).

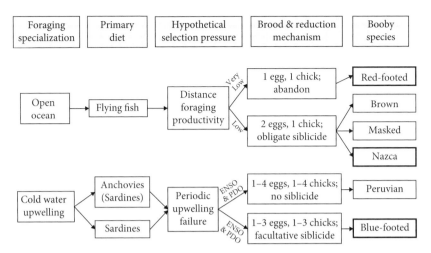

Figure 7.12. Correlations between foraging specialization, primary diet, brood size, and brood-reduction mechanisms in three native boobies of Galápagos (*boxes in bold*) plus the three other members of the Sula clade (*nonbolded*). Hypothesis 1 proposes that selection pressures arising from variation in food supply within their foraging ranges have favored mechanisms of brood size reduction, including abandonment and various forms of siblicide, among the three boobies native to Galápagos. Similar arguments and logic apply outside Galápagos to the other species shown here. *Source*: Data from Nelson (1978).

The aggressive behavior often draws blood, which may then attract hungry land birds who take blood meals, worsening the wounds and sometimes ultimately killing the chicks. As leading researchers stated, "NAV behavior is the direct or indirect cause of mortality of up to 24.6% of chicks, representing up to 41.6% of all deaths, in a given year" (Anderson et al. 2004, 960).

In the most comprehensive study to date, Anderson and colleagues (2004) logged a whopping 3,544 NAV events over the 1984–2002 field seasons. They found that nonbreeding males and females are equally likely to exhibit a NAV behavior, but because of a sex-ratio bias in adult Nazcas, a majority of NAVs are male. They found that the majority of NAV events were aggressive, with females slightly more often aggressive than males (60% vs. 55%), followed by 34% and 31%, respectively, that were "affiliative" events (combining the parental and courtship categories), and lastly, 6% and 14% that were sexual. Moreover, they found that female NAVs reached the highest levels of aggression more often than males (24.6% of 187 events versus 16.8% of 929 events, a statistically significant difference).

Nazca NAVs sometimes target other booby species, too. They target red-footed booby chicks in the trees, which is especially pitiful because those chicks do not

Figure 7.13. Aggressive NAV behavior among Nazca boobies. The attacking bird is a currently nonbreeding adult who is unrelated to its victim, a chick quietly sitting on its own nest. Females initiate a higher proportion of the most vicious attacks than males. *Source*: Photo courtesy of Jacquelyn Grace, Texas A&M, copyright 2009.

respond submissively and really get pummeled. Moreover, the Nazca marauders are enough to drive away entire nesting aggregations of blue-footed boobies. The latter prefer to nest at coastal sites, but if Nazca boobies are present, the Nazcas take the coastal sites and the blue-foots nest further inland, leaving a gap of some meters, reducing the NAV injury rate (Townsend et al. 2002). Curiously, most Nazca adults show NAV behavior at one time or another, and yet there still is no good explanation for the evolutionary emergence and persistence of this behavior. No one has yet identified a plausible fitness benefit (see Anderson 2016; Müller et al. 2008).

The NAV phenomenon is complex, incorporating a broad range of behaviors, including aggression as the most frequently observed form. For present purposes, let's focus on that aggression, to gain greater insight into the general aggressiveness of the Sulidae family, as discussed earlier. (Later, the full bundle of NAV behaviors are considered.)

Proximate Causes of NAV Behavior

In the study of animal behavior, it is useful—many authors would say crucial—to distinguish "proximate" causes, having to do with immediate, short-term reasons for behavior, from "ultimate" or long-term evolutionary reasons. Let us look at both in the context of NAV aggression, starting with some key proximate causes. Building on the finding that Nazca chicks show very high androgen levels, Anderson and team asked: Was the androgen involved in sibling aggression then associated with enduring hormonal changes in later life? Did chicks that had successfully carried out siblicide go on to be NAV aggressors at a higher frequency than nonsiblicidal individuals?

As a first test, they chose to compare NAV aggression frequency between Nazca colonies, which have more siblicide, and blue-foot colonies, which have less, using as controls the affiliative and sexual forms of NAV behavior. Unsurprisingly, they found a consistently higher percentage of aggressive behavior among Nazca boobies than blue-footed boobies, and lower levels of affiliative behavior. Aggressive NAV interactions do occur among blue-foots, but they are less common and milder. As predicted, the species with more siblicide exhibits more NAV aggression. Moreover, when the researchers looked at blue-foots living along the Peruvian coast, where the food supply is more favorable (thus favoring less siblicide), they detected a "far lower frequency" of NAV behavior (Müller et al. 2008, 3).[24]

In a second test, they looked at NAV behavior among Nazca boobies that had, and had not, carried out siblicide as chicks. They found significantly higher rates of NAV among the siblicidal Nazcas. The finding did not hold for affiliative or sexual NAV behaviors later in life, confirming that hatchling experience with siblicide has a special connection with NAV aggression in adulthood. One possibility, then, is that NAV aggression is somehow a delayed repeat of siblicidal behavior, carried out on a much older chick that happens not to be a sibling.

But there is more to it than that, as further experiments showed. One test asked, do victims of NAV at an early stage become perpetrators later, setting up a Nazca booby version of the proverbial "cycle of violence" (the "intergenerational transmission of adult–young maltreatment"; see, for example, Grace et al. 2011)? The answer, they found, was Yes: "In the present study, a bird's history as a target of NAVs proved to be [an even] stronger predictor of its adult behavior than was its history of siblicide" (Müller et al. 2011, 617). They suggested that

---

[24] The recorded differences were enormous: "When expressed as the number of NAV events per nonbreeder present, scaled to observation effort, Nazca [of Galápagos] and blue-footed boobies [of the Peruvian coast] performed 0.0725 (95% CI = 0.0588–0.0862) and 0.0017 (95% CI = 0.0014–0.0018) NAV events/nonbreeder/hour, respectively" (Müller et al. 2008, 3). That's almost a 43-fold difference.

"the intergenerational transmission of NAV behavior indicated by our present results [reflects a lasting developmental effect] of socially induced endocrine perturbations. We are focusing on the stress hormone corticosterone . . . as a probable candidate for physiological mediation of the cycle of violence in this species."[25] Among Nazca boobies, hormones—specifically androgen and possibly corticosterone—are clearly key in proximal causes of NAV behavior.

Evolutionary Causes of NAV Behavior

Keeping these proximate mechanisms in mind, we are back to the question: How can we explain the evolution of NAV aggression among Nazca and other boobies? Is there an adaptive benefit to the behavior? With all its hormonal priming and energy expense, how else could the behavior be sustained?

One intriguing hypothesis is that NAV aggression originated not via ordinary evolutionary mechanisms (e.g., natural selection, drift, etc.) but by a "social contagion" pathway. By this account, the evolved siblicide of Nazca boobies, supported by the underlying androgen mechanism, might "prime the pump." A spontaneous occurrence of NAV aggression by a single siblicide survivor in adulthood, or several, could trigger hormonal changes in their victims (also siblicide survivors), leading to their own NAV aggression later, and they in turn would affect others, such that NAV would spread like a contagion throughout a colony of birds. As noted by Müller et al. (2011, 617), it could also spread in this way to, and then among, blue-footed boobies, though perhaps more slowly since their androgen response would be reduced. It's a plausible scenario, and the androgen "pump" is certainly at the ready.

But what about the NAV aggression described here earlier among northern gannets? They lay one single egg and thus engage in no siblicide, yet they show aggressive NAV behavior at some of its highest levels. The social contagion explanation might conceivably work in the case of Galápagos boobies, but not in the case of siblicide-free gannets. Could both behaviors stem from some other process rooted deep in the phylogenetic history of the Sulidae?

It seems possible that NAV aggression could have evolved via some still-unrecognized evolutionary advantage. Rather than NAV being a simple outgrowth of behavioral and hormonal systems of siblicide, one could imagine it being the other way around: perhaps adult aggression evolved for its own reasons, and siblicide followed where advantageous, putting an existing androgen-aggression pathway to additional use.[26] By this logic, adult aggression toward

---

[25] Note that that the stress hormone implicated in the booby cycle of violence is a close evolutionary homolog of the hormone so important to the El Niño response of marine iguanas (Grace and Anderson 2014).

[26] Or surely the two behaviors—aggression by NAVs and siblicide by hatchlings—could also have coevolved, each contributing its incremental reproductive advantage to underlying genetic variants.

unrelated chicks could be fully homologous among northern gannets and boobies of the sulid clade who show this behavior (and specifically reduced by natural selection in those who do not). Nazca NAV behavior would amount to the everyday manifestation of an older mechanism that evolved for a specific fitness benefit. It would be yet one more booby trap.

My hypothesis is a bit outlandish and should not by any means be taken as an accepted understanding. But it suggests that NAV behavior could have a major fitness benefit during really extreme food scarcity. Could NAV aggression toward chicks appear to be "anomalous," dysgenic, and bizarre under a usual range of food supply conditions, but also favored in evolution because it saves adult boobies in unusual, desperate times? In short, could NAV behavior have evolved for benefits associated with survival cannibalism?

The impetus for this unusual suggestion comes from recent observations among brown boobies (*Sula leucogaster*), one of the wide-ranging tropical boobies. Having branched off roughly 3.1 MYA (Figure 7.5), the brown booby is a close phylogenetic relative of both the Nazca and blue-footed boobies of Galápagos. Indeed, it is similar enough to blue-footed boobies to produce viable hybrid offspring with them (another example of incomplete species; Taylor et al. 2013b), and its similarity with Nazca boobies includes obligate siblicide (Anderson 1990).[27] Interestingly, there are multiple reports of adult brown boobies' eating conspecific chicks (that is, chicks of their own species from other parents). One recent study (Neves et al. 2015) summarized four naturally occurring cases observed in the relatively small colony—fewer than 600 birds—of Brazil's Saint Peter and Saint Paul Archipelago in the Atlantic Ocean. One case was clearly survival cannibalism by "a female with a broken wing that ate a live chick when it was ejected from the nest by its parents" (p. 302). In another spontaneous case, a male and female competed for a chick freshly ejected from an adjoining nest, until the male swallowed it whole (Figure 7.14). Following these spontaneous cases, the authors ran an experiment: "We collected fresh chick carcasses resulting from natural mortality in their original nests or close to them. To assess propensity for cannibalism, we carried out seven experimental trials. We placed a [naturally caused] dead chick next to paired brown boobies defending a nest site without eggs or chicks" (p. 300).

These experimental trials generated two further cases where adults consumed brown booby chicks, and three cases where consumption was attempted but failed because the chick's carcass was too large to be eaten whole. In all five cases, adult females attempted or carried out the consumption, and several

---

[27] Tershy et al. (2000) presented arguments and data from San Pedro Mártir Island, Mexico, to suggest that at least some brown booby populations are also facultatively siblicidal, like the blue-footed booby.

Figure 7.14. Consumption of a chick within a colony of brown boobies (*Sula leucogaster*), on a Brazilian island. Here a breeding pair fight over a 3-day-old chick that was pushed out of the nest by the adjacent breeding female. The fight ends with the male swallowing the chick (*at right*). *Source*: Photo by G. T. Nunes from Neves et al. (2015, 302), copyrighted and used under Creative Commons License 4.0 (from https://creativecommons.org/licenses/by/4.0/). Brightness adjusted for reprinting.

cases included persistent aggressive pecking and shaking of larger chicks by adults, akin to Nazca NAV behavior. One could also add an older account of brown booby cannibalism mentioned in passing in Nelson's monograph (1978, 498). Plus, Neves et al. (2015, 302) noted a previous case of cannibalism in Galápagos: "A single time in the Nazca booby (*Sula granti*) in more than 15 years of research" (Humphries et al. 2006, 386).[28] Sulid cannibalism may be rare, but it does occur.

So, let me state the argument here as *Hypothesis 2*, that NAV aggression has evolved for the fitness benefits it confers to individual adults via cannibalism during life-or-death food-supply shortages. During duress from long-lasting El Niños, famine-inducing PDO swings, and other extreme pressures, NAV aggression would make available an additional source of food—independent, in the moment, of the depressed fish supply. The behavior and its hormonal priming mechanisms would be present today because they helped boobies and their ancestors, particularly adult females, survive drastic periodic reductions of food supply in the past.[29] A deep phylogenetic history of the behavior would

[28] Although a single instance is reported, Nazca chicks do commonly disappear "between nest checks in cases where ejection was not observed." While "the most likely interpretation [is] that the [second] chick was ejected from the nest and then taken by a predator" (Anderson 1989b, 364–65), it's quite possible that some have been cannibalized, as in the brown booby example of Figure 7.13.

[29] Neves et al. (2015) mentioned other contending hypotheses for the cannibalism they observed in brown boobies, including opportunistic feeding by females to restore energy during breeding, colony sanitation, and a density-dependent mechanism that, at high densities, eliminates adjacent nests. Further observations and experimental trials on site will help to narrow the list.

help explain parallels among brown, Nazca, and blue-footed boobies, and even the northern gannets. Moreover, an independent report of cannibalism by fully grown fledglings who consume little hatchlings in a colony of Peruvian pelicans (*Pelecanus thagus*) hints at an even deeper phylogenetic history of this behavior: "We suggest aggression toward, and cannibalism of, nestlings by fledglings are opportunistic behaviors, based on the opportunity of finding unattended nests, and are triggered by food deprivation, although hormonal mechanisms may also be involved" (Daigre et al. 2012, 603).

The hypothesis put forward here is testable, I believe, under the serious food-limiting conditions of "natural experiments" provided by El Niño and/or decadal PDO-style events. The predictions of Tompkins et al. (2017) suggest that climate change will continue to exacerbate the scarcity of Nazca prey—and by implication, of blue-footed booby prey—over the next decades. During a naturally occurring period of duress, one could attempt to keep tabs on all hatchlings in some subset of a colony of each species, with careful records of cause of death and apparent "disappearance"—the method that led Humphries et al. (2006) to their report. Motion- and heat-sensing nocturnal video would help to differentiate the proportion of chick disappearances related to predation and that related to cannibalism.

To this point, Hypothesis 2 focuses on NAV aggression per se, setting aside the affiliative and sexual forms of NAV behavior. However, the whole package might have a related evolutionary explanation: it seems possible that the other aspects of NAV behavior, which generally represent less risk to the perpetrator than aggressive behavior, could have evolved (without conscious understanding, of course) as a way to test the strength and resistance of potential quarry before committing to a fully aggressive assault. Alternatively, it may be that the separate behaviors have independent benefits of their own and became bundled over time by the evolutionary coordination of developmental or neural pathways. The remaining gaps in the story call for more work on these topics. Nelson's words in 1978 were prophetic: from the Sulids we can still "learn a great deal."

Looking back, this chapter offers a good example of "Anderson's rule"—a rule nicely summarized in a recent paper by the namesake and coauthors: "Food controls the breeding success of tropical seabirds (other environmental factors like predators and weather usually have negligible direct effects)" (Tompkins et al. 2017, 3). As shown by the three Galápagos boobies, it all comes down to food and its effects on attracting and keeping mates, nest maintenance versus nest desertion, brood-size regulation within the nest, migration, population resilience, and more. Indeed, two of the three Galápagos boobies—the blue-foot and the Nazca—show how specialization on remote islands can mean that breeding success hinges on a single prey type—namely, sardines. For these two boobies, sardines are a classic "limiting factor" or "limiting resource," as the terms are used in ecology, since an increase in sardine supply causes their increased growth, abundance, and/or distribution.

But there is even more to the control that food exerts. So much of what is shown in this chapter regarding blue-footed boobies' behavior, from parading to siblicide, from feet color to the fine-tuning of their color perception, relates directly to sardines and the carotenoids they offer. To the extent that Hypothesis 2 is valid, even the consumption of conspecifics may hinge on the availability of sardines in space and time. A basic message of this chapter is that all these diverse phenomena in the lives of the boobies appear to be linked to sardine supply. With so specialized a dietary focus on a single taxon of fish, it seems that the blue-footed boobies may be even more *bobo* than we initially thought!

Conservation Implications

This chapter, like its predecessors, concludes with a brief look at some salient conservation measures suggested by the exploration of boobies residing in Galápagos. To begin, there is an important lesson here about the IUCN Red List, the world's most comprehensive inventory of species conservation status. After careful study, the IUCN review (IUCN 2018b) concluded that the blue-footed booby warranted a Least Concern rating: "This species has a very large range, and hence does not approach the thresholds for Vulnerable under the range size criterion. . . . The population trend appears to be stable, and hence the species does not approach the thresholds for Vulnerable under the population trend criterion. . . . The population size is very large, and hence does not approach the thresholds for Vulnerable under the population size criterion. . . . For these reasons the species is evaluated as Least Concern."

Meanwhile, as we've seen, the Galápagos population of blue-foots was in precipitous decline, showing that the status of island natives with specialized feeding strategies can change quickly. But more than that, it shows that locally important populations of a species with otherwise wide distribution can be in serious decline without it affecting the overall rating (the IUCN report describes the decline in Galápagos in the "Threats" section, as part of the Least Concern rating). While it is certainly reasonable for the IUCN to keep its general rating for the reasons they describe, this case suggests that Galápagos might have its own Red List-style status report, especially for native species.[30] Or perhaps there could be

---

[30] The value of a Galápagos Red List was not lost on the local conservation community. As this chapter went into production, a list of this kind was published by a consortium of organizations (see Jiménez-Uzcátegui et al. 2019).

a sublist within the IUCN Red Book: a Galápagos-specific threat report to supplement the IUCN's broader, high-level assessments. It could be linked to the main IUCN page for any given species, signaling locally important population changes.

Second, there may well be a conservation "silver lining" to the evolved behavior of siblicide in the boobies of Galápagos and elsewhere. Siblicide suggests a simple and effective human intervention—one that could be used whenever there are convincing reasons to intervene in booby population dynamics. In the case of blue-foots' facultative siblicide, the option exists for Galápagos Park Rangers, or their deputies, to rescue smaller siblings before they are killed, and hand raise them off-site to a secure age when they could be placed back into the parental nest, or else to maturity and independence. Similarly, the initial siblicide survivors could also be rescued whenever short food supplies cause them later to be abandoned. In the case of Nazcas and obligate siblicide, the same intervention could work. It would be time-intensive, because one would have to act quickly, nest by nest, before the smaller chicks die. But there would be a big advantage: successfully hand-rearing "insurance" hatchlings could nearly double the per capita growth rate of the population—insurance, indeed, for a rapid population rebound. Something of a parallel intervention has been successfully used with poacher-orphaned macaw chicks in the Peruvian Amazon (see Brightsmith et al. 2005).[31]

When warranted, saving chicks from siblicide, or indeed from cannibalism, could conceivably buy valuable time for these locally endangered boobies, but it only works if the sardines return to Galápagos waters as a viable food supply. Careful analysis by Tompkins et al. (2017, 14) showed that neither blue-foots nor Nazcas "can maintain a stable population under the conditions of the Flying Fish Phase [the interval that began in 1997 with the SST change] at our Galápagos site. With a lipid content of < 1%, flying fish are effectively 'junk food' compared to sardines (lipid content 8%)." How will the sardines make it back?

Here, again (as in Chapter 2), we appreciate the links between Galápagos species and fisheries dynamics hundreds of miles away along the coast of South America. Once again, the pertinent social–ecological system for understanding the impact of human activities on boobies in Galápagos reaches all the way to the coastal fishers of Peru and Ecuador. How does their impact on coastal sardines in South America affect Galápagos? Of course, the sardines on the coast are also subject to the decadal swings of cold- and warm-water "regime" changes,

---

[31] I thank Stanford student Marika Jaeger for suggesting this intervention in her term paper for my Galápagos class, drawing from her experience as a summer researcher in Tambopata, Peru. More on the Macaw Project can be found at their website (http://vetmed.tamu.edu/macawproject).

in addition to the annual catch of the fishers. But current thinking is that "Galápagos schools rise and fall in association with sardines on the continental margin, on a scale of decades" (Anderson 2018, 14), owing to their capability for long-distance migration. What the fishers do in the Peruvian upwelling is destined to play a major role in booby conservation in Galápagos. The connection reminds us that local social–ecological systems are embedded inside larger ones, rather than existing in isolation.

The same researchers also found that if sardines return, the return may only be temporary. They noted that climate models for the Western Pacific now predict an SST rise of a whopping 4.5°C over the next 100 years, one of the highest changes anywhere on Earth (Tompkins et al. 2017, 2). Because this increment will mean water temperatures rise generally beyond the upper viable range for sardines, and well over their maximum spawning temperature, they may then be forever gone from the booby diets of Galápagos. Without them, Nazca and blue-footed species might well survive in other parts of their home ranges, but if climate predictions hold, almost certainly one or more long-standing icons of Galápagos will be a living icon there no more.

What can be done to avert such a tragic loss? Addressing climate change is certainly one part of the equation, as discussed in earlier chapters. Anthropogenic change contributes to ENSO and perhaps to decadal oscillations as well. It is one of the threats against which each of us can help, by reducing our carbon footprint and by pushing for supportive governmental action on carbon emissions in the countries where we live, especially the United States. Regulation of sardine fishing in coastal South America is surely another term in the equation, given the "migratory connection" to Galápagos (Tompkins et al. 2017, 15). Recently, a similar U.S. fishery in decline offered an example of what might be tried: in 2015, federal regulators called for early closure of commercial sardine fishing off Oregon, Washington, and California to prevent overfishing and damage to sardine stocks and their prey. The action was decisive and well warranted, given declining harvest trends. But it may also have been too late to avoid lasting reduction in the population.[32] However, even that outcome is an instructive warning for other sardine fisheries, like those of Peru and Ecuador.

---

[32] Since the U.S. closure, "the numbers of young fish added to the population from 2011 to 2017, known as recruitment, have been among the lowest in recent history," partly from natural ocean conditions, and preventing population rebound. Projections for 2020 indicate that sardine population size remains below the minimum cutoff level for reopening the commercial fishery. The fishery may soon be declared officially "overfished," which then requires a specific population rebuilding plan. This U.S. example attests to the importance of monitoring and regulation of coastal sardine fisheries (see https://swfsc.noaa.gov/news.aspx?ParentMenuId=39&id=23239).

As concerned global citizens, let us not be the *bobos* here, but instead let us support an effective, precautionary regulation of sardine fishing in the Peruvian upwelling. Let us also do our share, individually and collectively, to reduce greenhouse gas emissions and help stem the tide—the warm tide of El Niño, now more common and longer-lasting than before. Our prompt action could well give the blue-footed boobies of Galápagos something worth flaunting, once again, in their parades.

8

Not Earthbound Misfits After All

Galápagos seabirds . . . are renowned not for their great diversity of species but rather for their uniqueness, brought about by a broad and complex array of ecological, behavioral, and physiological adaptations.

Carlos Valle ("The Flightless Cormorant: The Evolution of Female Rule")[1]

This chapter examines two delightfully unique, flightless seabirds: the Galápagos cormorant, one of the world's most unusual organisms, and the Galápagos penguin, the only penguin to swim in waters of the Northern Hemisphere. As we explore their adaptations to the archipelago, three themes stand out: First, in pre-settlement Galápagos, neither species suffered great disadvantage because of flightlessness. Having no terrestrial predators allowed both species to nest on land near water's edge and specialize in diving for prey in the Cromwell upwelling. Both species survived and, from all indications, prospered during a million or more years of flightless life in Galápagos. Second, from divergent phylogenetic backgrounds, the two seabirds evolved a fascinating evolutionary convergence in Galápagos—not their flightlessness per se, because penguins were already flightless when they colonized this "little world." Instead, the birds exhibit striking similarities in their uniquely opportunistic mating practices. The third theme relates to their responsiveness to El Niño: Does the older flightless specialist, the penguin, have the advantage when El Niño comes to the archipelago and food supply falters, or does the cormorant, the seabird specifically retooled by evolution for conditions in Galápagos, show the greater resilience? Let us begin with the cormorant.

Galápagos Cormorant

The approximately 40 cormorant species of the world are well known for their successful hybrid lifestyle. In the skies, their flight is graceful and efficient. In the

---

[1] Opening quote is from Carlos A. Valle, scientist at the San Francisco University of Quito (USFQ) and leading researcher for the Galápagos cormorant. Valle's work informs much of this chapter.

Exuberant Life. William H. Durham, Oxford University Press (2021). © Oxford University Press.
DOI: 10.1093/oso/9780197531518.003.0008

waters, their dive is sleek and deliberate. They don't plunge like the boobies, but jump in an arc from the water's surface with a big gulp of air, relying on strong webbed feet to propel them to their prey. With those feet, their sleek design, their long necks, and their powerful beaks, they are formidable fishers. Such is their prowess that they are sometimes admired, sometimes feared, and sometimes even put to work—as in China and Japan, where traditional "cormorant fishing" is practiced (see Manzi and Coomes 2002). There, cormorants garner abundant catches, but the birds are outfitted with snares around their necks to ensure that most fish make it back, still in their throats, to their eager human caretakers. Elsewhere, cormorants' fishing prowess is despised because they are such able competitors, sometimes prompting retaliation.[2] If ever the local fish supply fails them, no problem: cormorants simply fly along the water's edge to a better spot, returning to their colony to rest, sleep, mate, and raise chicks. Nicely specialized for air and water, they are more awkward on land, walking with big feet and a wing-assisted hop.

Galápagos is the only place on Earth where resident cormorants have evolved to be completely flightless.[3] The endemic Galápagos cormorant, *Phalacrocorax harrisi,* has wings that are comically small for a large bird (Figure 8.1).[4] And they are large: males weigh up to 4.7 kg (10.4 lbs) and females up to 3.5 kg (7.7 lbs)—roughly 25% heavier than the next-heaviest cormorant species (Valle 2013), about the size of plump wild turkeys in the United States. As ornithologist Michael Harris put it: they are an "unmistakable, large dark bird with apparently functionless, tatty wings" (Harris 1974b, 73). We could fairly call them "Tatty and fatty."

The cormorant's gait is labored, more like a wobble, with pathetically little wing-assistance. Such clumsiness would never make it in a place with terrestrial predators; even in the Galápagos, cormorant nests are, at most, just a few awkward hops from the water. But lack of elegance on land is fully compensated in the water. With big streamlined bodies, large beaks, powerful webbed feet, and tiny wings wrapped close to their sides, they have evolved into diving specialists (Figure 8.2), agile and fast. Even the best human divers cannot keep up with them under water.

---

[2] In the estuary of the Columbia River of the United States, for example, federal agents sometimes shoot cormorant adults and poison eggs in nests, in efforts aimed at boosting the survival of salmon and steelhead (see Groc 2014; King 2013).

[3] Other flightless members of the order Suliformes (formerly Pelecaniformes) are known only from fossils; Olson (1980) argued that they went extinct with the evolution of predatory seals and porpoises.

[4] The genus name is from Greek for "bald raven;" the species name is from Captain Charles M. Harris, co-leader of the 1897–1898 expedition that collected the first specimen. For more on this expedition and others, see Larson (2001).

Figure 8.1. A male Galápagos cormorant dries his "tatty" wings. Note the extra-large body supported on strong legs and large, webbed feet. *Source*: Courtesy of Robert Siegel, MD, PhD.

Intriguingly, Darwin never mentioned this species. The *Beagle* passed through the heart of their current distribution along the Bolívar Channel (the small rectangle in Figure 8.3*A*), and Darwin spent one full day ashore near Tagus Cove (halfway up that rectangle), where their density along the coast is normally high. True, he spent most of the day inland exploring a geological feature now called Beagle Crater, just south of the cove, but he was in prime cormorant habitat getting there and getting back in a small skiff. How did he not see them? Could it be that he thought they were simply the plump chicks of already well-known Neotropic cormorants? Indeed, the adults do resemble overgrown chicks. Or did Darwin only see them in the water, with just the head and neck visible, so they might not have attracted his attention? Or did he not see them at all because their numbers were severely reduced at the time by an epidemic, an eruption, an El Niño event, or human hunters?[5]

I puzzled over this question for years because, during the *Beagle* voyage, Darwin detected and appreciated unusual phenomena much subtler than a large

[5] There is some evidence—such as the only reported hurricane in Los Angeles history—to suggest that 1835 was an El Niño year (see http://www.pbs.org/wgbh/nova/elNiño/reach/time.html). As this book went to press, Duffy (2020) published a thorough review of the possibilities for Darwin's oversight. Here I offer another explanation.

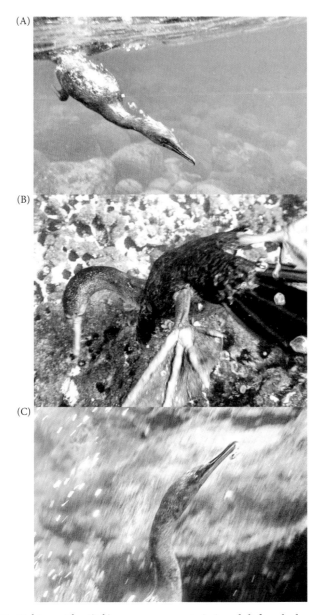

Figure 8.2. Fishing with a Galápagos cormorant. *A,* An adult female departs on a shallow dive after grabbing a breath at the surface, propelling herself with webbed feet toward the bottom, wings folded tightly along the body. *B,* Wide feet and strong legs speed the female toward the ocean floor, as plumage air bubbles out from the outer portion of her feathers. With agile neck and long beak, she probes her way along the nooks and crannies on the bottom until . . . *C,* Success! She grabs a bite-size blenny and heads for the surface to breathe again and swallow without taking in extra salt water. Studies of other heavy cormorant species on long dives show that these birds use their gulp of breath to adjust buoyancy (Cooke et al. 2010).

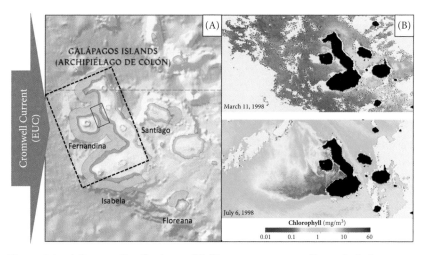

Figure 8.3. *A,* Present distributions of Galápagos cormorants (*large dashed black rectangle*) and Galápagos penguins (*dark* and *light orange* for breeding and nonbreeding areas, respectively) among the western islands of the archipelago, in relation to the Cromwell Current (or EUC, *large blue arrow*). High densities of both species occur today in the Bolívar Channel between Isabela and Fernandina (marked by the *small rectangle*). *B,* The impact of El Niño's warm waters (*top*) and La Niña's cold upwelling waters (*bottom*) on the photosynthetic activity of phytoplankton, the base of the ocean food web (as measured by chlorophyll concentrations). Warm El Niño waters stifle the special productivity of the upwelling region. *Sources: A,* BirdLife International and *Handbook of the Birds of the World* (2016, 2013). "*Spheniscus mendiculus.* The IUCN Red List of Threatened Species." Version 2019-3 (http://www.iucnredlist.org). Downloaded on March 8, 2020. *B,* Public domain, NASA's Goddard Space Flight Center, Ocean Color team and GeoEye, from Wikimedia.

flightless bird (Durham 2012). The first convincing explanation occurred to me while reading a report by Karnauskas et al. (2015) that the "cold pool" at the center of the Cromwell upwelling has been shifting northward and expanding in past decades. Because that upwelling is so rich and productive, and so crucial to the cormorant's food supply, perhaps the distribution of cormorants, too, has shifted northward since Darwin's visit. Extrapolating from Karnauskas et al.'s estimate of the rate of northward shift, it seems likely that when Darwin went ashore in northern Isabela, cormorant density would have been very low, maybe zero, although the area is the center of cormorant distribution today. Cormorant density at that time would have been higher down on the south coast of Isabela.

   Striking confirmation came from an 1813 observation by USS *Essex* captain David Porter (1823, 40–41). Porter said he and crew that year found "plenty of birds called shags"—a word used interchangeably with "cormorant"—at "Essex Point" on the southern coast of Isabela, where none are found today (roughly 80 km or 50 miles straight south of Tagus Cove). Alas, owing to wind conditions, in 1835 the *Beagle* sailed around that southern coast and Darwin was never ashore in the region (Estes et al. 2000, 352). Because of the cormorants' coast-hugging tendency, most likely he never saw them. The same is true for Galápagos penguins, also mentioned by Porter, but not by Darwin.[6] Penguins, too, keep close to the coast, and the *Beagle,* with Darwin aboard, must have sailed right past both populations. The inference also suggests that both species are quite sensitive to changes in ocean temperature—a topic examined more closely below. Collecting either species in 1835 would probably have changed the timing of Darwin's discovery of evolution, but I doubt it would have changed the outcome.

Today, we know the Galápagos cormorant descended from a volant ancestral cormorant who shared a common ancestor with two other American cormorants, the Neotropic cormorant and the double-crested cormorant, until 2.4 MYA.[7] Comparison with these phylogenetic relatives shows that the ancestral cormorant flew into the islands equipped with the "standard package" of cormorant features for their hybrid, flying–diving lifestyle. First, because all cormorants belong to the ornithological order Suliformes, they came with ancestral qualities common to that order—such as dark adult plumage, which helps with warming in sunlight after cold-water dives; webbed feet, which help with propulsion; and altricial young (helpless at birth), which require major parental feeding and protection. In addition, cormorants exhibit throat fluttering, a device for displays and for cooling, much like panting in dogs. The latter proved especially beneficial for thermoregulation in Galápagos, allowing nesting out in the open in full equatorial sun. Beyond that, we infer that ancestral cormorants arrived in Galápagos with all the features making cormorants highly successful fishers, such as hooked beaks that facilitate handling of prey, "wettable feathers" that allow water to displace air as cormorants dive, and reduced oil from the preen gland, adding to that wettability.[8] Wettability is important in cormorants

---

[6] Galápagos penguins were also mentioned by David Porter (1823, 41) at the same place and time as his reference to cormorants, but the penguins were not described in the scientific literature until Sundevall in 1871 (Frith 2016, 206).

[7] Burga et al. (2017) estimated the divergence date for these species at 2.37 MYA, a bit earlier than an older estimate of 2 MYA from Kennedy et al. (2009, 96). Related work by Burga et al. (2016, 2017) and Berger and Bejerano (2017) has identified key mutations behind flightlessness in the Galápagos cormorant.

[8] Cormorant feathers have an interior structure resembling that of more typical, fully waterproof feathers, with parallel barbules (small veins) joined by hooklets. However, on the outer edges, cormorant feathers are less structured and therefore more wettable. This dual arrangement is rare: in one study of 15 other water bird species, only the related European shag also exhibited a dual feather structure (Grémillet et al. 2005).

because air trapped in plumage generates buoyancy, which can only be countered by strong foot strokes that are energetically costly. The cormorants' wettable feathers result in half as much plumage air as in ducks of the same size, allowing them to forage longer and deeper. But long dives require time to dry out and warm up again, especially before temperatures drop at nightfall—hence the common spread-wing behavior of cormorants back on land (Figure 8.1). All this repertoire came with the ancestors.

When the ancestral cormorants flew into Galápagos as much as 2.4 MYA, the islands were very different from those of today. The big central island, which later broke up to become Santa Cruz, Pinzón, and Floreana (see Online Appendix B), is a likely candidate for an early cormorant nesting site because its position at 2 MYA was in the path of the Cromwell Current, which collides with the Galápagos platform, producing a nutrient-rich upwelling that sustains one of the most productive of all marine ecosystems (Figure 8.3B).[9] It put a rich and diverse diet literally steps away from the coastal nests of the cormorants, setting them up for the eventual evolution of flightlessness and other features.

At various points in the last million years, six volcanoes west of the big central island erupted to form today's Isabela Island. As their eruptions neared the ocean surface, they diverted the Cromwell upwelling of the time toward the south and west. The evolving cormorants would likely have followed the trail of cold water (and fish) to the southwest, eventually settling on Isabela's southern coast, where Porter found them. Over time, their distribution then shifted north, keeping them close to the center of the nutrient-rich upwelling—a good location for a large flightless bird . . . most of the time.

At this juncture, it is helpful to turn our attention to the Galápagos penguin, which, in contrast, arrived in the islands already flightless. Its ancestors most likely swam with the currents from the northern reaches of a related species in coastal Peru and Chile, and then evolution "adjusted" them for life in the isolated archipelago.

Galápagos Penguin

In contrast to the cormorant, typically regarded by visitors as fascinating but plump and clumsy, the Galápagos penguin (*Spheniscus mendiculus*) is seen as nimble, sleek, and cute (Figure 8.4). With an average adult height of roughly 50 cm (about 20 in.), and body mass around 2 kg (4.4 lbs), the penguins are among the world's smallest, and they are the only penguin found at such warm

---

[9] Drawing on the work of many paleoclimate investigators, Nathan and Leckie (2009) concluded that the Cromwell Current has upwelled in Galápagos for at least 6.50 MY.

Figure 8.4. Left, a pair of Galápagos penguins, *Spheniscus mendiculus*, with stripes on their faces and bellies. On fully webbed feet they stand only about 50 cm (20 in.) tall. Spots in their chest plumage may help in mate recognition (and are also useful to researchers). Right, the penguin with its back to us is molting (shedding old feathers), which takes about 2 weeks and happens twice annually. Since molting is asynchronous within the population, molting Galápagos penguins are visible for as much as 7 months of every year (Boersma et al. 2013). Their ragged appearance while molting may explain their species name: the Latin *mendiculus* means "indigent."

air temperatures. They are also super-fast swimmers, like darts in the water, belonging to the widely distributed genus *Spheniscus* (from Greek *sphēniskos*, "small wedge," in reference to their streamlined bodies and wings).[10] Like others in the genus, they have a striped appearance: their white belly, speckled with a few black feathers, is ringed with a solid stripe of black plumage, plus there's a second stripe of white feathers outlining the neck and face (Figure 8.4). Most penguins are not striped, and so the question logically arises, do the stripes serve a specific function, somehow enhancing the survival and reproduction of their *Spheniscus* bearers?

The question was explored experimentally by Wilson et al. (1987), who used a series of model penguins, with and without stripes, to study the reaction of the schooling small fish, like sardines and anchovies, who are the main *Spheniscus* prey species. These fish normally forage in loosely organized groups but react

[10] Other species of the genus include the African penguin (or "jackass penguin," because of its braying call), the Magellanic penguin of southeastern South America, and the Humboldt penguin of South America's southwest coast. The genus originated near Antarctica; ancestors of today's species followed circumpolar currents, eventually arriving at their present locations, as matches their phylogeny (Bertelli and Giannini 2005).

to predators by forming tight clusters in which individuals are hard to catch. In experimental trials, Wilson and colleagues found that the banding pattern of *Spheniscus* penguins scattered the clusters of fish both more frequently and for longer periods than the unstriped penguin coloration. In other words, moving stripes confuse the fish, breaking up their protective clusters and making them easier to catch . . . a fitting adaptation for penguins that feed on schools of small fish.

The stripes are a telltale feature of the *Spheniscus* genus, making the Galápagos penguins look like a smaller version of the Humboldt or Peruvian penguin from whom they diverged, we know from comparative genetic analysis, roughly 1 MYA (Gavryushkina et al. 2017). The Humboldt species, in turn, diverged from other, eastern *Spheniscus* penguins about 1.65 MYA, with ancestors before that time living in the Antarctic region. Given this origin, we expect Galápagos penguins to share quite a few features with Humboldt penguins, including similar adaptations to air and water temperatures. Because Humboldt penguins hunt in the cold waters of the Humboldt Current (whence their name) and form terrestrial colonies all the way to the warm tropics (at the northern end of their range in Peru), one expects them to show at least a few features not found in *Spheniscus* species residing further south. In fact, the Humboldt penguin's range has long put it into the heart of the ENSO zone where the rich Peruvian upwelling is periodically blocked by El Niño events. Thus, the Humboldt and Galápagos penguins have both lived at least 1 MY in different parts of the ENSO heartland. The latter's range today is shown in Figure 8.3A.[11]

As a result, the Humboldt and Galápagos penguins share a number of derived features not found in other *Spheniscus* species.[12] Among the features are:

1. They molt before breeding, not afterward like other penguins, allowing restoration of "unimpaired plumage" for insulation during brooding (Reilly 1994, 137).
2. They show highly variable breeding, with between zero and three clutches per year in accordance with the food supply—another case of "Anderson's rule."[13]

---

[11] Galápagos penguins also occur today in small numbers on Pinzón, the small island directly south of Santiago (not shaded in the map), evidence of the reach of rich Cromwell waters into the archipelago.

[12] One ENSO adaptation of Humboldt penguins has not been reported in its sibling species. During the 1997–1998 El Niño, penguins of northern Chile responded to the influx of warm water by swimming south as much as 116 km/day (70 miles/day), as far as 895 km (406 miles) south of their starting point, seeking cooler water and food (Culik et al. 2000).

[13] By Anderson's rule, food supply governs the breeding success of tropical seabirds. In a study of Humboldt penguins in Peru from 1993 to 1997, Paredes et al. (2002, 246) concluded that "Having as many clutches as possible when conditions are favorable appears to be a strategy used by Humboldt penguins to maximize their lifetime reproductive success within a productive but unpredictable environment" (p. 244).

3. They generally lay two eggs per clutch, with hatching asynchrony (Boersma et al. 2013, 269), such that eggs are laid 2 to 4 days apart, and then hatch in the same pattern.

4. When times are good, this clutching pattern affords two hatchlings; when food is scarce, it allows "brood size adjustment" to one hatchling or even zero (as in boobies).

These features provide reproductive flexibility, which is surely advantageous for a species living through recurrent ENSO events. One could almost think of ENSO as a switch for these species: La Niña turns ON both a strong Cromwell upwelling and penguin mating for the duration; El Niño turns them both OFF; and ENSO-neutral conditions turn ON the upwelling and mating, but for less time.

Adaptations of the Galápagos Penguin

Eventually, ancestors of the Galápagos penguin swam or drifted away from a mainland Humboldt colony and arrived in the islands—monogamous, flightless, specialized at hunting schools of small fish, with highly variable breeding and other adaptations to low latitudes. What further changes evolved in response to conditions in Galápagos? Careful analysis by researcher Dee Boersma and colleagues points to the following additional specializations.

First, Galápagos penguins show further refinements for thermoregulation in the face of extreme temperature ranges. Galápagos penguins average just half the weight of Humboldt penguins, and their small size dissipates heat better via a higher surface-area-to-volume ratio (an example of "Bergmann's law"—small size has advantages in warm places). In addition, the penguins make good use of behavioral thermoregulation, akin to marine iguanas, alternating between cooling off in the ocean and warming up on land. On land, they can cool down again by panting and extending their flippers like small radiators.

Second, Galápagos penguins have a unique molting pattern, which surely also helps with heat regulation. They are the only penguin to molt twice a year (Figure 8.4, right). They molt spring and fall, when food supply is normally good (molting is energy intensive), and when their old feathers are worn from sun damage or algae in their plumage. They thereby become "natty," on a par with the cormorant's "tatty and fatty." Twice-a-year molting restores plumage in both appearance and insulation, and they begin molting early: "Unlike any other penguin species, juveniles molt at 6 months to adult plumage" (Boersma et al. 2013, 291).

Third, "Galápagos penguins are the only member of their genus to occur exclusively in small colonies," or clusters, often with just three to ten penguins

(Steinfurth et al. 2007, 7). With such great thermal stress and equatorial ultra-violet radiation, they simply cannot "hang out" in large, open aggregations on beaches or grassy areas. Cool, shaded nesting locations, like miniature lava caves or tunnels, are crucial but limited in number, which spreads penguins out over long areas of coastline—contributing to their greater range than cormorants (Figure 8.3).

Fourth, while brooding, they are known for their "extremely aberrant coast-hugging behavior . . . which is shown by no other penguin species" (Steinfurth et al. 2007, 7). Half of all birds studied by Steinfurth and colleagues strayed less than 270 m (885 ft) from the shore, and no bird was recorded over 1,000 m (3,280 ft) offshore, even as they traveled between 1.1 and 23.5 km from their nests along the shore (mean distance traveled = 5.2 ± 4.9 km). Several factors may con-tribute to this unusual coast-hugging behavior. One such factor is proximity of the Cromwell upwelling—and thus penguin food supply—to shore, arising from the steep western edge of the Galápagos platform. Additionally, deeper water has a higher density of sharks, foraging fur sea lions, and other predators known to attack penguins. Finally, it may be easier and more energy-efficient to swim in inshore waters than out where currents are generally stronger.

The fifth adaptation is perhaps most important of all: Galápagos penguins are unique among all penguins in having no fixed breeding seasons. "They can breed in any month of the year, depending on food availability, up to 3 times per year" (Boersma et al. 2013, 289). Food availability, in turn, is related to sea-surface tem-perature (SST), with the result that breeding is triggered by SSTs of 24°C (75°F) and lower, when the Cromwell Current is strong and fish are abundant. As Reilly put it (1994, 136), "The lower the [water] temperature, the greater the breeding success." Another temperature-dependent variable may also affect penguin re-production: mosquito density and mosquito-borne disease. When water and air are both cool during La Niña events, mosquitoes are much less a problem than during El Niños, when, as we discovered with waved albatrosses, endemic and introduced mosquitoes can be a real threat. In the case of penguins, we have a direct measure: the birds show a high prevalence today of antibodies to malaria, a mosquito-borne disease (Palmer et al. 2013). We return to malaria later, but for now it suffices to say that, with cool temperatures, both food supply and mos-quito densities are better for penguin reproduction. In these good times, parental penguins even feed their fledglings who have grown and departed from the nest. Such "extended parental care" is rare among penguins (Boersma et al. 2017), but it surely speeds maturation during cool periods and thus helps fledglings get their adult plumage after only 6 months. Rapid maturation in the face of variable conditions can be a big boost to a penguin's evolutionary fitness.

With these five additional adaptations, the Galápagos penguin does remark-ably well for a bird of Antarctic origins living at the equator. While it's still sharply affected by El Niño, the resilient penguin is still in the game 1.1 million years after colonization, thanks to its special reproductive potential. That potential has evolved to be flexible, opportunistic, ever-ready, and efficient. It is theoretically possible for a female Galápagos penguin, under auspicious cool conditions, to lay three clutches of two eggs and hatch up to 6 young a year—an indication of reproductive resilience as impressive as any we've seen in these chapters. Will they be able to handle the ENSO-related challenges ahead?

Having sketched key local adaptations of Galápagos penguins, let us now do the same for Galápagos cormorants, who have lived in Galápagos 2.4 MY, more than twice as long as the penguin. How do their adaptations compare?

Adaptations of the Galápagos Cormorant

Leading researchers of the Galápagos cormorant, especially Carlos Valle (1994, 2013), Rory Wilson (Wilson et al. 2008), and the late Brad Livezey (1992), have filled in many of the details of this bird's unique lifestyle. It evolved away from the flying capabilities of its ancestors and became a fully dedicated seafloor hunter in the heart of the Cromwell upwelling. Indeed, Valle (1995, 612–13) argued that the cormorants' geographic range is small because they can only meet daily food requirements in the very center of the upwelling. This restricted range contributes to their being "the smallest naturally occurring marine bird population in the world," with only about 1,000 individuals in the best of times.

Living in this special setting, the cormorant evolved a list of unusual features— extra-big beak, large body mass, stunted wings, strong legs and feet, and a mating pattern found in no other cormorant—each of which enhanced survival and re- production, we infer, in its new niche. Consider, first, the cormorant's enlarged beak. It's a highly efficient tool for probing the nooks and crannies of the rocky seabed and for extracting marine prey (octopus, eel, blennies, and other small fish) from their hideouts (Figure 8.2). Under normal conditions, cormorants are able to feed themselves in a narrow band of water close to the coastlines, somewhat like the penguins, but always along the bottom and often at greater depths.[14] The cormorant beak is also a potent weapon, allowing the birds to rest and nest in the open, protecting themselves and progeny from aerial predators, especially the Galápagos hawk. One would be hard-pressed to disentangle these two benefits, probe and protector, but it hardly matters because the equipment

---

[14] As summarized by Duffie et al. (2009, 2104), "The majority of individuals stay within 2 km of their natal colony to breed. . . . The birds also seem reluctant to venture into the open water; they have rarely been recorded more than 200 m offshore." The researchers argue that populations on either side of the Bolívar Channel, just 5 km apart, "should be managed as at least two genetic populations to better preserve the species-level genetic diversity."

works so well for both. Livezey's (1992, 176) careful measurements showed that there's an impressive 40% increase in average adult beak length in the Galápagos cormorant compared to the closely related Neotropic cormorant.[15]

Second, consider the cormorant's extra-large body mass. In contrast to the penguin, which, coming from Antarctica, evolved a smaller body in response to warmer conditions in Galápagos, the cormorant faced the opposite challenge. Cormorants came from tropical origins and had a hybrid flying–diving lifestyle, so selection pressures favored the foraging efficiency of a larger body for diving deep in the cold waters of the archipelago. Indeed, the advantages of "big" would only have grown as selection pressures for flight relaxed. Such advantages would have included reduced buoyancy, higher oxygen storage capacity, and lowered heat loss (owing to a lower surface-area-to-volume ratio) in the colder water. In response to these conditions, and with no terrestrial predators from which to flee, the cormorant eventually evolved the largest average body size in its zoological family. In addition, just as a larger body mass in marine iguanas improves their odds of surviving El Niño, so it may be with Galápagos cormorants. Extra mass gives them more stores to live off during El Niño and may also enable them to dive deeper than usual for fish when El Niño first arrives.

As expected, the cormorant's large body mass enables it to carry out longer dives than its close relatives. The comparative study by Wilson et al. (2008, 642) found that "Galápagos cormorants are consistently among the [cormorant] species able to stay under water the longest for any given depth." They measured one dive, for example, that lasted 196 seconds (3.3 min) and reached the impressive depth of 73.2 m (240 ft).[16] What's more, their conclusion was based on the average duration of dives for males and females. Since males on average have a greater body mass than females, they derive even more benefit in terms of buoyancy, oxygen, and temperature, and have the capacity for longer, deeper dives. Valle (1994, 2013) studied the male–female difference among cormorants, which is called a sexual size dimorphism (SSD). He showed that their overall large size and unusually high SSD "are both novel traits that evolved in situ following

[15] As a partial control for overall scaling effects, Livezey (1992) found that the average height and width of skulls of adult Galápagos cormorants were 5% to 11% larger than in Neotropic cormorants. The average length of adult skulls was 18% to 20% more in the Galápagos species. Thus, bill length showed twice the difference of other measures.

[16] Another cormorant with long dives is the Kerguelen shag (*Phalacrocorax verrucosus*) of the southern Indian Ocean, quite possibly the closest thing to a "flying version" of the Galápagos cormorant. Averaging 2.3 kg (5 lbs, half the mass of adult Galápagos males), these birds reach 120 m in dives lasting up to 5 min, and they have big feet and leg muscles, even with the constraints of flight. Fittingly, their flights are "physiologically limited" to an average of only approximately 90 sec per flight, with rest between, and only 24 total flight minutes per day (Watanabe et al. 2011, 412). Why? They have evolved "short wings (to decrease air volume in the feathers and hence buoyancy)" and "small flight muscles (to allow for large leg muscles for underwater propulsion)," like their Galápagos counterparts, with their traits just not as extreme. It figures: they have no upwelling as strong and nutrient-rich as the Cromwell.

Figure 8.5. The sexual size dimorphism (SSD) in flightless cormorants is easy to see in the field and in this photo, being more than double the average SSD for other species of cormorants. The male is on top.

colonization of Galápagos" (2013, 145). The result, known since Snow's early investigation (1966, 289; see Figure 8.5), is that "the size difference between the sexes [is] greater than in other species of cormorants."[17] In addition to the advantages of large body size for diving, Valle argued, their high SSD offers the advantage of reduced foraging competition between sexes. Valle found that the extra weight of males allows them to dive deeper and longer and to bring back fish that are substantially larger and heavier on average than those caught by females (Valle 1994a, 40–41).

What about those tiny wings? Comparative anatomical work dating back to the early 1900s has shown that the Galápagos cormorants' wing bones are much shorter than those in other cormorants (Figure 8.6), and that the keeled breastbone, site of attachment for flight muscles, is also much reduced. By all indications, normal wing bone development is stunted in the cormorant because they stop growing prematurely.[18] The feathers, too, are stunted, both in length

---

[17] Valle (2013) measured the SSD in Galápagos cormorants, using the ratio of male to female weight in adults, and found it to be much more than in other cormorants, a whopping 1.37 vs. 1.16.

[18] Livezey (1992, 205) described the short wings of the Galápagos cormorants as "paedomorphic" (chicklike) because of the "premature termination of development" compared to volant birds of its same genus. Their flight feathers, which are also early to stop growing, measure roughly half the length of those in other birds.

Figure 8.6. A comparison of cormorants, showing that the flight-related bones of the wings and chest are greatly reduced in the Galápagos cormorant (*right*) compared to the double-crested cormorant, its closest relative. Measurements by Livezey (1992, 178) showed that the ulna—the heavier of the two bones at the *black arrows*—is 50% shorter in the Galápagos cormorant. Note the larger body size and rib cage of the Galápagos species, yet its keeled breastbone (*the lowest blue-colored bone*) is much smaller. As the site where flight muscles attach, this skeletal feature is a telltale sign of flightlessness. The beak is also slightly longer in the Galápagos species. *Source*: Katie Bertsche, adapted with corrections from Burga et al. 2017. *Science* 356: 921.

and number. As a result, the adult looks very much like an overgrown chick. It's another case of neoteny—the evolved retention of juvenile features in adults—akin to that of the rails.

To summarize, we infer a history something like this: tropical cormorants flew into the archipelago and settled alongside a uniquely rich upwelling; as years passed, mutation provided random variation in skeletal components, plumage, and musculature (among others); variants associated with bigger bodies and smaller wings were consistently better foragers, which translated into lasting reproductive advantages; generations came and went with those same advantages; a few million years passed, and voilà: *tatty and fatty!*

The cormorant's huge feet are another striking adaptation. Whereas its wings and flight muscles are chicklike (paedomorphic) and greatly underdeveloped, the legs and feet are overdeveloped (peramorphic) relative to those of other cormorants. They are already huge in newly hatched chicks, reminiscent of the large feet of puppies born to big adult dogs. Furthermore, Livezey (1992, 202) found that the shin bone is "disproportionately long," reflecting a sure advantage to long foot-strokes in aquatic locomotion. So we infer that disproportionately large hindlimbs and feet have evolved for their advantage in propelling big cormorants in underwater foraging.[19] In contrast, the penguins use their wings for propulsion and thus "fly" in the water.

Thus, in roughly 2.4 MY, the Galápagos cormorant morphed from a sleek, flying–diving hybrid like other cormorants, to a bird with an enlarged beak, an extra-heavy, dimorphic body, uniquely reduced wings, and enormous feet and legs. It may look comical to us, and its disjointed features may remind us of a child's "mix and match" book (a flippable collage of body parts from different animals). But in relation to its home environment, the Galápagos cormorant is a wonderfully specialized bundle of adaptations. In the richest part of the Cromwell upwelling, the cormorants' features enable successful survival and reproduction, just as Valle (1995) argued.

A Unique Mating Pattern in Galápagos Cormorants

What about the cormorant's unusual mating pattern? Never before seen in the whole order Suliformes, it is a pattern zoologists call "facultative sequential polyandry." Polyandry—the mating of one female with two or more males in the same breeding season—is very rare among birds. Valle (1994) reported that only about 5% of bird species are polygamous (having multiple mates), and even when polygamy evolves, 96% of the cases are polygyny, in which males mate with more than one female per season. Only 13 bird species exhibit polyandry. How did the islands' "creative force" bring about this unusual pattern? And was ENSO a key part of it, as we've seen in other unusual Galápagos adaptations?

As background to these questions, let's briefly consider the impact of ENSO on penguins living in the same area, starting with the warm-water phase. During El Niño, penguins lack their normally rich food supply, and their mortality is high—roughly 75% of adults died in the 1982–1983 event—and the penguins that survive show lasting effects. In 1982–1983, adult female survivors lost, on average, about 50% of their body weight (Boersma 1998). Fortunately, with the next La Niña, it all reverses: penguins fatten up and soon resume reproduction.

---

[19] A second selective advantage of their robust propulsion system is also possible, but unconfirmed: short bursts of speed may help them avoid marine predators, like sharks and predatory marine mammals.

Over a span of decades with a number of ENSO events, one therefore expects to find a negative correlation between penguin population growth rates and SST—high rates of growth are expected to associate with low temperatures. Such a correlation was handily confirmed by Vargas et al. (2006) using intermittent time series data from 1965 to 2005.[20]

Unsurprisingly, El Niño's high SSTs also negatively affect Galápagos cormorants. Tindle et al. (2013) showed that high SSTs are associated with fewer clutches per colony, and with fewer hatchlings per adult. At the population level, these changes would mean interannual decline in the cormorant population, just as in penguins. During La Niña conditions, the cormorants' situation is also similar to the penguins': Tindle et al. (2013) found more clutches per colony and more hatchlings per adult. How does that work?

The answer brings us back to polyandry, a breeding strategy "so unusual that it must surely be an adaptation to special circumstances" (Harris 1979, 143). It starts with inverted sex roles at the onset of mating: female cormorants initiate courtship in the water with elaborate displays, aggressively competing for males. It is truly a sight to see, with smaller birds chasing one another away from a larger one. Courtship continues on land with temporary "display nests" quickly assembled of local debris. Mate choice is in the hands, er stumpy wings, of the males. Once birds are mated, serious nesting begins rather uneventfully, looking very monogamous, and a lot of effort goes into building up a real nest of seaweed and local debris (Figure 8.7).

In due course, both sexes incubate a clutch of one to four eggs, and both continue parental duties when the eggs begin hatching 33 to 37 days later. Because the chicks are largely helpless and nests are out in the open, one parent must always sit on the nest for up to 40 days following hatching, to feed and protect the young from the hot sun, other cormorants looking for good nest sites, and Galápagos hawks. Meanwhile, the other parent forages along the coast to feed the young a graduated scale of prey, from tiny fish like sardines early on, to 15-cm (6-in.) fish at 30 days, to larger prey in time (Valle 1994a). This is a pattern also seen among albatross and booby parents with young hatchlings: one parent protects while the other forages, and larger prey are taken over time (Figure 8.8).

But then comes a surprise: at around 70 days from hatching, if SST is low and food remains abundant, the females do something no other cormorant does: they abandon the nest to the care of the male, and take off, but not to fly. The deserting females swim or wobble over to find a new mate, sometimes just a few meters away, often producing a second clutch within the same 8-month breeding

[20] Vargas et al. (2006, 110) represented SST by the "mean normalized sea-surface temperature (SST) anomalies for the period December–April [often the warmest El Niño period] that preceded each penguin count." But they tallied changes in the penguin population between counts up to 3 years apart—an awfully long interval for measuring the ENSO effects of a given year.

Figure 8.7. Once a male cormorant has accepted a female's overtures, nesting efforts look more serious and monogamous. Here, a male brings a nesting gift of bright red algae up the uneven lava several meters to the nest, where the gift is ceremonially received and carefully integrated.

Figure 8.8. Cormorant hatchling begs for food from its female parent (top), while the male parent preens nearby. To trigger regurgitation (bottom), the hatchling inserts its neck a mighty long way down the female's throat. At just about this stage of hatchling development, female parents sometimes abandon the nest, leaving the hatchling thereafter in the male parent's care. During favorable conditions, the females may mate again and produce a second clutch with a different male.

season. Dutiful by dint of natural selection, the males on the original nests continue to feed their fledglings for up to 5 months—the large fish they catch, relative
to what females catch, being a good match for the youngsters' growing appetites.

Note that females don't always desert the nest: the behavior is facultative, depending on conditions, like the blue-footed booby's siblicide. The key
determinants of desertion are a clutch size of one, which the male can adequately
provision alone, and abundant food, as during the La Niña phase of ENSO.
Females do not abandon the nest if there are two or more chicks or food is scarce;
instead, they stay to help raise the first clutch. However, if the water temperature
is cool, and cormorant females time it right, they can have at least one further
clutch that same year, maybe even two.

In this way, the female enhances her evolutionary fitness via polyandry, producing another clutch or two, so long as the upwelling persists. But there is one
more important condition: there must also exist nestless adult males to become
her second, or third, mates. Valle (2016, 164) reported that there's "an almost
permanent population of potential breeding [male partners] throughout the
year," because of nest failure—in turn related to inbreeding, egg infertility, and
even their oversize feet: large tottery males sometimes squash their own altricial
hatchlings.[21] Tindle et al. (2013, 134) confirmed that the process works for dis-
cerning females: during a strong La Niña cycle at their field site, "of 12 females
that raised juveniles, at least 11 deserted them, 10 were later found nest-building
or courting, and at least six [50%] laid second clutches." The cold, rich upwelling
provides female cormorants a special opportunity for added reproduction, and
evolution has clearly favored it. As in the case of penguins, moreover, the beneficial effect of lower temperatures may stem partly from a lower density of mosquitoes while the birds are nesting, with less nuisance and disease.

An Impressive Convergence

So here's what we've found: adaptation to ENSO in the heart of the Cromwell
upwelling has produced parallel opportunistic mating responses in two very different species. In the cormorants, when the upwelling is strong and rich, females
are able to mate via sequential polyandry as often as every 4 months. The net result is an opportunistic "boost" to their individual reproductive successes with,

---

[21] As Harris (1979, 136) noted, "These cormorants are extremely sedentary, never having been
observed more than 1 km outside the breeding range and only rarely more than 200 m from shore."
Inbreeding and its deleterious effects, including egg inviability, are a likely result. Tindle (1984,
161) suggested that localized inbreeding may have another consequence via kin selection. It is possible that "the male is selected to 'tolerate' his [closely related] mate's desertion." This is an intriguing
hypothesis that warrants careful testing and confirmation.

as a byproduct, a boost to population growth of the species. In the penguins, evolution favored a similar opportunistic boost, producing a clutch as often as every 4 months (three per year) under parallel conditions of cold water and abundant food supply. The penguin boost has reproductive advantages for both females and males and, as a byproduct, a positive boost for the species. The evolved opportunistic boost for both species comes into play at the same temperature point: below an SST of 24°C (75°F), which likely signals a strong upwelling. Yes, there are big differences in how the boost is generated—efficient brood generation by monogamous pairs versus polyandry by deserting females. Still, the convergence on temperature-dependent, opportunistic mating is impressive: a unique, distinguishing feature of Galápagos species in both cases.[22] And in both cases, relative freedom from mosquitoes and their diseases may be part of the benefits of cool temperatures.

Here, then, is an intriguing example of convergent evolution, all the more striking given the narrow geographic range where the species occur and their very different evolutionary histories. One species flew in from the warm tropics and began a new lifestyle in seclusion; the other swam in via the cold Humboldt Current, indirectly from Antarctica, with a long stopover on coastal South America. In Galápagos, the two species evolved in response to the same conditions of the Cromwell upwelling and ENSO, and today we see different but parallel opportunistic mating practices in both species. Triggered by low temperatures, it's surely one cool convergence.

Hypotheses

Building on this convergence, let me now offer three key hypotheses for this chapter. Given their similar dependence on the Cromwell upwelling, and their opportunistic mating responses to cold, prey-rich water, the two species can be expected to show similar demographic responses to ENSO cycles. We would expect to find, on the one hand, big population declines in both species during El Niño, because the warm waters stifle their food supply and the associated rains may also increase annoying and disease-transmitting mosquito populations. On the other hand, we would also expect major population rebounds during lasting La Niñas, because of bountiful food supply, few or no mosquitoes (in dry conditions), and the benefits of opportunistic mating. This is *Hypothesis 1,* the dual expectation of El Niño decreases and La Niña rebounds.

---

[22] Coastline hugging while foraging, noted above, is another impressive convergence, on which both species differ from their closest relatives. It's probably a strategy for predator avoidance, but that hypothesis remains to be tested.

But even with their hypothesized convergence, differences between the two species surely come into play. Because Galápagos cormorant body mass averages 1½ times that of penguins, and because cormorants have a greater diving range (to 70+ m or 230+ ft) and duration (to ~200 sec), they have two advantages when El Niño sets in: they can dive deeper toward cool water and the prey living there, and they have the body mass to help them "ride out" food deprivation for at least a little while. As a result, cormorants are expected to show somewhat slower rates of population decrease with El Niño events. That expectation is *Hypothesis 2a*. By the same token, *Hypothesis 2b* says that, because most penguin pairs raise two or more clutches during prolonged La Niña events, versus only a fraction of double clutches among female cormorants (less than one fifth of the adult females, by available estimates),[23] penguins are expected to show steeper rates of demographic recovery during La Niña. Putting both parts together, *Hypothesis 2* says that, although both species are seriously affected by ENSO events, the Galápagos cormorants are predicted to show somewhat less population flux, up as well as down, in the face of ENSO. In penguins, one expects more rapid change in both directions.

One last hypothesis brings us back to the topic of resilience. Because both species exhibit behavioral methods of mitigating warm-water and warm-air perturbations—for example, seeking shade (penguins) or gular fluttering (cormorants) as behavioral thermoregulation, and diving to extra depth as warm water moves in—it is reasonable to expect them to weather relatively mild El Niños with little or no change in population numbers. *Hypothesis 3a* proposes that both cormorants and penguins will show a measure of reproductive resilience, and thus little or no population decrease, in the face of mild warm ENSO phases. We expect it to take medium or stronger El Niños to produce notable population decline.

Similarly, we might also expect a plateau in both species' reproductive response to the cooling of ocean temperatures, as happens with La Niña conditions, effectively extending their ENSO-neutral response into weak La Niña conditions. Living in an upwelling zone, it might well be reproductively costly to respond to every influx of cool water as if it were the beginning of a lasting La Niña event, because some would surely be short-lived, fizzling out before prey fish reach renewed high densities. Let's call this *Hypothesis 3b* and test for it in manner parallel to *3a*: is there a range of mildly cool SSTs showing little or no change in

---

[23] During La Niñas, double clutching by female cormorants is far from universal, complicated by myriad individual contingencies. One study found that "11 of 61 females [18%] managed to rear two broods by deserting the first brood and nesting again that season with another male" (Tindle et al. 2013, 130). Like the shrinkage of marine iguanas, cormorant polyandry gives the impression of being an evolutionary "work in progress."

population sizes for cormorants and penguins? Or do they jump right in to reproduce whenever that 24°C threshold is crossed?

What would it take to provide strong evidence for or against each of these hypotheses? Ideally, one would have reliable data for both seabird populations and for local ENSO conditions for a long enough time interval to cover multiple declines and rises in order to indicate their relationship with ENSO. Remarkably, such data do exist in Galápagos, stemming from decades of local scientific concern for penguins, cormorants, and other seabirds, and I felt very fortunate in being able to assemble seabird census data extending back to the 1970s. I acknowledge with gratitude the painstaking efforts of personnel at the Galápagos National Park and Darwin Station to conduct annual censuses of both species each year—and at roughly the same time (late August or early September; see Mills and Vargas 1997), so I did not feel compelled to adjust or correct for other timing. With regard to measuring ENSO, there were a number of options, but the Oceanic Niño Index (ONI) seemed the best measure for purposes here because of its time depth (back to 1950), and because it has been one of the preferred measures used by NOAA (the U.S. National Oceanic and Atmospheric Administration). The ONI is built on a 3-month running mean of departures from average conditions (called "anomalies"), and further requires that the anomalies exceed +0.5°C or -0.5°C for at least 5 consecutive months to be labeled as El Niño or La Niña, respectively. The ONI has been devised for reliability.[24]

I made only one systematic change in ONI for this analysis: I turned the scale on its head, to show El Niño events as spikes downward—let's call them "stalactites"—and La Niña events as spikes upward. This depiction, reversing the normal top to bottom convention for the ONI axis, has a clear advantage in our context. It means we can expect swings in both ENSO spikes and seabird counts to move in parallel with each other—thus displayed, a confirmation of the hypothesis will look like a positive correlation of population and ENSO variables (but note that, in fact, the numbers will be negatively correlated—as ONI tracks SST increases, cormorant and penguin populations are expected to decrease). Using methods described in Online Appendix E2 and drawing on the talented assistance of statistician Dr. Lynn Gale, I offer preliminary tests of the three hypotheses as follows.

---

[24] Many ENSO measures, including ONI, are based on data obtained many kilometers from Galápagos, potentially adding time lags to the analysis because ENSO changes take time to spread across the Pacific. But ONI is the most complete data set available. Data from closer sites (see http://oceanmotion.org/html/resources/ssedv.htm) start too recently to allow full use of the cormorant and penguin data set assembled here.

Findings

Figure 8.9 shows how Galápagos cormorant and penguin populations responded to fluctuating ENSO conditions between 1970 and 2010, including major El Niño and La Niña events as measured by ONI. The plot shows the annual census counts of the birds, plotted as points (dots for cormorants, triangles for penguins), plus smoothed curves summarizing their trends over the years. Consider, first, the impact of El Niño's warm waters on both populations. The figure includes three of the four strongest El Niño years of recent history, 1972–1973, 1982–1983, and 1997–1998 (the 2015–2016 El Niño event is outside the time frame of my analysis), each marked by a dotted vertical line.

As expected, cormorant numbers plunged in association with the three strongest El Niños. The decrease is especially notable in 1982–1983, when the curve sank to roughly 625 individuals (and to 500 in the actual census count). The same is true for penguins: their curve also fell sharply in response to the two most recent "very strong" events (I was unable to locate reliable penguin data around the time of the 1972–1973 event). The next El Niño (1997–1998) was also associated with declines in the estimated populations of cormorants and penguins, to around 850 and 1,050, respectively. Extinction is a genuine threat from declines like these. There is also a loss of genetic diversity at each bottleneck (reminiscent of rails), exacerbating inbreeding issues and reducing genetic resources for disease resistance within each population.

Now consider the response of both populations to La Niña events, shown as spikes at the top of the figure. The cormorant curve rebounds after La Niña in 1974–1975, 1985, and especially 1998–2001. The penguins also rebound after the 1985 and 1998–2001 La Niñas. The populations respond similarly to warm and cold parts of ENSO. Hypothesis 1 is thus clearly supported by the data: ENSO affects cormorants and penguins strongly, and in parallel ways.

Figure 8.9 hints at another finding of potential importance: both cormorant and penguin curves on the left half of the plot (up to 1989) are generally well below the same curves on the right side (1993–2010). The magnitude of the difference can be appreciated from the lowest values reached for both seabirds during two major El Niño events: cormorant and penguin curves sank as low as 625 and 800 individuals, respectively, after the 1982–1983 El Niño, and to 850 and 1,050 after the 1997–1998 event. The latter event, it should be noted, had an even higher SST maximum, a key measure of El Niño severity. What might explain the relative deficit of over 200 individuals for each species in the earlier decade?

Many variables may be involved, including climate change (although unlikely, because cooler conditions of the earlier years would have been beneficial to both seabirds), differences in census techniques or data processing (also unlikely, as

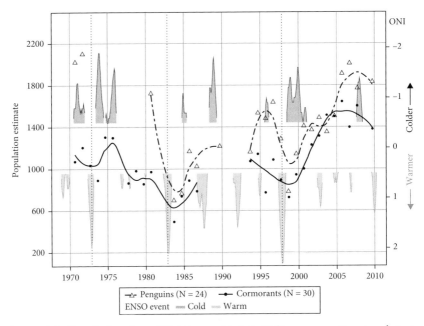

Figure 8.9. Population sizes of flightless seabirds in Galápagos over time in relation to ENSO conditions. Shown are the total numbers of cormorants (*smoothed, solid curve*) and penguins (*smoothed, broken curve*) estimated from census counts (*dots and triangles*, respectively) during the years 1970 to 2010. The figure also shows the warm waters of El Niño (*orange "stalactites"*) and cool waters of La Niña (*purple spikes*), using an inverted Oceanic Niño Index (ONI) axis at right (leaving blank the ENSO-neutral years to make the plot less confusing). These data support the hypothesis that both species decrease in numbers in response to El Niño events and rebound during La Niñas. As expected for species whose survival and reproduction are acutely influenced by ENSO, where there are pairs of census points for both species ($N = 23$), their rank order population sizes show strong correlation: Spearman's rho is 0.69, $p = 0.0004$. El Niños are rated "strong" above ONI = 1.5; similarly, La Niñas are rated "strong" below ONI = -1.5. Sample sizes vary because more cormorant censuses were taken during the study period. *Sources and methods*: See Online Appendix E.

techniques have been fairly constant), more disease exposure in the early period (again unlikely, for there have been increases in disease exposure in recent years), the impact of introduced animals (dogs, cats, rats, goats, etc., which surely play some role), and the impact of human fishing activity in the region (also surely involved—a lobster fishery boomed between 1970 and 1990, except for 2 years, 1978–1979, when the fishery crashed—see Chapter 9). Both of the last two variables surely played some role in the differences observed here.

More information is needed, but I suspect from the intensity of fishers' pursuit at the time that lobster fishing in the Cromwell upwelling area accounts for the lion's share of the difference between left and right sides of the figure. In the 1970s, Harris (1974a) reported that the Cromwell was "rich in [lobster] (*Palinurus* spp.) and there is a thriving commercial fishery. Up to the present these crustaceans have been caught by skindivers but a gradual fall-off of the catch, due to overexploitation, may lead to a change in fishing methods." Indeed, some species of lobsters were already being caught by underwater trammel nets and, "In the Galápagos these nets often [inadvertently] catch penguins, flightless cormorants, sea lions, and other endemic organisms," which drown (Holthuis and Loesch 1967, 221). In 1976, just 2 years after the Harris quote, the lobster catch of Galápagos, mostly from the Cromwell region, plunged from 600 tons to 60 tons as a result of overexploitation (Schiller et al. 2015, 439).

Possibly even more impactful than direct catch, the removal of significant numbers of lobster by fishers probably had an important "cascading effect" by reducing the usual lobster predation on sea urchins (Sonnenholzner et al. 2009; Ruttenberg 2001). The ensuing rapid growth of the urchin populations would decrease everything they eat, especially algae, to 20 m (66 ft) or more, potentially converting an entire productive rocky reef habitat to an "urchin barrens," with large negative effects up the food web to the food supply of penguins, cormorants, and many other organisms (Edgar et al. 2010). Among cormorants, the effect on food supply would be most pronounced for adult females, who routinely feed at the same shallow depths. Notably, in the cormorant sample on Fernandina studied from 1970 to 1979 by Tindle at al. (2013), the number of adult females fell from approximately 70 birds, a number that had been consistently higher than their male counterparts, to about 40, below male numbers, beginning in the same year that the lobster harvest plunged (1976).[25] Although it is difficult to generate an accurate estimate, the data in Figure 8.9 suggest that fishing in the region before 1990 could have reduced, directly and indirectly, populations of cormorants and penguins by as much as 150 individuals each.

We turn now to Hypothesis 2, which proposes that penguins show a sharper or more pronounced response to ENSO events than cormorants. That hypothesis, as well as Hypothesis 3 regarding resilience, calls for a different set of measures. For these hypotheses, I needed both to measure and to compare the rates of population change between censuses for the two species. Per capita rates of population change offer a clear advantage, because the population sizes of the two birds are clearly different. The per capita rate describes the rate of change

---

[25] These data are from Tindle et al. (2013, Figure 2B), using "derived numbers" of cormorants—that is, estimates for their study colony on Fernandina generated from a marked-recapture analysis of cormorants that the researchers banded along a wider section of the Bolívar Channel.

per year per bird, which allows us to compare their rates for testing Hypothesis 2 even with different total numbers. Because there's a less intuitive meaning to per capita rates of increase, and because there are more gaps in penguin data for the period from 1970 to 2010, I have put the full graphical analysis into Online Appendix E1, and here I simply summarize the key findings.

What I found was this: where there are comparable data following El Niño years, like 1997–1998 (very strong El Niño), 2002–2003 (weak to moderate El Niño) and 2006–2007 (weak El Niño), the cormorant per capita rates of change are less negative or even slightly positive compared to penguin declines (see Figure E1 in Online Appendix E). By the same token, when a La Niña event comes along, as in 1985 and 1998–2001, the penguin per capita rates of change reach slightly higher positive levels than do the cormorant rates. The differences are small (e.g., in the 1997–1998 event, cormorants fell to a growth rate of -0.1 cormorants per cormorant, while penguins reached -0.2 penguins per penguin), but they are in the predicted directions for both types of ENSO event. In short, the data are consistent with Hypothesis 2, that cormorant rates of change will be buffered on the way down during El Niños (Hypothesis 2a) and on the way up during La Niñas (Hypothesis 2b), but they do not provide what could be called strong support. On the other hand, given all the variables at play, it's impressive that there's a pattern at all.

As another test of Hypothesis 2, and as a test for Hypothesis 3—that both cormorants and penguins have evolved forms of resilience in the face of ENSO— I carried out a second kind of comparison. As in the previous test, I used per capita rates of population change to control for the effects of population size, but this time I used a *composite* ONI value for every census year. This composite value (cONI) conveniently summarizes in one number the ENSO conditions experienced by organisms *during that full census year* (i.e., the 12-month period prior to the tally; see Online Appendix E2). It is scaled so that cONI from 1.0 to 1.5 represents a year of "moderate" El Niño conditions, with 1.5 to 2.0 for a year of "strong" El Niño, and similarly negative values of cONI correspond to years with moderate and strong La Niña conditions. Again, using smoothed curves, I explored the responses of cormorant and penguin per capita growth rates to ENSO conditions as measured by cONI (details in Online Appendix E2).

Results of this third test are shown in Figure 8.10. Consider, first, the penguin response represented by the broken curve.[26] As expected, that curve has an overall negative ("downhill") slope, moving from positive rates of per capita change at negative cONI values—La Niña conditions—to negative

[26] Note that the data points shown in Figure 8.9 are those belonging to cormorants, not penguins. The broken curve is the proper smoothed curve for the penguin census data, which are shown in Figure E2 of Online Appendix E.

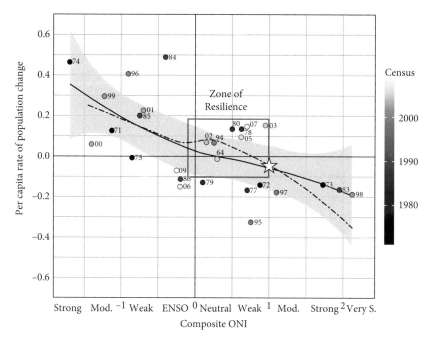

Figure 8.10. Per capita rates of population change in Galápagos flightless seabirds in relation to ENSO conditions for the interval 1970 to 2010 (La Niña, *left*, below 0.0, to El Niño, *right*). With the advantage of controlling for different population sizes, these smoothed curves summarize trends for cormorants (*solid curve*) and penguins (*broken curve*) as a function of the ENSO indicator, "Composite ONI" (cONI), for the year before each census date (cormorant data are shown by small color-coded circles with two digits for the year). As expected, both species show negative rates of per capita increase for cONI greater than 1.0—indicating moderate or stronger El Niño conditions—and both show positive rates of per capita increase when the ocean is cooler, below cONI = 0. Both curves show broad zones of resilience to El Niño (*blue box*), and only dip to seriously negative values above cONI = 1.0 (*yellow star*), confirming Hypothesis 3a. The overall similarity of the two curves (support for Hypothesis 1) is highlighted by their similar slopes and magnitudes. The penguin curve lies within the 0.95 confidence band of the cormorant curve (*shaded zone*). Consistent with Hypothesis 2, the cormorant curve is a little flatter over most of the cONI range, showing a slightly more buffered response to ENSO swings. *Source*: See Online Appendix E.

rates of change at positive cONI values—El Niño. The overall curve nicely summarizes the penguin population's responses to ENSO—it increases when waters are cool and decreases when warm. But the curve is not symmetrical on the left and right sides of cONI = 0.0 (neutral ENSO conditions). First, notice that the tail of the curve on the positive side of the cONI scale, representing population decline with El Niños, is steeper than the tail on the negative side of cONI, reflecting population growth with La Niñas. In other words, penguin populations decrease rapidly when the composite El Niño experience is in the range of moderate to strong. The overall negative slope to the curve echoes the rapid population response in Figure 8.9: the penguin population quickly shrinks under El Niño conditions.

Second, notice also that the broken curve for penguins has a relatively flat zone (a "bump") in the middle of Figure 8.10, between cONI = 0 and cONI = 1.0. This zone—indicating relatively constant per capita growth rates over the interval out to moderate El Niño conditions—provides support for the "zone of resilience" predicted by Hypothesis 3a. It confirms the efficacy of penguin mechanisms, many of them behavioral, for coping with mildly warm ocean temperatures.[27] The implication is that penguins "get by" quite well up to a cONI value of about 1.0 (moderate El Niño conditions). Their population continues growing slightly (more births than deaths over the year) even when the ocean is moderately warm. Such a response most likely reflects their various thermoregulatory behaviors and perhaps a tendency to dive deeper than usual for prey. Yes, penguins are re-silient in the face of mild El Niños.

Now look at the smoothed solid curve in Figure 8.10, representing per capita growth for Galápagos cormorants (amid corresponding cormorant census tallies marked with their years) by cONI. Because there are more cormorant census tallies than penguin tallies in the total data set, the cormorant curve extends slightly beyond that of penguins. The solid cormorant curve is similar in overall shape and slope to the broken penguin curve, again confirming Hypothesis 1: the two species do show highly similar responses to ENSO. However, the cor-morant curve is shallower above cONI = 1.0, in the moderate to very strong El Niño range, than that of the penguins. The implication is that cormorants are, yes, slightly but consistently more buffered in their response to the warmer waves of ENSO. Again, the data are in the direction predicted by Hypothesis 2a, al-though not conclusively: the penguin curve lies fully within the 0.95 confidence band of the cormorant curve (the shaded zone).

Evidence for Hypothesis 2b is more mixed, owing to a cross-over of the curves on the cool-water side of the plot, at cONI values less than –1.0. There, with cool SSTs, the two populations show generally the same per capita rates of change, with just the slightest hint of a higher rate among cormorants. Because we lack penguin census data for the La Niña years of the 1970s, it is

---

[27] The curve of declining per capita growth in penguins deflects at the blue box in the figure. To the left of the box, it looks as though the curve is headed for the point 0.0—that is, headed for nearly zero per capita growth rate at ENSO-neutral conditions, and from there to negative per capita growth rates for positive values of cONI. Instead, with the deflection, the per capita growth curve for penguins remains positive all the way out to cONI = 0.8.

difficult to interpret this cross-over, which could be an artifact of holes in the penguin data for those years. Or could it point to the special population boost for cormorants from sequential facultative polyandry? We'll need more data to know for sure.

Figure 8.10 supports Hypothesis 3a, that both species exhibit a measure of resilience to mild El Niño conditions. The data for both species remain above seriously negative growth rates for an impressive range of warm SSTs, out to cONI = 1.0 (moderate El Niños, shown by the star). It is impressive that, with all the complexities of time-series data for these populations, there emerges a flattening of the curves for both species. Both species fare reasonably well as the sea warms from ENSO-neutral conditions into mild El Niños. With the passing of years and accumulation of additional data, it will be important to reconfirm this finding and to see if the resilience zone shifts at all upward or to the right. But for now, even this preliminary test is encouraging: cormorants and penguins do appear to have ways of coping with mild El Niño conditions, enough to prevent their numbers from decreasing drastically at those times.

The test of the symmetrical Hypothesis 3b of resilience to cool SSTs is also interesting. In response to La Niña conditions, one might predict that evolution would favor a "sit and wait" strategy for mildly cool SSTs (cONI values from 0.0 out to –1.0 or so) as noted above, and thus not turning quickly to mating with every cool SST. If so, one would expect to see that per capita rates of population increase would remain fairly constant for a range to the left of cONI = 0.0, symmetrical with the evidence for El Niño resilience to the right. This suggestion is *not* supported by the data—per capita growth rates grow steadily as the sea cools, starting just below cONI = 0.0. In other words, penguins and cormorants do show resilience to the harmful El Niño phase of ENSO, but they respond immediately to the positive reproductive potential of the cold La Niña phase. For birds of cold-water ancestry—the penguins—this may not be surprising. But for the cormorants, coming from warm-water ancestry, to have evolved a sensitive cold-water response suggests that the emergence of cold water is itself reliable and stable, and so, too, the food supply. The plot suggests that there are fitness advantages to mating just as soon as SST drops below 24°C (75°F).

In conclusion, this analysis adds a key adaptation to the toolkits of cormorants and penguins for life in Galápagos. It adds to each a convergent, opportunistic mating pattern triggered by a sensitive ocean temperature "switch." Valle (2016, 162) definitely had it right in the opening quote: this is "uniqueness brought about a broad and complex array" of adaptations. The net result here is impressive reproductive resilience, allowing the birds to rebound from intermittent El Niño setbacks with population increases just as soon as the water again turns cool and productive. Both species also show impressive tolerance for mild amounts of sea-surface warming. They are really quite remarkable in this regard, avoiding serious per capita population decline during mild El Niño conditions all the way to cONI values near 1.0. No wonder both species have survived a million or more years of flightless life in Galápagos. They are certainly not "earth-bound misfits!"[28]

Conservation Implications

Let's conclude with what these findings imply for cormorant and penguin conservation. First, consider the issue of fishing in the Cromwell upwelling, a zone both extra-productive for fishers *and* extra-hazardous for the small, localized populations of flightless Galápagos endemics. Figure 8.9 hints at the direct and indirect impact of fishing on these focal populations. It hardly matters which has the greater magnitude—direct bycatch in fisher's nets or indirect impact via changing food supply. Dating back to at least the 1970s, human activity has impacted the survival and reproduction of flightless seabirds in both these ways. We've seen such an impact before, in albatrosses, tortoises, and boobies, among others, but the convergent reproductive flexibility of the two species here is different. Their evolved resilience to ENSO in the archipelago helps them rebound from population losses, including anthropogenic losses, whenever SST is low. The dips and squiggles of population plots like Figure 8.9 may be disconcerting, but they also provide hope: these are resilient species.

According to Wagner and Boersma (2011), even volant pursuit-diving birds (that is, flying birds that actively pursue their prey under water) are heavily affected by fishing activity in their ranges. All the more reason to be concerned for the birds in Galápagos with no option to fly. Fortunately, by the early 2000s, fishing in the Cromwell upwelling region—both legal and illegal—waned as stocks were depleted, consistent with the rebounds seen in Figure 8.9. But there remains a great need for surveillance (noted also by Crawford et al. 2017), since fishing activity persists there. A top priority for conservation is surely an effective monitored zonation system that designates low-impact artisanal fishing areas, with technologies that do not harm seabirds, and "no take zones" in and around the center of the Cromwell upwelling. As Chapter 9 discusses, such a zoning system was agreed upon in the year 2000, was implemented in 2005, was revised in 2014, and is increasingly monitored today.[29]

Second, there's the matter of introduced species. Given the restrictive ranges and sedentary lifestyles of flightless seabirds, the terrestrial predators that

[28] Expression borrowed with appreciation from the Pink Floyd (1986) song, "Learning to Fly."
[29] For more on zonation in the Bolívar Channel ecosystem, see Moity (2018). A ranger outpost and a speed boat have improved monitoring in recent years.

humans have brought to the islands—especially dogs, cats, and rats—are a grave threat. Take cormorants first. Having evolved to be appropriately tatty and fatty for the world of pre-settlement Galápagos, and to nest out in the open, they are now maximally vulnerable to land-based predators, with only their longer-than-average cormorant beaks for defense. Penguins may be slightly better off, given their propensity to seek shelter in lava caves and tunnels, which are surely easier to defend. But their beaks are a much smaller weapon against hungry mammalian predators. Clearing all terrestrial predators from islands on the frontline of the Cromwell upwelling, and keeping it that way, is an important conservation priority. Project Isabela was a crucial step in that direction, eliminating the plague of dogs from western Isabela, but, alas, not cats and rats. Fernandina is of extra importance to penguin and cormorant conservation because of its added degree of isolation and refuge. At this writing, happily, it still has no introduced mammalian predators.

Third, both flightless birds are readily susceptible to introduced diseases. Yes, they live in relatively remote areas, which would seem to offer a degree of protection, and yes, they live in dispersed small colonies, with fairly low densities even within their restricted ranges, which may impede the spread of various contagions. But neither of these defenses is sure protection against introduced, vector-borne infirmities. In addition, both species have unusually low genetic diversity (due to founder's effect at colonization, plus repeated El Niño diebacks), which hampers disease response.[30] As described in Chapter 2, the black salt marsh mosquito pesters waved albatrosses when they mate, but they suffer mosquitoes only a few months of the year, albeit key months for reproduction. In contrast, cormorants and penguins are a year-round captive audience.

Worse still, two additional mosquito species have been introduced to Galápagos, *Culex quinquefasciatus* and *Aedes aegypti*. Both are a menace: the former are vectors for avian pox viruses, a filarial parasite (a skinny wormlike nematode), and various malaria parasites of the genus *Plasmodium;* the latter are vectors for serious human illnesses, including dengue and Zika. Evidence for the filarial parasite traces to 1985, and it has since been estimated to have a prevalence as high as 59.5% in cormorants and 21.7% in penguins (Santiago-Alarcon and Merkel 2018). Evidence of *Plasmodium* infections in Galápagos penguins was found in 2009 (Palmer et al. 2013, 770). Subsequent research, reported in the same paper, found that 97% of a large sample of penguins (181 birds from eight sites) have antibodies to the parasite, indicating prior infection.

---

[30] When diversity was compared between Galápagos and Humboldt penguins, for example, within the genes for a group of immunologically active proteins (of the MHC complex), the Galápagos penguin had only 15% as many variable sites (Bollmer et al. 2007 and references therein). Genetic diversity is similarly low in the Galápagos cormorant compared to its closest relatives (Burga et al. 2017).

Fortunately, penguins show few symptoms of malaria and the parasite does not complete its life cycle in penguin hosts (Parker 2016, 181). But even if it simply weakens them, causing low additional mortality (up to 5%) after El Niño events, projective models show that *Plasmodium* has "the potential to drastically reduce the probability of persistence of the Galápagos penguin population over the next 100 years" (Meile et al. 2013, 440). Three additional groups of invertebrates are also a potential menace: lice, biting midges, and louse flies. Vector control, especially of the *Culex* population, is thus an important conservation priority, but how can it be accomplished? Although much remains to be learned about these diseases, no doubt they already contribute to the ENSO-related demographic fluctuations in Figures 8.9 and 8.10.

Last, we must also consider exacerbations of El Niño. Even with a certain capacity for coping with failed upwellings and food scarcity, penguins and cormorants remain highly vulnerable to prolongations and higher frequencies of El Niño events, both of which have been linked to anthropogenic climate change (Chapter 2). One study has already concluded that, if El Niño events do not change but continue at roughly 5% probability per year of a strong El Niño and 20% probability of a weak El Niño, the Galápagos penguin population "has a 30% probability of extinction within 100 years" (Vargas et al. 2007, 143). According to the model, the extinction probability jumps to 100% when frequencies of strong El Niño events are 12% or more (regardless of weak El Niño frequency).[31] What a tragic fate that would be for these stunning, world-record examples of island evolution. But therein lies further motivation for finding ways to turn the ever-more-integrated social-ecological system from a potential trap to a positive protector. Given the ENSO vulnerability of these two species, the question is, how?

The findings of this chapter suggest several strategies. First, consider the "housing shortage" for penguins. On land, Galápagos penguins need shaded, cool shelters for protection from tropical sun and heat. In such refugia they make nests and brood their young out of sight of aerial predators. But suitable natural sites near the Cromwell upwelling are limited in number; which is why the population is so spread out as noted earlier. During ENSO-neutral times, the penguins manage quite well, as the graphs in this chapter attest. But during El Niño events, adequately cool, protected dwellings can make the difference between life and death.

Recognizing this key limiting resource, Boersma and her group have undertaken a campaign to build penguin "condos"—lava lean-to's, caves, and other

---

[31] Similar models proposed independently by Putnam (2014, 51) indicate that, if current El Niño trends continue, "Without some form or a mix of types of intervention, the penguin's chances of extinction are 20% over the next 100 years. . . . Climate change alone could increase that probability to 60%."

protective lava structures—within the penguins' normal range. Since 2010, they have built 120 nest sites on or near Isabela, Fernandina, and Bartolomé. Over that interval, an estimated 24% of the total active nests (nests containing eggs or chicks) have been in the experimental condos. Near the center of the upwelling, condos account for about 43% of penguin breeding activity.[32] The effort has proven a major success: it is only a matter of time before graphs like Figure 8.9 will show a significant upswing as a result of the condo-construction efforts. Alas, condos are not appropriate for cormorants, who seek open, flat coastal areas for their nests—so exposed that they sometimes delay feeding their hatchlings when kleptoparasitic frigatebirds are circling nearby. Probably the best protection we can offer nesting cormorants, and surely a boost to penguins as well, is freedom from the predators that humans have introduced. This is an important priority.

Consider now the question of food supply during El Niño events. Maybe the time has come to consider providing penguins and cormorants with supplemental fish during El Niño periods, at least for adults (who can then feed their chicks or not). Such interventions have normally been ruled out in Galápagos; but with growing evidence of human impact on the ENSO cycle, perhaps it may be time to reconsider that option. Using curves like those presented here, and data on fish consumption during good times, one could estimate the size of emergency food supplements required to bring both species through SST changes of a given magnitude. One could well tailor such interventions to the added degrees of SST change and reproductive impairment that human activity contributes to the natural ENSO cycle. Alternatively, if population rebounds should lag, humans could step in to promote population growth via captive breeding programs, run by GNPD and CDRS, much as they do for tortoises (see Chapter 3). For penguins, there is precedent elsewhere: following a serious oil spill in 2000, chicks of the phylogenetically related African penguin (*Spheniscus demersus*) were hand-raised. Investigators found that, on average, survival and breeding success of the experimental chicks were no different from those of naturally reared chicks (Barham et al. 2008).

But another intervention might also be considered, one that could be assessed for feasibility during a La Niña period, when both species are normally thriving. Since neither cormorants nor penguins navigate far from shore on their own, one could implement an experimental relocation of a sample of one or the other species, or both, to temporary refuge on an island less affected by El Niño. Ideally, this would be an uninhabited, predator-free island of the Eastern Pacific that experiences similar ENSO-neutral SSTs but smaller SST changes during El Niño events than do most Galápagos islands, and that has a rocky reef with dense

---

[32] See https://www.Galápagos.org/conservation/our-work/ecosystem-restoration/increasing-the-Galápagos-penguin-population/ on the heroic penguin condo project of Boersma and colleagues.

and stable prey populations close to shore. Alas, only a couple of islands fit the bill—both relics of the Galápagos hot spot,[33] and neither fully escapes the warm waters of El Niño events. Instead, perhaps one or more artificial, floating islands could be tried, complete with coastal rocky reefs, engineered to the scale of needs for cormorants or penguins, or both. Conceivably, the islands could be moored somewhere mid-range for both species, and thus could be naturally colonized during non-El-Niño years by a diverse Bolívar Channel rocky reef community. As a serious El Niño approached, the islands could be towed outside the zone of warmest El Niño waters and moored perhaps near a seamount, or perhaps the Costa Rican Thermal Dome (a cold-water upwelling in the open ocean off Costa Rica) to provide rich food webs akin to the Cromwell upwelling. Experimental trials would obviously be critical before implementation.[34]

But if an experimental trial proved successful, then a less intrusive, short-term El Niño "work around" might be possible. When prolonged ocean warming is forecast for the archipelago, a sample of Galápagos cormorants or penguins, or both, could shelter on one or more of the substitute islands and remain there for the duration of warm waters. When neutral or La Niña conditions return, the decision could then be made to bring back the birds, and the islands, to the Cromwell region, or not. It *is* a stretch, but relocation efforts have been successful for other seabird species (e.g., short-tailed albatrosses; see Chapter 2). And better to stretch *our* efforts than the limits of *their* survival. Such measures may be needed to "buy time" for these amazing, highly endangered species. But let us never lose sight of the most important, overarching goal: doing our best to reign in the anthropogenic contributions to climate change and thus ENSO. Like penguins and cormorants, we have a problem we cannot fly away from—will we show appropriate and timely resilience?

The main arguments of this chapter are summarized in Figure 8.11, showing key adaptations of penguins and cormorants to the initial challenges of pre-settlement Galápagos. The middle of the figure displays the impressive

---

[33] One possibility is Colombia's Malpelo Island. According to Naranjo et al. (2006), no other cormorant or penguin species are currently using the island, though it has suitable rocky reefs around its shores. Alternatively, Cocos Island of Costa Rica could be considered. In either place, a protected landing area near shore would need to be found or constructed, with open areas for cormorant nests and secluded lava nooks for penguins. The birds would need careful monitoring, regarding feeding success, marine predator attacks, disease, and health status, etc. Though neither island is fully outside the zone of El Niño influence, nearby eddies of current partially alleviate the local scarcity of fish during El Niño times (Corredor-Acosta et al. 2011). If this strategy is successful, the stewardship advantage is the continued reproductive autonomy of the birds.

[34] Artificial islands are another project, akin to the Payment for Conservation Services in Chapter 4, that could be funded by a sustainability assessment paid by foreign visitors at the time of entry, in addition to, and separate from, the Galápagos National Park entrance fee. Perhaps one or more retired cruise ships could serve as bases, with added lava-like topography provided by a shell of plastics—all the better if recycled from the great Pacific gyre.

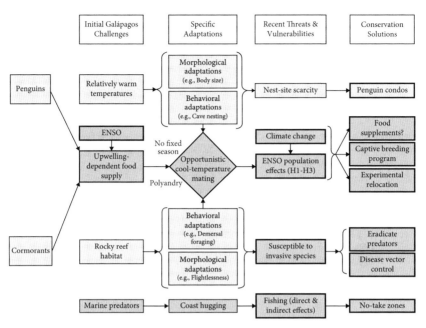

Figure 8.11. Schematic summary of key arguments of Chapter 8. The separate evolutionary adaptations of penguins and cormorants to challenges of pre-settlement Galápagos are *lightly shaded*, and their convergent adaptations to the Cromwell upwelling and marine predators are *darkly shaded*. The population effects of ENSO, explored in Hypotheses 1 to 3 (indicated here as "H1-H3"), suggest several conservation efforts, as do the effects of invasive species, fishing, and nest-site scarcity for penguins. Boxes caused or affected by human activity are *bolded*.

evolutionary convergence of the two species—a convergence on opportunistic mating strategies as an adaptation to the ENSO cycle. In penguins, the opportunism is achieved by the absence of a regular breeding season. In cormorants, opportunistic mating follows from nest abandonment by females under favorable conditions and their polyandrous mating in the same season with at least one other male. In both species, opportunistic mating is triggered by the cool SSTs that mark the onset of La Niña conditions and abundant fish densities. As discussed, the three main hypotheses of the chapter were each subjected to preliminary tests against census counts and ENSO measures, leading to the conservation implications shown in the figure.

Preceding chapters focus on the beautifully unique terrestrial organisms of Galápagos, their evolutionary response to local environments, and their resulting conservation challenges today. The next chapter considers a humble marine organism common to a wide range of coastal tropical America. It's an enduring case of direct exploitation, where humans are still today the resource users, a case where humans' actions directly impact the survival and reproduction of the organism in Galápagos, and where they indirectly affect a full food web of other organisms, including cormorants and penguins. As becomes clear, it offers a good look at the present and future challenges for human stewardship in this most amazing "little world."

9

Fishing in a Common Pool

*Historically, the Galápagos Islands are famous as a living laboratory
for the study of evolution and ecological processes. [We argue] for the
added recognition of the Galápagos Marine Reserve (GMR) as a living
laboratory for the study of cutting-edge social processes.*
                              Heylings and Bravo ("Evaluating Governance")

The period from the 1980s to 2010s was one of rapid change in Galápagos, well
beyond what most residents or observers ever imagined. The decades brought
exponential tourism growth, increased immigration, unprecedented urban expansion, and new waves of introduced plants and animals. A crisis in two fisheries, explored in this chapter, prompted unprecedented social unrest in the
1990s and early 2000s, leading to repeated protest occupations of the National
Park Headquarters and the Darwin Station, to giant tortoises killed and scientists
held hostage, and to the shooting of a park guard. It was the worst of times.
    Circumstances got so bad that, on June 26, 2007, Galápagos was officially
posted on UNESCO's "List of World Heritage in Danger," a status given to World
Heritage sites imperiled by any combination of natural and human causes.
UNESCO authorities explained that

> The principal factor leading to the inscription of the property [as] World
> Heritage in Danger arises from the breakdown of its ecological isolation due to
> the increasing movement of people and goods between the islands and the con-
> tinent, facilitating the introduction of alien species which threaten species na-
> tive to the Galápagos. This is fueled by poor governance, leading to inadequate
> regional planning and unsustainable tourism development. Illegal and exces-
> sive fishing pressures in the Galápagos Marine Reserve (GMR) were another
> factor contributing to the Danger listing. (Strahm and Patry 2010, 6)

The breakdown of isolation, one of the key ingredients in the secret recipe of
Galápagos, is explored in Chapter 4. This chapter explores the other problems
mentioned by UNESCO: unsustainable tourism, illegal and unsustainable
fishing, and poor governance. The discussion looks at how the issues developed,
at various efforts to resolve them, and at where they stand today. The chapter

Exuberant Life. William H. Durham, Oxford University Press (2021). © Oxford University Press.
DOI: 10.1093/oso/9780197531518.003.0009

examines the lessons learned from the turmoil, and how the lessons have helped shape the ongoing governance, planning, and tourism in the archipelago.

Two things are most impressive about this final part of the story. First, the fishing problems that served as key triggering events were focused in the same Cromwell upwelling area that is home to the Galápagos cormorants and penguins. Fishing activity had a major impact on the Bolívar Channel ecosystem, greatly reducing the populations of important species in the food web (Toral-Granda and Martinez 2004; Ruiz and Wolff 2011; Eddy et al. 2019). Making matters worse, a major El Niño event (1997–1998) came along right in the middle of the fishing crisis, acting like a multiplier on fishing's impact (Wolff et al. 2012). As shown in Chapter 8, the negative impact of the two together on cormorants and penguins brought them close to the population lows of the 1980s, and to renewed risk of extinction.

Second, it is particularly striking that, when major problems surfaced in the archipelago, so did local efforts at resolution. In this time of global climate change and major global biodiversity losses, we may often think of environmental problems as huge and distant, beyond the reach and capabilities of local people and local institutions. This chapter is a good antidote to that view. It shows, instead, that serious, far-reaching environmental problems have surfaced in Galápagos, generating deep social divisions and disputes, sometimes getting seriously out of hand. However, it also shows that dedicated local efforts at communication, negotiation, and implementation work wonders toward finding solutions. We may generally think of Galápagos as a kind of showcase for learning about ecology, evolution, and other natural phenomena but, as the opening quote points out, the islands have also become a fascinating and appropriate context for understanding important social processes.[1] It may not be an exaggeration to say that Galápagos brings all kinds of processes into sharp relief, including some that may prove influential for the future of our planet.

Let us begin with a brief look at the humble organism central to this story.

The Brown Sea Cucumber

One of the most innocuous of all key marine organisms of Galápagos is the brown sea cucumber, *Isostichopus fuscus* (Figure 9.1). With a live weight of just over 0.2 kg (half a pound) and a full-grown length of about 20 cm (8 in.), the lumpy brown sea cucumber (BSC) lives at a depth of 2 to 30 m in the rocky reef

---

[1] As in the opening quote of this chapter, people sometimes refer to Galápagos as a "living laboratory" or "natural laboratory." To avoid implying that its natural processes are under purposeful human control—which is far from the reality—I feel that "showcase" is a better metaphor. For related discussion, see Quiroga (2009, 2014), Hennessy (2018), and Hennessy and McCleary (2011).

Figure 9.1. The brown sea cucumber, *Isostichopus fuscus,* is a marine organism valued in Asia for food and medicinal uses. Once common along the Pacific Coast from Mexico to Peru, these echinoderms reached their highest densities along the coastlines of Galápagos prior to the 1990s. For scale, the small fish at right is roughly 5 cm (2 in.) long.

communities that ring each island of the archipelago. Called *pepinos* (in Spanish, literally "cucumbers"), they are slow-moving bottom dwellers and detritivores (general detritus scavengers), subsisting on the array of organic matter that falls to the sea floor.

As a member of the phylum Echinodermata ("spiny skins"), sea cucumbers reproduce in much the same way as sea stars, sea urchins, brittle stars, and crinoids—via broadcast spawning. Adult males and females simply release their gametes into the water, in a roughly synchronized release so that densities are high in the same place at the same time, allowing fertilization to take place through the collision of gametes. In many locations, the BSC takes its cues for spawning from seasonal changes in water temperature, likely linked with food supply (Toral-Granda and Martinez 2007, 2095). In Galápagos and coastal Ecuador, the BSC has evolved a particular adaptation, apparently because ocean currents provide adequate nutrients for year-round spawning: males and females release gametes at sunset on the same night, within a few days of the new moon (Mercier et al. 2004, 2007), which they detect via small light-sensitive eyespots on their tube feet. For this mechanism to promote reproductive success, and thereby replenish local populations, there must be a certain density of mature adults in any given locale. Below that density, reproduction often fails completely because the gametes rarely find one another—an example of the Allee effect.[2]

Though they may not seem very impressive, BSCs play a number of important roles in the marine ecosystems they inhabit. For one thing, these detritivores are like "vacuum cleaners" of the sea. They ingest debris from the ocean floor, excreting feces rich in organic matter that feed algae and sea grasses and thus help with the productivity and energy flow of marine ecosystems. In addition to debris, the sea cucumbers hold in check the spread of harmful brown mats of cyanobacteria in the rocky tidal zone (Beltram et al. 2019). BSC feces are also generally rich in calcium carbonate, which makes them basic on the pH scale, to the point where some authors suggest they may help locally to combat the effects of ocean acidification (Wolfe and Byrne 2016). Their activities and excrement are important enough to local food chains, says Chantal Blanton, former Director of the Darwin Foundation, that "If sea cucumber [are overfished], you can say goodbye to the fish, the penguins and the flightless cormorants" (quoted in Brooke 1993, brackets added; see also Valle 1995, 601).

The second important role of sea cucumbers is as prey species, especially as larva and juveniles, for diverse predators, including many species of finfish, other echinoderms (such as sea stars), and crustaceans. Sea cucumbers are also important food items for humans, especially in East Asia, where sea cucumbers are a significant part of the diet, valued for their vitamins, minerals, and general vitality promotion. Extracts of some species have been shown to have antihypertensive effects and to slow the growth of cancer cells, among other bioactive properties (Bordbar et al. 2011; Pangestuti and Arfin 2018).[3] Sea cucumbers are also valued in East Asia as an aphrodisiac and male eroticism booster, for which there is precious little laboratory support, though it surely props up their commercial value. Together, the various uses of sea cucumbers, including the BSC, which commands one of the highest prices of all sea cucumbers, fuel a market demand that results in annual catch rates of approximately 15,000 metric tons (33 million pounds!) of sea cucumbers, all species included (Purcell et al. 2013, 36).

No wonder sea cucumbers are readily depleted from in-shore fisheries in the tropics. That's an enormous, almost unfathomable harvest, which leaves many areas of ocean floor without their native detritivores. Figure 9.2 shows just how much damage has been done. It represents the impact of fishing across the range

[2] This is one example of the more general Allee effect—a positive relationship between population density and the fitness of an organism (Stephens et al. 1999).

[3] Janakiram et al. (2015, 2910) observed, "Sea cucumbers consist of vitamins, minerals, cerebrocides, peptides, and lectins, and also contain [many] unique molecules, . . . [which, in sufficient quantities,] possess anti-microbial, anti-oxidant, anti-angiogenic, anti-inflammatory, immunomodulatory, and anti-tumoral properties. . . . As supplements, these sea cucumber extracts have been shown to suppress inflammation and increase innate immune responses."

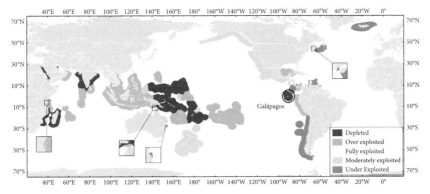

Figure 9.2. Patterns in the global harvest of sea cucumber species in 69 fisheries: the most depleted areas, where "catches, and/or stocks, are well below historical levels, irrespective of the amount of fishing effort exerted" (*shown in red*), and the areas of over-exploitation, areas "exploited at above a level that is believed to be sustainable in the long term" and thus at a high risk of depletion or collapse (*shown in orange*). Galápagos (*black circle*) is labeled "depleted"—it is the most overharvested area of the Americas. *Source*: Republished with permission of S. Purcell from Purcell et al. 2013. "Sea Cucumber Fisheries: Global Analysis of Stocks, Management Measures and Drivers of Overfishing." *Fish and Fisheries* 14, no. 1: 34–59. License conveyed through Copyright Clearance Center, Inc.

of the BSC, which is confined to the Pacific coast of Central and South America, and a similar impact on dozens of other sea cucumber species. The figure confirms that large areas of ocean around Zanzibar and Madagascar have been depleted of their sea cucumbers, as have the coastlines of India and the Pacific shores of Indonesia. The Galápagos archipelago, once famed for the highest natural BSC densities of all the Americas, is now the most seriously depleted region of the Western Hemisphere. In recent decades, at least seven species of these innocuous little echinoderms from around the world, including *I. fuscus*, have been pushed by Asian markets to the IUCN status Endangered.

How has this happened, in Galápagos of all places? One would sooner expect the BSC to enjoy a safe haven in Galápagos, given its status as an Ecuadorian National Park, a Biosphere Reserve, home of the Charles Darwin Research Station, and a UNESCO World Heritage site.

An Exotic Introduced Organism

The tale of the BSC would be neither complete nor fully understandable without a subplot dealing with yet another organism: the invasive, terrestrial mammal we call *Homo sapiens*. Our species is key to understanding BSC depletion, and not just by human consumption, as important as that is, or even by capture at the hands of human fishers, which directly produced the depletion. The story of the BSC is interwoven with another pronounced dynamic in the islands involving people: namely, tourism. During the years when BSC harvest in Galápagos was reaching record levels, there were also record numbers of human visitors to the islands, with such financial impact that even simultaneous bumper crops of harvested BSCs did not hold a candle to the revenue from tourism.

As the UNESCO quote at the beginning of this chapter so aptly recognized, uncontrolled growth in tourism to Galápagos and its indirect effects, via immigration and stimulation of local commerce, became a big contributor to the "World Heritage in Danger." Data from GNPD show that tourism continued to grow rapidly right up to the start of the coronavirus epidemic, maintaining a trend dating back to the 1980s (Figure 9.3; see also Durham 2008). A total of 275,800 tourists bought visitor permits in 2018, up 14% from 2017, and up

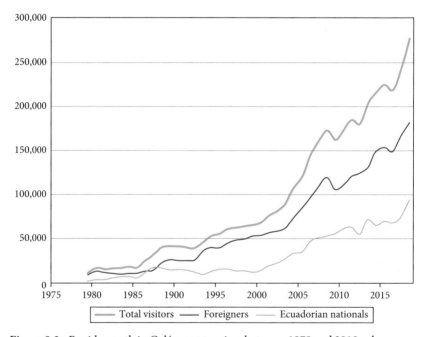

Figure 9.3. Rapid growth in Galápagos tourism between 1979 and 2018, when a total of over 275,800 visitors arrived. Foreign visitors (*in blue*) increased again in 2019, but totals of both foreigners and Ecuadorian nationals (*orange*) fell to near zero for much of 2020 (not shown) with the onset of the coronavirus pandemic. *Source*: Data from Galápagos National Park Directorate (http://www.Galápagos.gob. ec/en/visitor-statistics/).

a whopping 59% from the previous 10 years. No doubt the annual totals have fallen sharply with the pandemic; however, they are sure to rise again, but ideally to a lower, more sustainable level.

Continued exponential increase in tourists visiting a small collection of island ecosystems is sure to cause problems sooner or later. Before the pandemic, tourism was only very loosely regulated in Galápagos (Tullis 2016) by a limit to the number of berths in boat-based tourism,[4] and by building permits for hotels and lodges in the case of land-based tourism, also called "island hopping." As Figure 9.3 shows, these measures had not managed to control, or even to slow, the overall rates of increase in visitors. Although rates of boat-based tourism had leveled off, and actually declined by 9,000 visitors between 2007 and 2015, the overall tourism numbers continued to grow, fueled by an increase of 73,000 island-hopping visitors over the same interval (DPNG 2016). In island hopping, visitors stay in hotels on one of the four main inhabited islands (the fifth, Baltra, has no hotels), making day trips with guides by land or small boats to nearby GNP visitor sites.

It is not the wear and tear of tourists themselves on the ecosystems of Galápagos that is the main concern, although that changed somewhat with the boom in land-based tourism. Until recently, concern focused on the indirect effects of tourism. As summarized schematically in Figure 9.4A, tourism had its greatest impact through stimulating economic activity in the islands, which, in turn, prompted immigration from mainland Ecuador, where average incomes have been less than one-third of those in Galápagos (Villacis and Carillo 2013, 76).[5] The immigrants have been diverse, representing differences in class, race, and ethnicity, as described in Chapter 4. They also brought along a diverse biota of introductions, including pets, house and garden plants, and, in many cases, agricultural flora and fauna. The immigrants' modification of habitats in Galápagos, for agriculture and settlement, profoundly changed terrestrial ecological conditions in many areas, enabling invasive species to outcompete Galápagos natives. Consequently, tourism began to undermine the

[4] The term "boat-based" tourism is used here instead of "cruise-based" because the GNPD prohibits visitation by boats with over 100 passengers. In boat-based tourism, berths are regulated by permits called *cupos*, issued by the Park. Epler (2007, 11) noted that "As of July 2006, the tourist fleet's capacity was 1,805 berths. Forty-six individuals, companies or families owned the 80 tour vessels operating in Galápagos. Of these, 26 owners (57% of the total) owned one vessel each [and are thus small-scale operators] and cumulatively controlled 1/3 of all tourism permits." At the other end of the scale, "14 vessels and 14% of all *cupos* are held by three [large companies], one of which owns six vessels."

[5] A model by Taylor et al. (2003) makes an explicit link between tourism and immigration. In their model, a 10% increase in total tourism in Galápagos prompts a 5.3% increase in immigration to Santa Cruz, the largest human population in the islands. If immigration there were blocked, island wages would then increase about 8%.

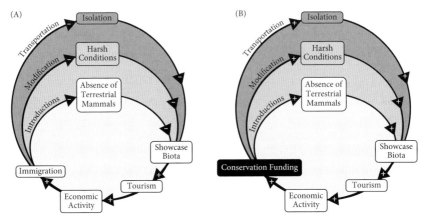

Figure 9.4. Simplified schema for the direct and indirect effects of tourism in Galápagos. *A,* For much of its history, Galápagos tourism unwittingly undermined the very showcase biota that brought visitors to the islands. Tourism fueled growth in economic activity and immigration that fed back negatively on the valued biota via introductions (e.g., new mammalian predators and competitors), habitat modification (as land areas were converted to human uses, urban and rural), and transportation (which greatly reduced the islands' isolation). *B,* There is a genuine potential for tourism to forge, instead, a positive relationship with the habitats and organisms of Galápagos. Through increased entrance fees and a recommended sustainability assessment, tourism and economic activity can generate new levels of conservation funding, which can then promote positive environmental feedback by helping to restore *Scalesia* forests, for example, to control or eliminate introduced mammals, to reduce pollution, and to promote environmental education. In this simplified model, tourism becomes an ongoing positive influence on both showcase biota and local livelihoods. (For a more complex model of tourism and its effects, see Espin et al. 2019.)

very same showcase biota that attracted tourists in the first place.[6] The dwindling *Scalesia* forests of the humid zones are a tragic example. Making matters worse, the tourism revenue that drove the feedback loop was "leaky," meaning that much of the tourism earnings—from 61% of all boat berths, for example—accrued among mainland Ecuadorian and foreign owners, not among Galápagos residents (Taylor et al. 2009).

---

[6] Nash (2009, 108) offered this summary: "Open immigration quickly depletes both natural and social capital. Add more tourists and residents, and the need grows for more housing, food, fresh water, diesel fuel—thousands of tons of a thousand different things every month. Everything multiplies, including the dread scourges of island biogeography: invasive alien species. Uninspected traffic [now a thing of the past] is a conveyor belt, carrying invasives" at record rates.

But the flow of immigrants toward opportunities in the Galápagos tourism sector was only one flow of arrivals. Another flow—actually a big *wave*—of major impact was the influx of fishers who came to Galápagos during the 1990s. This second wave had far-reaching implications for a key social-ecological subsystem of the archipelago.

Fishing in the Common Pool

The story of Galápagos fishing can be told as the history of a traditional "open-access" fishery (unrestricted and unregulated) that worked relatively well, without overt conflict, for decades after renewed human settlement efforts in the 1920s. As noted in Chapter 4, Ecuadorian law "provided all colonists with the possibility of receiving twenty hectares each of free land . . . and the right to hunt and fish freely on all uninhabited islands" (Lundh, 1999, Chapter 12). Although life was not easy, settlers fished as they pleased, with relatively little adverse impact: there were few fishers and abundant coastline. Farming was more impactful (see Chapter 6), since the best agricultural conditions exist in the same highland areas as the giant daisy forests, and farmers felt compelled to clear them for crops and cattle. Little thought was given at the time to agroforestry shaded by *Scalesia*.

But another difference between fishing and farming became increasingly relevant as the years passed. Ocean fish are a good example of what conservationists call common-pool resources (CPRs).[7] When unregulated, CPRs can sustain only very low rates of usage, as in the case of low-density harvesters each taking small harvests of the resource. With more harvesters, bigger harvests, and no restrictions, pursuit of individual benefits generally causes depletion for all— an outcome once erroneously called the "tragedy of the commons" (Hardin 1968) but now more accurately regarded as the "tragedy of open access" to the resources (Feeny et al. 1990). Such a tragedy is especially pronounced with resources subject to the Allee effect, like BSC, which require a minimum density for reproduction. Under open access, there is little to stop users from hunting such a resource below the minimum density required to keep the population from extinction.

---

[7] CPRs have two properties especially relevant to their management. First, they have high subtractability (or depletability): once a given resource unit (i.e., a fish) is procured, it cannot then be used by others, even if the resource is renewable at the population level. Subtractability can also be high with land for farming (and other so-called private goods). Second, high subtractability is accompanied by low excludability: it is difficult to keep other users away from the resource, since it occurs in a free-ranging population that is potentially available to all. With CPRs, it takes institutions, rules, and enforcement to restrain users and prevent a resource "free for all." In contrast, the land worked by farmers shows high excludability via boundaries and fences, and thus is not a CPR.

The central problem addressed in this chapter can now be brought into focus: How does a social–ecological subsystem like the sea cucumber fishery of Galápagos move from open access to some form of regulated use, avoiding tragedy? With ongoing changes in market demand, in fishing technology, and in numbers of fishers, will institutions, rules, and enforcement be implemented in time to prevent overfishing and collapse of this and other important fisheries?

Change Comes to Galápagos Fisheries

At first, open-access fisheries worked well with the sparse settler population. But fishing became much more impactful in the last half of the 1900s owing to a series of changes. First was the arrival of industrial fishing in neighboring Pacific waters, both U.S.-based tuna fishers in the 1940s and Ecuadorian commercial fishers in the 1950s and 1960s. The success of these efforts almost certainly conveyed to later fishing fleets that, although near-shore fisheries officially became property of the Ecuadorian government with the creation of the GNP in 1959, there was minimal monitoring or enforcement of rules for these resources, even quite close to the islands.

The second change was both demographic and technological, when a good number of fishers from the coast of South America came to Galápagos in pursuit of lobster: "According to natives and older migrants in Isabela, the first large in-migration of coastal fishermen to the Galápagos coincided with the lucrative boom of lobster fishing from 1982 to 1984.[8] These migrant fishermen introduced new fishing techniques that were already being used on the [mainland] coast, such as the trident, which is used for spearing lobster, and the air compressor, which is used for breathing during diving [now known, rather humorously, as 'hookah' fishing]" (Bremner and Perez 2002, 308). The immediate effect of this immigration, seen clearly in Figure 9.5A, was twofold—a larger population of active fishers, and deeper, more extensive lobster harvesting, a commercial fishery that grew significantly in the 1960s and 1970s (Figure 9.5B). Eventually, the lobster fishery collapsed: a daytime pursuit in the 1970s, with daily yields of 10 kg per diver, lobster is today a difficult nighttime pursuit, with maybe half that yield (Schiller et al. 2015; Alex Hearn email to author, August 24, 2016). Partly in response, the Ecuadorian government declared a new Galápagos Marine Resources Reserve (GMRR) in 1986, to prevent marine abuses and over-exploitation of the kind. But the GMRR was poorly planned and implemented,

---

[8] This is the boom period in lobster fishing that impacted the Galapagos cormorant and penguin populations, as discussed in Figure 8.8.

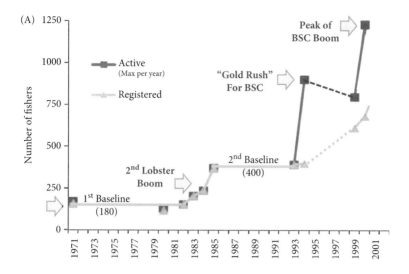

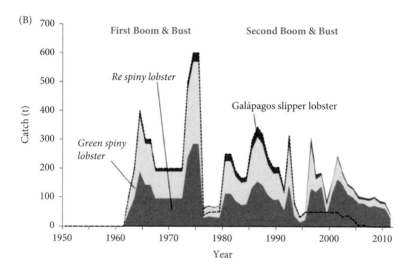

Figure 9.5. Galápagos fisheries in profile. *A,* The number of active fishers increased significantly via immigration between 1980 and 2001, prompted by lobster and sea cucumber booms. Numbers peaked at 1,250 in 2000, and then decreased to 400 active fishers by 2012 (Engie 2015). *B,* Booms and busts of the Galápagos lobster harvest between 1950 and 2010. Harvests of green spiny lobster (*Panulirus gracilis*), red spiny lobster (*P. penicillatus*), and the Galápagos slipper lobster (*Scyllarides astori*) are shaded. The second boom and bust includes several "mini" boom-and-bust cycles within it, one related to the 1982–1983 El Niño. *Sources*: *A*, Adapted from Castrejón (2013, 137), in *Galápagos Report 2011–2012*. Puerto Ayora, Galápagos; *B*, Republished with permission of L. Schiller from Schiller et al. 2015. "The Demise of Darwin's Fishes: Evidence of Fishing Down and Illegal Shark Finning in the Galápagos Islands." *Aquatic Conservation: Marine and Freshwater Ecosystems*, 2, no. 3: 439. License conveyed through Copyright Clearance Center, Inc.

remaining little more than a name for the following 5 years (Reck 2014, see also Barragan-Paladines and Chuenpagdee 2017).

The third change was the most dramatic, bringing us back to *el pepino*. In 1991, against the backdrop of continuing growth in Asian sea cucumber demand (Purcell et al. 2014), zealous overfishing on the mainland coast brought collapse to the continental fishery (and many others, Anderson et al. 2011). The "roving bandits" of aggressive middlemen and marketers behind the collapse were not deterred: they led an impressive, unregulated immigration of fishers of diverse identities (Ecuadorian and other), classes (boat-owners, laborers), and technologies (from small fiberglass speed boats, *fibras*, to semi-industrial fishing boats, often of Asian ownership) to Galápagos beginning that same year.[9] With a tradition of de facto open access in Galápagos fisheries, there followed an uncontrolled "gold rush" of sea cucumber harvesting, supplemented at times by lobster harvesting from rebounding local populations. The number of active fishers jumped in one year from roughly 400 to over 900 (as in Figure 9.5A), and on to 1,250.

Meanwhile, the Ecuadorian government was moving forward with its intention of protecting the archipelago's marine life, via the GNPS, much as it was protecting terrestrial life. In 1992, just as the sea cucumber rush began, the first management plan was issued for the GMRR, increasing the GNPS's jurisdictional role in protecting the marine resources of the islands.[10] Before long, these events pitted Park personnel against middlemen and fishers, now a tumultuous blend of "old" fishers who felt legitimized by their traditional access, and "new" fishers, mostly *pepineros* (sea cucumber harvesters) who had recently come to Galápagos from coastal Ecuador, with methods and marketing new to the islands. Because the Darwin Research Station (CDRS) had, by this point, worked for decades with the GNPS on terrestrial conservation, the Station and its personnel were also part of the opposition in the eyes of the fishers.

The ensuing 24/7 scramble for sea cucumbers produced a period of overt conflict unprecedented in the human history of Galápagos. The conflict laid bare some key social fault lines, present for many years in the archipelago but largely latent before the impact of the immigrants. For nearly 20 years, it pitted fishers,

[9] As described by Castrejón and Defeo (2015), the "roving bandit" middlemen, typically from Asia, sponsored fishers from mainland Ecuador to move to Galápagos and exploit *I. fuscus*. As BSC exploitation grew, resident small-scale fishers were also drawn into the industry. As a result, "a resource that had not been traditionally exploited by the local population became unexpectedly the most lucrative fishery of the Galápagos and, most importantly, a pervasive partnership was created between roving bandits and fishers" (616), commonly involving loans from middlemen for technology improvements. Consequently, when fisheries were later closed for conservation, fishers were unable to repay their loans, increasing pressure to fish illegally.

[10] The GMRR management plan set up an interinstitutional "Control and Vigilance Commission" with representation from Fisheries, the Navy and the GNP Director (Reck 2014). With the surge in the sea cucumber fishery, GNPS eventually took on a greater role in management and jurisdiction.

middlemen, and buyers against state authorities and conservationists, resulting in lasting social strife and unrest, damage to the sea cucumber fishery, and harm to lobster populations as well. This period of Galápagos history has proven pivotal for understanding the social–ecological systems of the islands, past and present, so let us now return to that framework.

Background to the Conflict

Considered in terms of the SES of resource users (diverse fishers) and resource units (BSCs and lobsters), the heart of the conflict was an attempt by the government to change the governance system from open access to a set of rules restricting harvest in space and time. There were surely good reasons for a change, including preventing overexploitation of sea cucumber and lobster populations, thus protecting long-term livelihoods of the fishers themselves. In addition, the interactions in this SES (Figure 9.6) from the removal of BSC and lobster virtually guarantee negative impact upon other Galápagos organisms, including the penguin and flightless cormorant, long a part of the Park's terrestrial mandate. The role of external drivers is especially pronounced: the high price for resource units ensured the profitability of their capture. The depletion of mainland fisheries prompted hundreds of diverse users to move to the archipelago.

It was not so much the rule changes per se as the process behind the changes in the 1990s that gave impetus to the conflict, and later made possible its resolution. So it is useful to focus for a moment on the topic of rule-changing. As Ostrom and colleagues originally envisioned the SES framework, the "rules in use" of a common-pool resource could stem from a central authority like the state, or emerge on their own from the norms of cooperative action on the part of resource users (norms being locally approved standards of behavior). Ostrom (2005, 130), for example, reviewed studies of conditions that promote cooperative rule observance among resource users and conditions that thwart such behavior. She concluded that "a social norm, especially in a setting where there is communication between the parties, can work as well or nearly as well at generating cooperative behavior as an externally imposed set of rules and system of monitoring and sanctioning. Moreover, norms seem to have a certain staying power in encouraging a growth of the desire for cooperative behavior over time, while cooperation that is primarily there due to external imposed and enforced rules can disappear very quickly."

Ostrom inferred that local norms "have a certain staying power" within their community that contrasts with imposed rules, which tend to be ephemeral without heavy sanctions. Let me call this conclusion *Ostrom's rule*: imposed regulations for cooperation lack the staying power of regulations based on

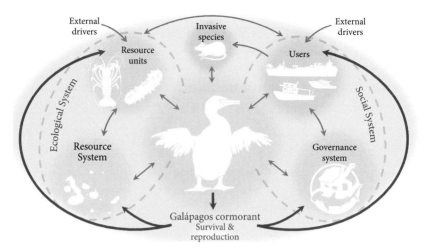

Figure 9.6. An SES model of the Galápagos sea cucumber and lobster fisheries, showing their impact on the survival and reproduction of flightless cormorants, as an example. Note the diversity of fishers, represented by widely varying boats, whose activities are affected by external drivers such as high market prices. Note also the diverse impacts of fishing, including direct impacts on birds injured or killed by fishing gear and propellers, and the effects of competition on cormorants' food supply. Indirect impacts include harm from invasive species (rats shown, plus dogs, cats, mosquitoes, and pathogens) that may escape on shore from fishers' activities. There is also an indirect cascade of effects from reductions in sea cucumber and lobster populations, given their ecosystem services in the Cromwell upwelling (e.g., lobster predation limits herbivorous sea urchin numbers, maintaining biodiversity and health of the reef, on which cormorants depend). *Source*: Adapted from Ostrom (2009).

community norms. Ostrom made clear that fishers can, and do, regulate themselves under conditions where they are free to choose *and* able to benefit from the management of their CPR as a regulated common property or "commons" (Cox et al. 2010). Sometimes it is not restraints themselves that fishers resist, but the way in which they emerge.

Through an anthropological lens, Ostrom's topic here is the cultural evolution of a fishery's rules in use, whether by the free and unfettered emergence of norms from the fishers or by the imposition of rules from powerful external agents. A few years earlier, in a related, more general, theoretical work (Durham 1991), I proposed that imposition and choice are the two main processes of cultural change in human societies, and I argued the more general case that the products of choice within a social group have a staying power that imposed phenomena can only achieve with coercive force from outside. The role of imposition in cultural change interested me quite apart from Galápagos, but I noted early on that governmental imposition was a key component of the sea cucumber conflict. It looked as though events playing out in Galápagos might offer a "test case" for some of my arguments, so I followed the ensuing saga from the 1990s to today. As it unfolded, the events of the conflict were sufficiently clear to allow a preliminary test of the following hypothesis—one that will help us understand years of strife over Galápagos fisheries.

Hypothesis

These arguments form a testable hypothesis for the sea cucumber conflict: regulations that are externally imposed upon resource users harvesting CPRs under a tradition of open access will be unstable owing to noncompliance, defiance, or both—unless noncompliance and defiance are suppressed for a period of time through force by the imposers. In the Galápagos context, this prediction can be recast as the hypothesis that imposition fosters resistance: peaceful and effective long-term solutions to fisheries management will require rules and regulations that are, in some meaningful way, the product of choice among fishers. Let me propose a preliminary test of this hypothesis, drawing on the history of regulations for managing the sea cucumber and lobster fisheries of the GMR.

"The Sea Cucumber War"

In 1992, when the Ecuadorian government sought to impose order and to regulate the lobster and sea cucumber fisheries, there followed a series of major protests by both long-time resident Galápagos fishers and the hundreds of *pepineros* newly arrived from the continent. Conflict began after the government issued two bans intended to prevent overharvesting: a 2-year ban on sea cucumber harvest, and a 7-year ban on lobster harvest. Soon thereafter, fishing cooperatives were formed on three islands, in large part to protest government closures. The next 6 years constitute the first phase of conflict.

Although parallel sea cucumber gold rushes took place elsewhere in the world (Anderson et al. 2011; Crona et al. 2015), this period was especially intense and heated in Galápagos. The local and immigrant fishers reacted immediately to the top-down impositions intended to curb their use of long-time open-access resources. Their livelihoods, almost entirely dependent on fishing (since there had been little effort to include them in tourism development), were directly threatened. By the same token, it is also understandable that the GNPS would act firmly to enforce conservation measures: for nearly 30 years, they had carried out a serious terrestrial conservation mandate in the archipelago, based in national law and abundantly reinforced by the international visibility of Galápagos. Their efforts in the fishing crisis were a logical extension. In addition, many of the iconic organisms that nest terrestrially—cormorants, marine iguanas, and penguins among them—are dependent on near-shore marine resources, which were then threatened. On islands, there's a logical overlap between land and sea conservation anyway. But what the GNPS saw as logical, even overdue, came across as heavy-handed imposition to *pepineros*.

A synopsis of key fishing-related events for the years 1992 to 2016, Figure 9.7, helps highlight the overall pattern. As shown in the top section, Phase I (1992–1997), each time GNPS officials tried to limit the season or close one or other fishery, even for a period of partial recovery, the newly organized fishermen rose in protest.[11] The conflict, dubbed "Pepino War" (Merlen 1993; Valle 1994b), deeply divided the human inhabitants of Galápagos.

This pattern of impositions and resistance during the first 6 years of the conflict provides a revealing, if short, preliminary test of the chapter's hypothesis. Imposed rules for compliance had virtually no staying power, nor the clout to exact compliance, instead triggering protest and resistance at every turn, as seen in the figure. The conflict grew more serious every year, showing little sign of resolution, with two additional factors feeding the fishers' reactions. First, through long-standing working relations with the Park, CDRS—well known for hosting outside researchers—inadvertently added a Global North/South dimension to fishers' resentment. There was a feeling of haves vs. have-nots, rich vs. poor, and developed vs. developing worlds in the conflict. Second, when wealthy boat-owners from the tourism sector sided with the Park and Darwin Station against the fishers, fearing loss of tourism revenues from the protests, tensions inevitably developed along lines of social class. As Besserer (2019, 51) put it, "local compliance with reasonable environmental regulation was low, as a rebellious act against perceived economic oppression."

From the start, the conflict was deep and difficult, attesting to the volatility of imposition. An even more telling second test of this chapter's hypothesis was provided by events that followed.

---

[11] For insightful accounts of events itemized in Figure 9.7, see Nicholls (2006, Ch. 6) and Meltzoff (2013, Ch. 2). For a journalist's account of the whole turbulent period, see Bassett (2009). For details of the 1995 fishers' occupation of the GNPS headquarters and blockage of CDRS, including videos from 1994 and 1995, see the website of James Young (https://james-young.org/wp/Galápagos-islands/).

Forming a Co-Management System

By 1996, the imposition of state authority, even with good intentions (believing them to be necessary for the long-term sustainability of *pepineros'* revenue streams), had backfired into escalating protests by sea cucumber and lobster fishers. But three key events that year opened a pathway toward a resolution. First, Galápagos was made an official "Biological Reserve" in 1996 by the Protected Areas Authority of Ecuador (INEFAN). That declaration gave GNPS a clear mandate to coordinate revision of the Reserve's management plan. The designation allowed the Park to consider anew how best to protect the Reserve's marine resources.

Second, a social scientist from Harvard University and former Galápagos Peace Corps volunteer, Theodore Macdonald, was invited to carry out a conflict analysis of the turmoil, to meet with various stakeholder groups, and to make recommendations to the GNPS and CDRS for a peaceful, long-lasting solution. Macdonald found that,

> Much of the current management dilemma rests on the [fishing] residents' *sense* of rights to the resource. The [fishers] of the Galápagos have a sense of marginality and, with it, resentment, as a result of government policies. Rules are perceived as alien, imposed, and inappropriate. . . . [To] a large extent, it is the *manner* in which rules have been made that frustrates and angers local people, *not simply the rules themselves.* That manner has been exclusion and it has led to a strong sense of marginality. The resultant feeling, in turn, leads the residents to act as if the land and marine resources were located in areas of 'open access' . . . despite clear state claims and related rules. (Macdonald 1997, 3-11; emphasis added)

Macdonald's observation that a good part of the issue was "the manner in which the rules have been made" speaks to the sensitivity of fishers to rule changes by imposition—consistent with the hypothesis here. Macdonald offered a far-reaching practical suggestion: although the coastlines and fisheries were all government property by law, he recommended "rethinking the entire situation *as if it were* managed as 'common property' [i.e., co-owned by the various stakeholders including fishers] and building recommendations from there" (Macdonald 1997, 3; emphasis added). In short, he suggested giving all players a genuine and equal hand in the game.

Third, a 1996 UNESCO delegation to Ecuador indicated that, with the conflict and violence of the sea cucumber "war," this World Heritage Site could well be declared "in Danger of Extinction" if drastic changes were not made. The president of Ecuador quickly indicated a willingness to consider urgent measures, including a special law to revise governance of fisheries in Galápagos if need be.

Year(s)	Changes in Fisheries Governance	Fisher Responses
	Phase I	
1992	• GNPS given jurisdictional role under 1st management plan for GMRR • Presidential decree closes BSC fishery	• Immigration of hundreds of fishers, BSC "gold rush" begins • Protest of fishery closure: 15 tortoises killed on Isabela
1993	• GMRR opens BSC fishery; 3-6M BSCs taken	• Bustling business with BSC middlemen
1994	• GMRR opens BSC fishery with 0.55M quota; closes early after 6-10M BSCs are harvested	• Protests with 81 tortoises killed and arson-set fires on Isabela
1995	• GMRR issues "precautionary closure" of BSC fishery	• Protests: fishers strike, damage to GNPS and CDRS buildings
1996	• Increased jurisdiction for GNPS in GMRR; begins joint monitoring with local fishers • BSC fishery closure continues, first arrests & confiscations for illegal BSC fishing	• Protests continue, fishers seize GNPs office • Nine tortoises killed on Santa Cruz
1997	• BSC fishery closure continues, along with illegal harvesting • GNPS and CDRS create *Grupo Núcleo* discussions with diverse stakeholders	• Fishers seize CDRS; GNPS ranger is shot at illegal BSC camp • Fisher representatives invited to *Grupo Núcleo* discussions
	Phase II	
1998	• 1998 Special Law converts GMRR to a larger GMR, with participatory co-management • Industrial finishing outlawed; artisan fishing only by registered coop members	• Fishers gain one representative in each PMB and INA • Fishers included in planning discussions, including calendar
1999	• New GMR management plan, with principles of adaptive management & zoning • PMB sets annual calendar plus quota for BSC fishery	• Fisher representatives agree to new management principles • Fishers exceed quota, express feeling outnumbered in PMB
2000	• PMB issues lobster limit of 50 tons; raised to 80 tons following protests	• Protests damage GNPD and CDRS offices; tortoises taken hostage
2001	• PMB sets annual calendar & quota for BSC fishery	• Fishers do not exceed quota for first time
2002	• IMA closes Fishing Registry; 5-year moratorium on new fishers; no quota issued for BSC	• Fishers strike 11 days on Santa Cruz; still make record BSC harvest
2003	• PMB issues liberal BSC quota	• Second-highest BSC harvest
2004	• PMB issues BSC quota and announces ban for 2005-06 • Environment Minister talks with fishers about alternative livelihoods	• Fishers siege CDRS, hold 30 scientists hostage, withdraw from PMB for 9 months
2005	• Longline fishing banned in GMR • Official recognition of failure of co-management system	• "Recreational fishing" with tourists begins under authorization • Some fishers paid for work in conservation & invasives removal

Figure 9.7. Timeline of trouble in the lucrative sea cucumber and lobster fisheries of Galápagos. The table summarizes changes in the governance system (*left*) and responses of resource users (*right*). Instances of imposition are shown in light shading on the *left*; protests are shown in dark shading on the *right*. Three phases are shown: Phase I, the sea cucumber gold rush of 1992–97; Phase II, early participatory co-management years of 1998–2010; and Phase III, consultative co-management years, 2011–2016. *Source*: Inspired by Table 8.1 in Castrejón et al. (2014, 162–63) with additional information from Ospina (2006, 229), Celata and Sanna (2012, 984). Years 2012 and 2013 omitted for brevity.

2006	• BSC density low, fishery closed; GMR implements coastal marine zoning • New Technical Team starts work on fisheries management plan	• No organized protest • Fishers included in Technical Team for fisheries management
2007	• BSC fishery opens briefly; UN declares Galápagos "World Heritage in Danger"	• No organized protest
2008	• New Ecuadorian Constitution gives legal rights to nature, advocates *sumac kawsay* ("good living") • New governance structure, the Governing Council, replaces INGALA	• No organized protest
2009	• GNPD approves fisheries plan, Capítulo Pesca, with more local knowledge and fisher input than before	• Fishers agree to minimum density for BSC harvest
2010	• BSC fishery closed; UNESCO removes Galápagos from list of "World Heritage in Danger"	• In protest, 6 goats deliberately reintroduced to Santiago Island
Phase III		
2011	• BSC fishery opens for 60 days, but is closed again from 2012 to 2015	• No organized protest
2014	• GNPD issues new Management Plan for Galápagos protected areas (terrestrial and marine)	• New participatory management rules for spiny lobster fisery
2015	• 2015 Special Law abolishes PMB & IMA, creates new fishery planning tool, Calendario Pesquero • Only 1 monitored site of 14 shows BSC density over approved threshold	• New, direct discussion between GNPD & fishers • No organized protest: BSC fishery opened for 50 days
2016	• GNPD presents revised zoning scheme to fishing coops for feedback • Presidential decree adds major new Marine Sanctuary on top of zoning	• Fishers on San Cristóbal react to imposed new zoning system • Protest on San Cristóbal spreads to Santa Cruz; zoning delayed

Figure 9.7   Continued

Sensing an opportunity to build a negotiated solution drawing on Macdonald's proposal, the GNPS and the CDRS held a participatory planning workshop in June, 1997, bringing together in official conversation, for the first time, the main stakeholder groups involved in the fishing conflict. That group in turn designated a smaller *Grupo Núcleo* to continue the discussions. The *Grupo Núcleo* took seriously the "unique opportunity to draft an appropriate legal framework and to revise the management plan" for a new GMRR participatory management system (Heylings and Cruz 1998, 2).

In due course, the *Grupo Núcleo* was successful at two major goals: first, they reached a consensus, with compromises, on a general set of principles for participatory management of the Marine Reserve. Second, they agreed that the Reserve should be a national protected area featuring the GNPD as the jurisdictional authority,[12] with the following additional provisos. They proposed that fishing rights in the Reserve be exclusive to members of local artisanal fishing cooperatives, that all Reserve stakeholders be well defined and limited in number, and that all stakeholders have genuine access to decision-making power. The goal agreed upon was not some weaker "consultative" participation, but full cooperative co-management between the state and stakeholder groups.

Thus, a new process of rule change by choice, rather than imposition, seemed well on its way, with genuine, broad support from the fishers and other constituencies. With these accomplishments in hand, the *Grupo Núcleo* orchestrated lobbying of the Ecuadorian National Assembly, in the face of opposition by powerful industrial fishers from the northern coast of Ecuador, and and succeeded, marking the start of Phase II in the timeline (Figure 9.7).

The 1998 Special Law

In March, the 1998 Special Law, or *LOREG 1998* (its Spanish acronym), was approved by the National Assembly and President.[13] It officially redefined the GMRR as the Galápagos Marine Reserve (GMR), with boundaries extended to 40 nautical miles from the islands, and the proviso of no industrial fishing within the boundaries. As the *Grupo Núcleo* had wanted, it also outlined a new GMR and fisheries governance system based on co-management principles. It stipulated two supervisory committees, one local and one national, both inclusive of stakeholder participation (Figure 9.8). The first was called the Participatory Management Board (PMB; also known as the "*Junta*"), with local, individual representatives from five direct GMR stakeholders: the Artisanal Fishers Cooperative Association, the Galápagos Chamber of Tourism, the Association of Naturalist Guides, CDRS, and the newly renamed Galápagos National Park Directorate (GNPD). In the PMB, the Special Law held, all decisions must be taken by consensus. The second, more powerful decision-making body was the Inter-Institutional Management Authority (IMA), headed by the national Minister of the Environment, with government officials and single stakeholder

---

[12] From founding in 1959 to 1998, the Park was administered by the *Servicio Parque Nacional Galápagos* (SPNG), translated here as the Galápagos National Park Service (GNPS), under the *Instituto Ecuatoriano Forestal y de Areas Naturales y Vida Silvestre* (INEFAN). In 1998, as part of the Special Law, it was renamed the *Dirección del Parque Nacional Galápagos* (DNPG), translated here as Galápagos National Park Directorate (GNPD), and reassigned to the *Ministerio del Ambiente* (Environmental Ministry) of Ecuador).

[13] The full name is "Ley Orgánica de Régimen Especial para la Conservación y Desarrollo Sustentable de la Provincia de Galápagos."

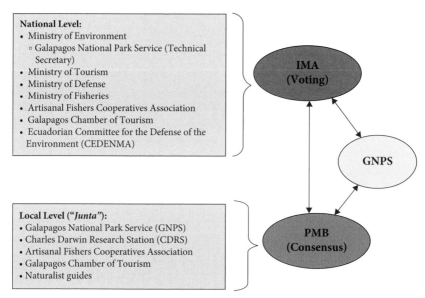

National Level:
- Ministry of Environment
  - Galapagos National Park Service (Technical Secretary)
- Ministry of Tourism
- Ministry of Defense
- Ministry of Fisheries
- Artisanal Fishers Cooperatives Association
- Galapagos Chamber of Tourism
- Ecuadorian Committee for the Defense of the Environment (CEDENMA)

IMA (Voting)

GNPS

Local Level ("*Junta*"):
- Galapagos National Park Service (GNPS)
- Charles Darwin Research Station (CDRS)
- Artisanal Fishers Cooperatives Association
- Galapagos Chamber of Tourism
- Naturalist guides

PMB (Consensus)

Figure 9.8. The participatory governance system for the Galápagos Marine Reserve (GMR) created by the Special Law of 1998, consisting of two supervisory committees, one local and one national, both inclusive of stakeholder involvement. The local committee was the Participatory Management Board (PMB; or "Junta"), with representatives from five GMR stakeholder groups (*box at bottom*). The more powerful, national decision-making body, the Inter-Institutional Management Authority (IMA), was headed by the Minister of the Environment, with both government officials and single stakeholder representatives (*box at top*). *Source*: Reproduced with minor modifications and permission of V. Toral-Granda from Altamirano et al. 2005. "The Application of the Adaptive Principle to the Management and Conservation of *Isostichopus fuscus* in the Galápagos Marine Reserve." *FAO Fisheries Technical Paper*, pp. 247–58. Permission also from The Food and Agriculture Organization, Rome.

representatives. At this top level, decisions would require voting. Altogether, it was a bold and forward-looking arrangement.

By 1999, participatory processes for the new Marine Reserve were underway, and negotiations soon produced three important results. (1) The new Management Plan for the Marine Reserve featured an impressive list of principles on which the participating parties agreed (Castro 2005, 25). Noteworthy among them was the principle of "integration," which held that every decision must be a joint decision of integrated management; there would be no playing to special interests. (2) By 2000, there was further in-principle agreement on a zoning scheme for Galápagos coastlines, with designated areas for fishing, tourism, and conservation (so-called no-take zones; NTZs).[14] (3) There was agreement on the need for a seasonal calendar each year for the sea cucumber and lobster fisheries, including negotiated annual opening dates. The co-management system was evidently off to a good start: imposition had abated and regulations elected by choice in a locally based *Junta* seemed to be taking hold among stakeholders. Events of 1998–2000 were thus consistent with Ostrom's rule and the main hypothesis of this chapter.

Fisheries Management

In 1999, after lengthy discussion, participating parties agreed to reopen the sea cucumber fishery for 60 days, and to collaboratively define a similar calendar for each successive year. As shown in Figure 9.9*A*, nearly 4.5 million sea cucumbers were harvested the first year (not including illegal take outside the calendar), 5 million in 2000, and 8.3 million in 2002. The annual fishing calendar continued, but the next 3 years saw steady declines in annual harvest, suggesting that earlier years, especially the bumper harvest of 2002, had overtaxed the standing stock of the echinoderm—not the outcome the management system was supposed to produce. The inference of damage was further strengthened by consistent declines after 2002 in catch per unit effort (CPUE) reported by the fishers (Defeo and Castilla 2012). Under fishing pressure, Galápagos *pepinos* were getting harder and harder to find.

The sea cucumber density plot (Figure 9.9*B*) constitutes another good example of the tightly linked human and nature interactions that comprise the SES of this case. Before 2002, every harvest was followed by a healthy rebound in sea cucumber stock density. Had careful monitoring and full collaboration been realized, those years might have provided a basis for a program of stewardship by adaptive management—that is, stewardship with a self-correcting management system to prevent overharvest and to maintain a healthy stock even as it allowed an annual harvest. The impact of El Niño could also have been factored into such a management plan, with close monitoring and feedback, leading one day to

---

[14] Agreement was reached on no-take status for an impressive 18% of the coastal area. However, immediate surveys in 2000–2001 found pro-fishing and pro-tourism bias in the allocation of areas (Edgar et al. 2004). Fishing zones, for example, had two to three times higher BSC and spiny lobster densities than did the NTZs. This bias appeared to help secure fishers' approval of zonation. Eleven years later, Buglass et al. (2018, 204) studied the economically important red spiny lobster, *Panulirus pencillatus*, and found "no appreciable effect on lobster sizes or abundances inside the NTZs when compared with adjacent fished zones." Partly to blame are the locations and small sizes of many NTZs plus confusions and use conflicts that arise because zonation is not linear along coastline, but covers areas 2 nm deep (Moity 2018).

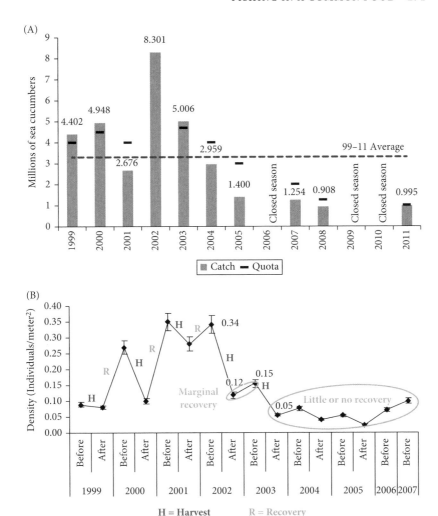

Figure 9.9. The collapse of the Galápagos brown sea cucumber (BSC) fishery under heavy exploitation, 1999-2011. *A,* Annual totals for BSC catch (*numerals*) and quota (*black lines*) from 1999 to 2011. Quotas were set by the participatory management system (no quota was established in 2002). Harvest spikes in 2002 and 2003 were followed by years of low and declining harvest, indicating damage to the sea cucumber stock. *Dashed line* indicates average total catch across the same time period. *B,* Annual average densities of BSC before and after harvest, 1999-2008 (a slightly different time range than in *A*). The zigzag pattern of human harvest (H) and natural recovery (R) of the BSC illustrates the tightly linked interactions in this social-ecological system. It also shows how quickly human impact can exceed the reproductive resilience of a focal species: the harvests of 2002 and 2003 exceeded the capacity of the sea cucumbers to recover. *Sources*: *A,* From Reyes et al. 2013.

Figure 9.9. Continued
Galápagos Report 2011–2012, p. 145. Puerto Ayora, Galápagos; *B,* Reproduced
with minor modification and permission of V. Toral-Granda from Toral-Granda.
2008. "Sea Cucumbers. A Global Review on Fishery and Trade." *SPC Beche de Mer
Information Bulletin*, 28, p. 4. Permission also from The Food and Agriculture
Organization, Rome.

---

a perpetually sustainable system. The challenges of such a notion are in monitoring the densities and collaboratively regulating harvest to prevent damage to
the stock.

Alas, the years following first efforts by the Special Law's new governance
system show pointedly what can happen when effective management fails. The
harvest of 2002 brought the stock density down so low that the rebound was only
a fraction of the harvest, meaning the 2003 harvest started at a very low sea cucumber density and only lowered it further (Figure 9.9*B*). The next couple of
years confirm the damage to the stock—there was precious little rebound, local
densities were evidently below the Allee threshold, and the fishery was in a state
of collapse.

Not surprisingly, parallel declines were also seen in the second most important fishery, spiny lobster.[15] Seeing declining lobster harvest data in 2000, the
PMB overruled fishers' desires and set a seasonal limit of 50 tons. The new participatory management process itself had come to the point of imposing a limit on
the fishers: sensing imposition despite the rhetoric of participation, the fishers
took to the streets again in protest (Figure 9.7; see also summary in Ferber 2000).
The national government and navy special forces stepped in, and a settlement
was reached with an increase to 80 tons of lobster that year. The fishers also asked
for the right to carry out longline fishing in the GMR (a technology discussed
in Chapter 2). An interinstitutional panel was formed to consider the potential
impacts of the request, realizing that sharks were among the target species fishers
sought to harvest from the GMR in this manner.

In 2004, after reviewing new monitoring data, the PMB announced a 60-day
opening of the sea cucumber fishery, with a generous harvest limit of 4 million,
but a sobering rider: because BSC harvests and CPUEs were falling, the fishery
would then be closed for recovery for all of 2005 and 2006. Fishers saw the prospect
clearly: the year 2004 might go OK, but then they faced 2 years with zero legal sea
cucumber income. Sensing another imposed and uncompensated closure, fishers
reacted strongly in protest (Figure 9.10), blocked roads, seized the Darwin Station,
held hostage 30 scientists and untold numbers of tortoises, and threatened to release

---

[15] Here, "spiny lobster" refers to both the red and green spiny lobster species that can be found
near Galápagos *(P. penicillatus* and *P. gracilis,* respectively).

Figure 9.10. Protests in 2004 at the Charles Darwin Research Station. *A*, Fishers protest the coming closure of the sea cucumber fishery. *B*, A counterprotest by Galápagos National Park employees. Even with the early participatory management in place, fishers continued to perceive that regulation of their sea cucumber and lobster harvests was imposed by the GNPD and responded in protest as they had for over a decade. *Source:* Photos from CDF archives.

goats on several islands (at what was the peak of Project Isabela eradication activity). When fishers withdrew in protest from the PMB, Ecuador's Minister of the Environment offered to begin a series of direct meetings with fisher representatives to discuss solutions. Despite the representatives' deeply held identity as fishers, the new harvest levels were bad enough that representatives began asking about new career options, or "conversions" into pursuits like ecotourism.

By 2005, both the PMB and IMA officially acknowledged that the co-management system was not working. The IMA approved the concept of recreational fishing, giving fishers a path into tourism by allowing them to sell daily fishing trips to visitors, during which they could showcase their traditional techniques and knowledge. At the same time, on the advice of the special inter-institutional panel, the IMA prohibited all longline fishing in the GMR. There remained widespread disenchantment with the co-management system; quietly, on the side, a technical commission with an inclusive set of members was set up to negotiate a new management plan for key fisheries.[16] In retrospect, this commission was a genuine breakthrough. In the smaller and less formal arena, fishers enjoyed more influence; they actively participated in discussions of regulation and monitoring, offered input from their accumulated local knowledge of fisheries, and helped to set the future research agenda for both BSC and lobster.[17] This was the first forum in which fishers had meaningful influence over the decision process. It was also during a time that local and indigenous perspectives were given increased legitimacy nationally, via a new Constitution in 2008 that promoted an indigenous concept of "good living" for all Ecuadorians (to which we return below) and significantly streamlined the Provincial government of Galápagos with an impressively simple "Governing Council" (*Consejo de Gobierno*).[18] After 3 years of work, the fishing commission reached a new set of agreements—called the Fishing Chapter (*Capítulo Pesca*)—designed to be integrated as a literal chapter in the 2009 Management Plan for the Marine Reserve. By 2011, there were no new regulations imposed by the government, and fishers' protests subsided, marking the start of Phase III (as shown in Figure 9.7).

[16] Observers reported that better efforts were made to hear and incorporate the perspectives of the fishers themselves, including their opinions on needed fisheries research. When the plan was signed in 2009, it was "the first plan unanimously approved by all fishing sector representatives" in a decade (Castrejón et al. 2014, 174).

[17] More recent studies have confirmed the accuracy and value of fisher's local knowledge, including knowledge of cold-water refugia used by fish during El Niño events (Cavole et al. 2020) and of coves used by Galapagos sea lions for collective hunting of yellowfin tuna (Paez-Rosas et al. 2019). By all indications, fishers' knowledge has value for the long-term goals of sustainability in the archipelago, all the more reason to honor their voice in conservation.

[18] The Governing Council replaced the much larger National Galapagos Institute (INGALA) that proved too complex and unwieldy, much like the PMB and IMA.

After So Promising a Start, What Happened?

A decade after the 1998 Special Law, the sea cucumber and lobster fisheries were in shambles—so much so that the sea cucumber harvest was closed again in 2008—and the whole co-management system was in question. Certainly, there were many positive aspects of the 1998 regime: it had been designed through a participatory process, with a clear legal basis; it promoted dialogue among all sectors, with institutional support by the GNPD; it required decisions by consensus at the level of the PMB; and it prompted cross-cutting agreements on principles of operation, coastal zoning, and fishing calendar dates. But on closer inspection, there remained problems in decision-making that challenged the very participatory nature of the process. Analysts before me have highlighted key issues (see Castro 2005; Celata and Sanna 2012; Hearn 2008, Ospina 2006; Stacey and Fuks 2007; Castrejón and Charles 2013), so I draw here upon their insights. As often happens in conflictive situations, all sides seem to have felt their positions victimized to some extent by circumstances and the behaviors of others.

Starting with the fishers' point of view, one can distinguish several major issues. First was the matter of representation. In the PMB, decisions were taken by consensus, with only one representative per sector—that person having the responsibility for information-sharing with their constituency. One problem, fishers felt, was that one leader could not easily represent the divergent interests of four different fishing cooperatives across three islands, with a total membership of over 1,000 during the gold rush days.

Second, fishers felt the co-management institutions, IMA and PMB, did not work as expected—as a kind of negotiation arena where their interests and preferences would be freely heard and discussed. Instead, within these institutions, "rational arguments [had] a . . . dominant and hierarchical base, which is neither questioned nor explicit: there [was] a superiority of science over empirical knowledge. Science [was] on one side of the table" (Stacey and Fuks 2007, 73, brackets added). By arguing from the power of science, conservation and tourism representatives became more powerful actors in the decision-making process, able to bend the outcome to their favor.

Finally, and perhaps most importantly, the fishers perceived the cards as stacked against them from the start. They felt their representation was structurally disadvantaged in every PMB meeting by a more closely aligned vision among the other four representatives, each with a vested interest in conservation and rigorous enforcement of protective measures. The fishers clearly felt the imbalance early on, and more strongly when PMB decisions about the BSC and lobster harvests started to feel like a new form of imposition. When quotas and closures had to be decided, and fishers could not sway the consensus, they used more confrontational tactics to get what they wanted (Figure 9.7, Phase II). Even when, in 2008, the CDRS was legally converted from a voting to advisory member of the PMB, as fishers demanded in 2004, they still saw a 3-to-1 disadvantage in the participatory process. In the fishers' view, there was still imposition and it was still unacceptable. At first, many observers were surprised that the early participatory management phase—from 1998 to 2010—was marked by violent protest much as in Phase I. But in retrospect the protests highlighted the fishers' perception of continuing imposition despite the rhetoric of inclusion. Ironically, Phase II supports the hypothesis even though the institutional setting was designed to be participatory. In both of these phases, Ostrom's rule was alive and well.

The point of view of the conservationists was understandably different and nearly opposite the fishers' perspective.[19] Conservationists often felt themselves "ganged-up against" in the PMB when the tourism sector supported the fishers to prevent their disrupting tourism and making Galápagos seem unwelcoming or unsafe to visitors. At times, conservationists also felt cornered into peculiar positions by arguments to compromise with fishers. One of these was "Give them *pepino* [i.e., give in to their demand for more] so they don't keep asking for [rights to] longline" (Alex Hearn email to author, August 24, 2016), a "deal" that could have had serious implications for both resource management and tourism. As a result, there was significant jockeying in the PMB, especially since diverse members of the group had no training in consensus-building. Meetings lasted many hours and compromise was difficult: "I remember a 16-hour meeting in a room that had no air conditioning, no food, [and] plenty of mosquitoes. . . . I [could] have strangled my grandmother by the end of it had she been in that room. Not enough emphasis was placed on creating consensus scenarios BEFORE the meetings" (Alex Hearn email to author, August 24, 2016).

Ironically, many conservationists felt that the fishers held sway over local politics. Like other provinces of Ecuador, Galápagos then had two Members of Parliament (MPs), but unlike in other provinces, the Galápagos MPs were elected by fewer than 10,000 voters, of whom over 2,000 were fishers and their spouses. Numbers alone thus gave Galápagos fishers disproportionate national political influence at the time, the more so because of their history of activism. In the conservationists' view, local politicians would thus "court [fishers] constantly, stoking conflict for their own personal benefit" (Alex Hearn email to author, August 24, 2016, brackets added). Repeated fishers' demonstrations were the most obvious symptom of a system deeply flawed from the start. Lu at al. (2013, 93) put it well: "Over 170 years after Darwin's visit, it appears that these volcanic islands are still teaching us about the nature of struggle, albeit of a different sort."

---

[19] I am grateful to Alex Hearn, biologist of the University of San Francisco in Quito (USFQ) and participant in many of these meetings, who shared his perspectives with me on the fisheries crisis of the 1990s and 2000s.

New Directions in Fisheries Management

The 1998 system fell under its own weight. In June, 2015, Ecuador's government passed a new "Special Law" for Galápagos (with Spanish acronym, *LOREG 2015*; see Ortega Pacheco and Bustos 2016, Section 4; Neira 2016). The new law eliminated the complexities of the IMA and PMB, replacing them with direct four-way discussions among key parties—representing the GNPD, the Governing Council, the corresponding Ministry (of Agriculture, Cattle, Fishing and Aquaculture), and the fishing community—regarding fishing calendars and harvest limits. Such discussions came close to putting key stakeholders on the same level (Ramírez-González et al. 2018, Figure 2). The law converted the PMB to an advisory council and also created a citizens' Consultatory Council, open to the public, for input to Marine Reserve administration (Article 22). The GMR thus retains a form of participatory co-management, inclusive of fishers, but now simply "consultative" in Ostrom's terms (2009) and thus more informal, less bureaucratic, and much less hierarchical. The overall reaction of fishers has been noteworthy: "Despite these changes, the participation of fishermen in the [new, less formal co-management proved to be] permanent and very active" (Ramírez-González et al. 2018, 23).

Key to the functioning of this new regime, in my estimation, was the precedent set by work on the Fishing Chapter (*Capítulo Pesca*)—in some ways the crowning achievement of the earlier, more complicated system. It provided a genuinely participatory "workshop" atmosphere, more accepting of fishers' input than the formal PMB and IMA structures. Among other things, the Fishing Chapter described management rules and fishery status indicators that used a scheme of "stoplight colors" familiar to all stakeholders: green indicating a "healthy" fishery, yellow, "in recovery," and red, "critical." This color system is part of the Technical Report to the new "Fishing Calendar" (*Calendario Pesquero)* that, as of the 2015 Special Law, replaced the Fishing Chapter (see DPNG 2016, *Anexo* 1). As with its predecessor, the drafting of the Technical Report was participatory, with no fewer than six scheduled workshops, each listing the artisanal fishing sector as the first of the four groups involved.

The objectives for specific fisheries within the Calendar for 2016–2021 read like a compendium of lessons learned in the previous two decades. For example, note these objectives for the spiny lobster fishery, especially the last:

Biological: To recover the abundance of lobster populations, ensuring a healthy population structure.

Economic: To improve the economic profitability of the lobster fishery [in the] long term.

Social: To contribute to the improvement of the life quality of fishermen and their families.

Governance: To strengthen the participation of the artisanal fishing sector, and the capacity of the GNPD and other relevant institutions, in the co-management of the lobster resource. (in Ramírez-González et al. 2018, 21; emphasis added)

The corresponding objectives for BSC fishery management are nearly identical. However, under "methods of management," the Calendar indicates that the BSC fishery is closed for the full 5-year period (*"veda de cinco años"*) because participatory censuses continue to report densities below the threshold standard (11 individuals per 100 square meters).[20] At the Calendar signing ceremony in October 2016, Ecuador's Minister of the Environment said, "we are giving the world an example that, yes, we can contribute to the improvement of the quality of life of our [fishing] compatriots at the same time as we conserve natural resources with productive sustainable activities."[21]

Lessons Learned

A major lesson from the first decades of strife brings us back to Ostrom's rule: imposed state regulation of fishing was neither stable nor sustainable owing to the fishers' steadfast resistance. This hypothesis was supported in Phase I, before 1998, by a perfect concordance: 5 of the 6 "gold rush" years featured imposed restrictions from governance, and the same 5 years featured massive organized protest. The hypothesis also held up in Phase II, from 1998 to 2010, when imposition became a subtler product of a system designed for participatory co-management. In that phase, imposition was prominent in 6 of 12 years, a majority of which (4 of 6) again sparked fishers' protest. The two exceptions occurred in 2005 and 2006, when specific concessions were given to the fishers (i.e., jobs in conservation in 2005, and inclusion in the technical team in 2006) and they did not organize large protests.

An additional, unexpected test case surfaced in Phase III, which began in 2011. Generally peaceful relations characterized the years 2011 to 2015, suggesting that fishers felt their new role to be effective. That sentiment prevailed even when disagreements surfaced during meetings in January 2016 over proposed revisions to the coastal zoning system (conceived in 2000 and implemented in 2006). But a few months later, protest erupted once again when fishers saw that, in the interval since their meetings, without consultation, major new changes had been made to the zoning plan as officially decreed.

---

[20] However, it also says that "if the sea cucumber resource recovers before the five years are up," and thus meets the threshold standard, "the fishery will be opened starting on June 1 [of that year]" (DNPG 2016, 12; my translation).

[21] From the Ministry of the Environment's website (http://www.ambiente.gob.ec/mae-firma-calendario-pesquero-para-cinco-anos-en-Galápagos-como-medida-desarrollo-sostenible/), posted October 25, 2016.

For example, the January 2016 proposal for a small NTZ around Darwin and Wolf Islands in the north morphed, under advice from an international NGO and ignoring fishers, into a large "Marine Sanctuary" (40,000 square km, 15,444 square miles), created by presidential decree. As described in more detail in Burbano et al. (2020, 8), "In early March 2016, the leaders of the [fishing] co-operative organized meetings with all their members to gather support against what they called an 'imposed zoning system'" that eliminated their access to key areas. Following a familiar pattern, fishers' protests followed on San Cristobal and Santa Cruz. "The hardest part," said one fisher, "is to be told what has already been decided" (quoted in Burbano et al. 2020, 9). Fishers' protests opened the way to adjusting the boundaries in their favor, but not the total area of the sanctuary, and delayed its implementation.

Looking back on these events of recent decades, the conclusion seems inescapable: imposition bred resistance during all three phases (I, II, and III) of the timeline, even with three very different structures of governance. The whole episode speaks to the validity of Ostrom's rule. We also see that fishers do comply with new regulations when they have exercised meaningful choice over what the regulations will be.

Another noteworthy lesson concerns the effectiveness of efforts at co-management. Comparing average catch per unit effort (CPUE) for lobster and sea cucumber—a measure as meaningful for fishers as for analysts—one sees a striking contrast over the focal years of this analysis, 1992–2012 (Figure 9.11). As shown in Figure 9.11A, the CPUE figures for lobster show a clear pattern of improvement during co-managed regulations, following an uneven initial period of implementation (reflecting, in part, compromises to fishers' demands). But, in Figure 9.11B, the sea cucumber CPUE figures do not improve over the co-management years, even when stern measures were enacted, such as short harvest seasons and complete closures—an indication that co-management worked for lobster but not sea cucumber. Why this difference?

Three contributory factors were important. First, the durability of dried sea cucumbers provided marketing options not available for more perishable fresh lobster. Second, that durability went hand in hand with the willingness of middlemen to pay ever higher prices as Galápagos sea cucumbers became scarcer—sometimes surpassing $1.50 per cucumber in the early 2000s and reaching a whopping $4.00 each by 2011 (Defeo and Castilla 2012). These conditions, together with fishers' sense of unjust imposition, fueled substantial illegal sea cucumber harvest during co-management implementation.[22]

Third, high prices and illegal harvest combined to bring sea cucumbers to record low densities on the ocean floor. In most heavily fished areas, the densities were below the estimated Allee threshold—the "magic number" of 11 BSCs per 100.

---

[22] Schiller et al. (2013) estimated that 3,000 tons of *I. fuscus* were taken illegally between 1950 and 2011, with the highest concentration of illegal catch between 1994 and 1999.

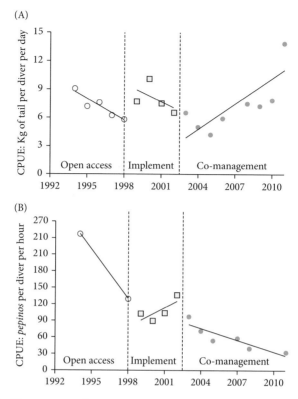

Figure 9.11. The transition from open access to co-management in Galápagos fisheries for lobster (*A*) and sea cucumber (*B*), as reflected in catch per unit effort (CPUE). Using market-appropriate units, the *y*-axes measure changes in the average yield for lobster (1994– 2012) and sea cucumber (1994–2011). The period labeled "Open access" runs from 1992 (the start of the "sea cucumber wars") to the Special Law of 1998; "Implement" refers to the implementation of the PMB and IMA, 1998–2002; and "Co-management" refers to the main years (2003–2012) of participatory co-management under the 1998 law. Linear trend lines are shown for each interval. *A,* Transition to co-management appears to have worked for lobsters; CPUE (and presumably lobster population) rebounded during co-management 2002–2012 and continued thereafter (Ramírez-González et al. 2018). *B,* Transition to co-management did not work for brown sea cucumbers, as CPUE—after slight improvement during implementation—continued to decline out to 2011, where, to this day, they remain too scarce to harvest (with a brief exception in 2015). See text for explanation. *Source*: Data from Defeo et al. (2016) plus sea cucumber data for 1994 from Shepherd et al. (2004), assuming an average of 4.5 hours per diver per day.

The result was fishery collapse, with echinoderm densities so low that knowledgeable observers wondered if, in many areas, they would ever recover. As recently as 2015, they BSCs remained below that the threshold in all but one of the fourteen 14 sites monitored throughout the archipelago; surprisingly the fishery was opened for a short period in 2015, and then closed for up to 5 years. The "roving bandit" middlemen, meanwhile, simply picked up and moved on to other opportunities.

Another key lesson learned was the need for surveillance and the capacity to intercede with speed in cases of illegal harvest or sale within the GMR. Following Ecuadorian legislation in 2007, the GMR now hosts a sophisticated marine surveillance system to monitor the locations of industrial ships, including those of middlemen, in or near protected waters, and the results have been impressive. Figure 9.12, for example, shows the improved GMR protection with respect to pelagic tuna fishing during the focal decades of this chapter.

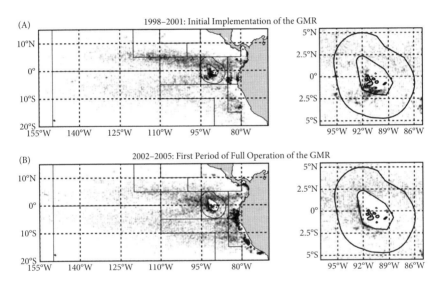

Figure 9.12. Improving protection of the Galápagos Marine Reserve (GMR) 1998–2009. Maps depict the changing intensity of industrial tuna fishing, which was prohibited within the GMR (*innermost polygon*) by the 1998 Special Law. *A*, In the first years of the GMR, protection improved for only a small region in the center of the archipelago. *B*, Protection improved greatly by 2001–2005. Each gray dot represents the setting of one tuna net for yellowfin, skipjack, or bigeye tuna by the main category of tuna-fishing boats (class 6 purse-seine Ecuadorian vessels) allowed to fish within the Exclusive Economic Zone (EEZ; *polygon outside the GMR*). *Source*: Republished with permission of A. Hearn from Bucaram et al. 2018. "Assessing Fishing Effects Inside and Outside an MPA: The Impact of the Galápagos Marine Reserve on the Industrial Pelagic Tuna Fisheries During the First Decade of Operation." *Marine Policy* 87: 212–25. License conveyed through Copyright Clearance Center, Inc.

Finally, notable lessons have also been learned from the transition to the New Special Law in 2015. Close inspection of the new governance rules for Galápagos fisheries reveals additional support for the hypothesis that the products of cultural change via choice have a staying power not enjoyed by those from imposition. What has endured from the tumultuous BSC gold rush are a number of products of genuinely participatory decision-making where fishers could see their clear influence: the collaborative monitoring system, the stop-light indicators, the face-to-face workshops, and even participation of outside NGOs in Calendar planning.

It is fascinating that a place that has taught us so much about the genetic evolution of local endemics now offers examples and lessons of local *cultural* evolution. Although it remains to be seen how effective and long-lasting the newest co-management system will be, evidence already suggests that genuine participation and decision-sharing with the fishers have a greater chance of working in the long run. In addition, today there is a healthy new emphasis on collaborative research agendas for Galápagos (see Izurieta et al. 2018), yet another initiative with precedents tracing back to the Fishing Chapter. Finally, the new, 10-year integrated management plan for the protected areas of Galápagos—the National Park and the Marine Reserve together (see DPNG 2014)—seeks "for the first time in the history of Galápagos, a directionality in its management, highlights the benefits of services from [its] ecosystems, and the need to articulate its sustainability [between institutions]. This management proposal aims at generating positive changes in the resident population, while implementing with social and environmental responsibility, what all Ecuadorians seek: *el buen vivir* ("good living")" (DPNG 2014, 16; my translation).[23] This unambiguous emphasis on management for sustainability, with a high quality of life for all inhabitants, is a worthy lesson from Galápagos for all of planet Earth.

Conservation Implications

A key question remains: Are the changes in fisheries management from 1992 to present enough to prevent another "gold rush" for the same or other species, should the BSC miraculously recover? Some experts are pessimistic:

> What processes are in place to ensure [today] that the same mistakes are not made? I would argue that [little] has changed—there are still 1,000 registered

---

[23] From the Spanish *buen vivir*, in turn from the indigenous Kichwa, *sumac kawsay*, which "implies both material and spiritual well-being and is identified with a life of fullness, ... equilibrium, and harmony with the various dimensions of the human being" (Caria and Domínguez 2016, 19).

fishers, of which only a third are full-time fishers. There is still common access for all fishers to all resources, and there is still no mechanism to determine [let alone enforce] a quota. And finally, and probably more importantly, there is still no mechanism to engage with the merchants [aka middlemen] who, at the end of the day, are the most important stakeholders whether we like it or not, because they drove all the prices. . . . There is nothing to suggest that this would not happen all over again. (Alex Hearn email to author, August 24, 2016; brackets added)

Hearn's arguments have much validity, and they draw further attention to key issues. For instance, his suggestion of "engaging" with the BSC marketing system aligns with broader arguments today for "coupling production-based management systems [for sea cucumber] with processing and export regulations" (Baker-Médard and Ohl 2019, 1). And there is certainly great need for a quota system, to prevent the rush to maximize catch whenever an open season is declared. A quota system would benefit the long-term sustainability of both fisheries. But it is also time to face the reality that it would be difficult to sustain 1,000 active fishers, even with a good quota system, in this little archipelago.

A good first step to accompany a quota system would be to work with fishers and fishing cooperatives to estimate, across a variable range of ocean conditions, how many fishers Galápagos can sustain, while respecting the zoning system, the density management of the Fishing Calendar, affordable technology, a new quota, and "good living." At the same time, renewed efforts are needed in capacity-building, to prepare fishers for attractive positions in other sectors of the economy. Sea cucumbers have taught us that fishing, while important for local food supply and livelihoods, requires regulation for sustainability, which in turn works best when it comes from the active input of stakeholders. The same is true for tourism, to which we now return.

Tourism Revisited

One of the great hopes in Galápagos is that tourism, particularly bona fide ecotourism—with reduced environmental footprint, enhanced benefits for island residents, including fishers, and steadfast support for conservation—will prove capable of reversing the negative feedback cycle introduced at the start of this chapter (Figure 9.4A). If successful, tourism would then move from being one of the key causes of "World Heritage in Danger" to being a major part of the solution (Figure 9.4B). Ecotourism has that potential, I am convinced (and see Stronza et al. 2019), by at least four pathways, none being exclusive of the others.

First, tourism has the capacity to offer supplemental and alternative employment to fishers, a fact not lost on practitioners themselves. In the midst of the tumult of 2004, fishers appealed to Ecuador's Minister of the Environment for ways to help them benefit from tourism's burgeoning success in the islands, which had largely excluded them to that point. Still, today, programmatic effort is needed to facilitate fishers' capacity-building in tourism, and to repurpose their boats and trucks for tourism purposes (see Walsh et al. 2019). We have seen how the PMB/IMA system approved the idea of recreational fishing, whereby fishers can offer fishing tours to visitors, with opportunities to share local knowledge of marine ecology and natural history. Unfortunately, licensing of recreational fishing did not go smoothly (Engie 2015), and today "experiential fishing is no longer the focus, but rather a means for conducting operations that are very similar to [land-based] tours" (Erazo et al. 2017, 98). There remains a need for practical tourism training, environmental education, and language instruction (English is crucial for success in tourism), for fishers and their children. It is one of the best options in the eyes of many fishers.

Second, ecotourism can have a stimulating effect on local public education, helping to shape locally meaningful curricula and incentivizing students of all ages to learn more about the special place where they live (see Cruz et al. 2017a). Between meeting the employment needs of ecotourism, conservation efforts and the broader educational goal of sustainability in Galápagos, the potential is there for the islands to become a recognized leader in environmental education. Educational reforms at the national level in Ecuador since 2006, plus renewed commitment locally by NGOs like CDF, Galápagos Conservancy, and the Scalesia Foundation, are making great strides in this direction (see Creamer and Cabot 2019). Public environmental education of children and their parents remains one of the highest priorities for a sustainable Galápagos, as does strengthening expertise in ecology and evolution among Galápagos teachers (Cotner et al. 2016). Tourism reinforces the value of these efforts.

Third, tourism has the potential to generate larger and more reliable revenue for local communities and for conservation. The most obvious means to this end would be simply to increase the Park's visitor fees for international visitors, keeping existing provisions for local distribution of resulting revenue or revising them as warranted.[24] This measure makes good practical sense. Today, a 7-day visit to Galápagos, a common trip duration and a good standard for comparison, has an admission cost of $100 USD per visit (i.e., under $15 USD per day). Most observers agree that this fee is simply way too low, particularly when compared with a week's entrance fees at other world-class destinations (Figure 9.13). It also undervalues what visitors are willing to contribute toward conservation, as

---

[24] Benitez (2001, 13) noted, "Prior to implementation of the [1998 Special] law, an average of only 30% of visitor use fee income reverted to the budget of the GNPS, while the remainder went to INEFAN" (Ecuadorian Institute of Forests, Protected Areas, and Wildlife). The Special Law required much more to stay locally: 50% to GNPD, 25% to municipal governments, 20% to the Governing Council of Galápagos, and 5% to parochial governments.

Destination	Price (USD)	7-day Price (USD)
Machu Picchu (Perú)	$92 per adult per day[†]	$644
Mt. Kilimanjaro (Tanzania)	$70 per adult per day	$490
Serengeti National Park (Tanzania)	$60 per adult per day	$420
Ngorongoro Conservation Area (TZ)	$60 per adult per day	$420
Kruger National Park (S. Africa)	$23 per adult per day	$161
Galápagos National Park (Ecuador)	$100 per adult per visit	$100
Angkor Wat (Cambodia)	$37 per adult per day	$72[††]
Torres del Paine (Chile)	$32 per adult per visit[†††]	$32

Figure 9.13. Galápagos National Park has a very low visitor fee ($100) compared to many popular ecotourism sites around the world (shown are fees per person for a week's visit). Increasing the GNP fee would be one good way to increase funding for conservation efforts, as would a supplemental sustainability assessment (see text) and stimulating visitors' philanthropy (Ardoin et al. 2016). *Source*: Websites for respective destinations, accessed in June 2018, with currency conversions also 2018. †Price for both morning and afternoon time blocks. ††At Angkor Wat there is a special reduced rate for 7-day visits. †††High-season price per adult per visit.

revealed in willingness-to-pay surveys carried out by Oleas (2008) and Schep et al. (2014). The surveys suggested that the GNPD could *double or triple* its entrance fees with little or no adverse effect on international visitation (provided it includes some program of discounted rates for disadvantaged visitors). The matter warrants fresh assessment in the post-COVID world, but the general point remains: fees for international visitors should not be substantially less per day than at other top-ranking ecotourism destinations. Also helpful would be a graduated scale for the fees, making longer visits less expensive per day, thus encouraging visitors to stay longer and learn more from their experience, plus contribute more to the local economy.[25] As part of this effort, boat-based visits could be encouraged via the fee structure to become "hybrid tours," with at least a night or two on land, thus reducing "leakage" (revenue flow off islands) and promoting local spending.[26]

---

[25] In discussions leading to the 2015 Special Law, a "sliding scale" entrance fee was considered, encouraging longer visits to the islands (see Nicholls 2014, 147), but the proposal was not in the law as passed. More recent discussions factor in the length of stay in mainland Ecuador, as a way to spread Galápagos tourism revenue more broadly. A 2019 proposal would have increased the entrance fee to $200 for those visiting the islands with at least three nights on the mainland and to $400 for those spending only a night or two on the mainland. Following the pandemic, these figures may need to be adjusted, while maintaining the incentive for longer stays.

[26] A related idea would save tourism to certain locations for visitors who make a special contribution of effort or funding to conservation. Locations could include visitor sites not currently included in normal visitor circuits, perhaps Pinta, Pinzón, and Alcedo Volcano on Isabela. In order to visit such locations, visitors would contribute either time and effort, or additional conservation philanthropy, beforehand.

In addition to an increase in visitor fees, I propose a separate and smaller "sustainability assessment" for foreign visitors to the islands, for two purposes. First, the assessment would provide a measure of direct tourism support for the CDF (and its research station), which derives no support from entrance fees but is, instead, funded anew each year, and not always reliably, by donations, grants, and a small endowment (see Online Appendix A). Envisioned from the start by UNESCO as the main way for international support to assist with conservation in this World Heritage site, CDF has been a crucial, independent partner to the GNPD for research and conservation efforts since the Park opened more than 60 years ago. Moreover, CDF is uniquely positioned to play a leadership role in the transition to a sustainable Galápagos, including the Foundation's emerging strength in social sciences, but it sorely needs financial volume and stability to make this possible. Such an assessment seems to me a fitting and fair way to institutionalize international support for CDF at this key juncture in Galápagos history.

The sustainability assessment could also support timely special projects, such as the "payment for conservation services" effort mentioned earlier, funding farmers, fishers, and other locals for specific GNPD-supervised conservation efforts. The value of such a program lies in motivating and educating an array of citizens, not just scientists and specialists, to play active roles in efforts like the planting and care of tree *Scalesias* on farm properties, for example. It's a step toward "democratizing" conservation in the islands and spreading the conservation payroll more widely. The assessment could also make possible conservation efforts requiring special equipment or technologies, like the floating refugium suggested in Chapter 8 for penguins and cormorants. Even at post-pandemic visitation rates much lower than those of the last decade, a modest and separate assessment of, say, $20 per foreign visitor could generate annually a million dollars or more for each of these purposes. The program would also motivate recipients to share with visitors the good work being done with funding from their assessments.

Fourth and finally, tourism is an increasingly important source of philanthropic donations for conservation and community improvements. As shown in Figure 9.3B, such donations have the potential to turn around the systemic feedback of tourism from negative to positive. Even with revised entrance fees and a supplemental assessment, the role of travelers' philanthropy is crucial to support the various initiatives, terrestrial and marine, of the Darwin Foundation and affiliated NGOs, including the environmental educational programs discussed above. A major component of such philanthropy will always be direct contributions from donors (individuals and foundations) to causes they support. But there is a growing second component that pools philanthropic contributions from visitors on group tours. A number of tour operators are already exemplary models for philanthropic "reinvestment" in Galápagos (Powell and Ham 2008; Ardoin et al. 2016), soliciting donations from their guests and/or returning a fraction of their revenues to local causes and worthy NGOs. The practice is admirable, appropriate, and very important to the local quest for sustainability.

Paradoxically, these four pathways warrant action while another profound change is made simultaneously: effective regulation of tourist numbers to the archipelago. The time has come for meaningful limits. An effective system to cap the total, with limits on numbers of boat-based and land-based visitors, including a lottery-like component to ensure equitable access, is essential for the future of Galápagos.[27] Yes, the price of average admission can certainly be raised for international tourists, which will help, as will the sustainability assessment proposed above. Yes, new tourism products should be offered, including those favoring the skills and knowledge of fishers, farmers, and their families. And yes, added consideration must be given to need-based and educational discounts to continue visits by a broad spectrum of visitors to this special place. But now is the time—especially in the wake of the COVID-19 pandemic—to work out new, effective, and equitable limits on visitation.[28] Not surprisingly, limits are also recommended by a recent UNESCO "monitoring mission" to the islands (Lethier and Bueno 2018, 8).

A related point warrants emphasis: although total tourism growth was steep in the decades before the pandemic, growth in responsible educational tourism lagged behind—if anything, it could use help (see Ortega Pacheco and Bustos 2016, 34–35, on Ecuadorian policies to this end). The time has come to reemphasize the importance of visitors' stepping up to help Galápagos, through what we might call collective conservation, akin to crowd sourcing, as a fundraising strategy. Imagine the benefits to Galápagos if individual visitors were to take small-scale responsible actions before, during, and after their visits that would then be multiplied across thousands of visitors.[29] Collective conservation

---

[27] Galápagos has a long history of ineffective limits to tourism, tracing back to the first National Park "master plan" in 1974. As summarized by Moore and Cameron (2019, 144), "That plan recommended limiting tourism to 12,000 tourists per year. Subsequent plans raised that limit to 14,700 (1978), 25,000 (1982), and 41,767 (1984), and more than 200,000 (2017)." The economic benefits of more visitors, although seductive, can be easily matched or exceeded by modest increases in entrance fees for foreigners.

[28] Before the pandemic, an impressive effort was made to explore feasible limits for tourism using a "participatory systems dynamics model," described in Walsh and Mena (2016) and Pizzitutti et al. (2017). The process and results, described in nontechnical terms by Tullis (2016), generated alternative scenarios of varying levels of risk, one of which produced stable revenues into the future by limiting tourism to 242,000 annual visitors. Although the Ecuadorian government invited the effort, it was never acted upon. It would be useful to repeat the analysis in the post-COVID-19 period, using appropriate new assumptions and allowing tourism to build up anew from low levels.

[29] A "collective conservation" website for Galápagos would help, with concrete steps that visitors could undertake as part of a responsible visit to the archipelago (of value, perhaps, to other World Heritage destinations, too).

actions could include reading and study before travel (to enhance the learning experience, and to know best practices for reducing stress to native organisms); choosing responsible tour companies and hotel operators whenever possible, such as Smart Voyager Certified providers;[30] arranging land-sea "hybrid tours" whenever possible to promote local earnings; and considering appropriate acts of travelers' philanthropy to bolster overall conservation and sustainability efforts in the islands. Responsible educational tourism, boosted by the collective conservation actions of individual visitors, is surely one of the most sustainable livelihood options for Galápagos.[31]

A chance to fully consider such measures provided a proverbial "silver lining" to the dreadful 2020 coronavirus pandemic. The virus hit Ecuador hard, including an imported outbreak in Galápagos, bringing tourism and daily life to a standstill in a matter of days. While it certainly had sobering impact on the people and economy of the islands, the pause also opened up some time and space for discussion and reflection outside the annual flood of tourists. It sharpened attention to the matter of food sovereignty in the islands, for example, because supplies ran low when cargo shipping was temporarily interrupted. It became clear that more thinking and action are needed to sustain local food production and distribution and to reduce the overall dependence on imports. The lull also provided opportunities. One was to see the effect on flora and fauna of the sudden and enduring disappearance of people from visitor sites. Places that, on a given day before the pandemic, may have had 100 human visitors by noon, and another 100 in the late afternoon, suddenly had not one. Is our repeated daily presence really as benign as we like to think? Now we can answer that question. Another was the opportunity to take stock and consider, without the press of normal operations, desired changes in tourism, like those raised here, and more. Once local COVID-19 transmission was curbed, the islands had the rare chance to "reboot" tourism in positive new directions. Let us hope that the decisions going forward will make wise use of the opportunity arising from the pause.

In the long run, there are two paramount goals for Galápagos tourism. One is for visitors to learn for themselves, first hand, a sample of the awe-inspiring

[30] Smart Voyager, initiated in 1998, is a program of certification for sustainable tourism in Galápagos and South America. Certification considers "rigorous environmental, social, and security norms, that, when completed, guarantee a decrease in the environmental impacts that a tourist operation may generate, ensuring direct benefits to the local population and an active participation by the tourist in the conservation of natural resources" (see http://www.ccd.ec/turismo).

[31] This is not to argue against occupational diversity: indeed, too much dependency on tourism makes the economy vulnerable to recessions, epidemics, international conflicts, and more. As argued earlier, both fishing (this chapter) and agriculture (Chapter 4) are important for local food sovereignty, not to mention the importance of livelihoods in conservation and sustainability research, as exemplified by the GNPD and CDF, and also the Galapagos Science Center (see Online Appendix A). Going forward, thought should be given to additional, complementary financial "engines" for Galápagos, perhaps including an international center for sustainability science.

lessons Galápagos offers, and to spread those lessons far and wide as we seek ways of living sustainably on planet Earth. Tourism is a mutually beneficial way to share the test bed of our future that is Galápagos. The second goal is for eco-tourism to provide a sustainable, strong revenue stream for the local economy, one that can contribute positively both to conservation of the showcase biota and to sustainable livelihoods for the people who live and work there. To my way of thinking, a sustainable social-ecological system in Galápagos is in the interest of all humanity. It is in our collective interest to conserve the flora, fauna, and habitats of an archipelago that continues to teach us so very much about life on Earth. And it is definitely in our interest to have an instructive, positive example of the ways human beings can be part of sustainable social-ecological systems so that many more can be devised on the planet while there is still time. The world today needs good examples, and Galápagos has the potential to be a front-runner. It is important that we individually and collectively support efforts toward conservation and sustainability in this special place.

10
One Big Social–Ecological System

Just as the Galápagos became a model for Darwin, so it is a model for
anyone who cares about our future. . . . The Galápagos really matters
because what happens in these islands is an honest, unfalsifiable look
at the future that faces our children, our grandchildren and our species.
Henry Nicholls (*The Galápagos: A Natural History*)

This book aims to make three points above all others. First, it shows that understanding an organism's evolutionary history is a fundamental tool for working toward its conservation. The organism's evolutionary history reveals the organism's unique features and strategies for coping in the isolated Galápagos archipelago of old, and it helps us to understand the organism's challenges as it confronts the globally integrated Galápagos of today. It often provides surprises and suggestions we haven't yet considered, such as the potential benefits of the joint conservation of tree *Scalesias* and tree finches, or a new explanation for peculiar behaviors in Nazca and blue-footed boobies and how that can benefit their conservation.

Second, the book shows convincingly that the survival and reproduction of Galápagos organisms are affected today by human activity. We must accept that Galápagos is no longer a collection of pristine natural ecosystems, and it has not been so since about 1535. The "little world within itself" that Darwin visited is now one big social-ecological system, or better, a nested set of SESs. Yes, Galápagos retains an estimated 95% of its original terrestrial biota, but they are joined by over 35,000 diverse human inhabitants, hundreds of introduced species, incessantly interacting social and ecological subsystems, and intimate daily connections with the forces of globalization, including two to three hundred thousand annual visitors in the years before the coronavirus pandemic. As a result, human activity is implicated, for example, in the struggles of waved albatrosses to survive and reproduce. Humans are involved in the population fluctuations of penguins, cormorants, and many other species. Of course, our impact is not all negative, as attested by the tortoise populations painstakingly bred under supervision to replenish their numbers on Pinzón, Española, Santa Fé, and (soon) Floreana. The point remains that the natural history of Galápagos species is now a *social* and *ecological history*. We're part of it all.

Exuberant Life. William H. Durham, Oxford University Press (2021). © Oxford University Press.
DOI: 10.1093/oso/9780197531518.003.0010

Third, the chapters in this book document a sustainability urgency. Chapters show that Galápagos rails have recently gone under, or nearly, on two islands, and that there has been lasting damage to the sea cucumber fishery, where sea cucumber densities may take decades to recover in some areas—if ever. The text explains that marine iguanas have grown larger since the early 1900s, but probably not fast enough to keep pace with the changing pulse of El Niño in this time of accelerating climate change. And the text argues that social issues of the archipelago —including alternative employment options for fishers, limits on tourism growth, improved food sovereignty, and compelling needs for environmental education—call for urgent attention. It's clear we must act quickly but wisely, putting to work the tools of evolutionary analysis, the insights from SESs, and all the economic resources that tourism and philanthropy can provide, before there is more irreparable damage to the unique World Heritage site that is Galápagos. Moreover, we cannot expect the hard-working Galápagos conservation community to handle it all alone, cleaning up after over 400 years of global open access to the islands. If our caring is commensurate with how much we value and learn from these islands, we must take action now. We can each do our part. All is not lost, but it is high time for us to pitch in.

Take-Home Messages

The following general themes stand out from the previous chapters, and I list them even as I also emphasize their tentative nature. Galápagos tells us plainly that neither organisms nor knowledge remain indefinitely as we saw them at one time. Here, again, I focus deliberately on themes pertaining to organisms of the terrestrial realm, to those at the interface of land and sea, and to our human impact. I leave to others the wonderful world of Galápagos marine life, with its abundant similarities to, and differences from, what this book describes.

Key features of terrestrial Galápagos derive from its hot spot origins in the eastern equatorial Pacific amid moving oceanic plates. Today's islands are simply the current "frame" of a volcanic movie going back more than 17 million years in the same remote area of the Pacific Ocean. The changing volcanic islands of the region have always been isolated from the continents by approximately 1,000 km and, in most cases, isolated from one another, making each a separate petri dish for colonizing organisms. The islands come and go independently in time, giving each a unique size, age, altitude, proximity, and position in relation to ocean conditions in the region. Some are low, generally dry and desertlike; others are high, with abundant moisture at altitude; all of them offer unusual challenges.

The convergence of trade winds at the equator bathes the islands in a unique system of air and water currents of profound ecological impact. Normally, the main wind and ocean currents flow swiftly east to west, providing a "conveyor belt" for continental colonizers to reach the islands. The currents bring with them differences in water and air temperature, aquatic micronutrients, and terrestrial moisture. One of them, the Cromwell Current, flows well below the surface, west to east, producing a cold upwelling against the western edge of the archipelago, supporting a uniquely rich food web for a mid-ocean location. But every so often, things reverse: the surface currents move west to east, warm water settles around the islands, the Cromwell upwelling and productivity are stifled, and a deluge of rain falls on the islands. Galápagos is virtually the center of the El Nino-Southern Oscillation (ENSO), adding periodic shock waves to its already serious environmental challenges. Putting it all together, this setup has made Galápagos an unparalleled "natural experiment" for terrestrial life. The fascinating organisms we see are simply the results so far.

Colonizing organisms came to Galápagos from many points of origin, by diverse means, and at widely disparate times. Aided by the winds and currents, colonizing organisms came to Galápagos from American continental areas as far north as Baja California and as far south as Tierra del Fuego. A few extra-hardy travelers made it all the way from Asia, riding a more northerly current from west to east. Of the total, we infer that some floated or rafted out, others flew or hitched a 'free' flight, and a few blew out on strong winds. The small pool of organisms who successfully colonized the islands was thus a highly selective subsample of hardy continental organisms, and surely even selective again of those who started out. Similarly, estimated arrival times are widely spread out over the last 13 MY, becoming more concentrated in the last 4 MY.[1] If there truly have been islands in the area of today's Galápagos for over 17 MY, as geological evidence suggests (see Werner and Hoernle 2003), why weren't there more early colonizations?[2] This is a topic that only further research can illuminate.

Colonizing organisms encountered further harsh conditions as they began life in Galápagos. The conditions included swift ocean currents, fresh lava flows, hot sandy beaches, steep and irregular basaltic cliffs, and thoroughly dry, desertlike conditions in low-lying areas. Galápagos does have humid zones on islands reaching above 300 m (1,000 ft) or so, where there is both seasonal rainfall and cloud interception. That means each moist area is surrounded by a natural "moat" of lowland desert habitat, in addition to the obvious "moat" of ocean, making

---

[1] Some species—like ancestors of the leaf-toed geckos and waved albatrosses—arrived early, before 13 and 5 MYA, respectively; others arrived as recently as 126,000 YA (Galápagos hawks) and 84,000 YA (Galápagos martins). For insightful reviews of colonization, see Parker (2018, Chapters 2 and 3).

[2] Possible explanations abound, including (a) a major catastrophe that wiped out terrestrial life of the time, such as a simultaneous eruption, or a complete subsidence, of all former islands, and/or (b) different patterns in currents of wind and water, impeding crossing from the continent or providing less upwelling-based food for those that made it.

them "botanically one of the most isolated regions in the world" (Kastdalen 1988, 8). Galápagos thus came to be colonized by a selective group of unusually tough organisms, among them distance-flying birds, desert-adapted reptiles, durable insects and spiders, and hardy xeric plants. No terrestrial mammals made it, other than a few hardy rice rat species (Castañeda-Rico et al. 2019) and an endemic subspecies of the eastern red bat (Jiménez-Uzcátegui and Snell 2016).

Successful colonists were thoroughly reshaped for life in this new little world. The demanding ecological conditions of Galápagos, including its cadence of diverse and unpredictable perturbations—such as El Niño floods, La Niña droughts, eruptions and lava flows, fires, island subsidence, and the periodic arrival of new colonizing species—heavily impacted the survival and reproduction of the early inhabitants, as indeed they still do today. What may seem like very small differences among individuals—a few millimeters in beak dimensions or in tail flatness—have translated into survival and/or reproduction differentials under such conditions, and hence into fitting illustrations of evolution by natural selection. Iguanas remain an outstanding example, stemming from a single colonization and diversifying to one sea-foraging and three land-foraging species, each a testimonial to the harsh conditions of life in the archipelago. Galápagos is thus a showcase for unusual adaptations produced by Darwin's "main means of modification" (1859, 6) operating through demanding and ever-changing local conditions. It's a restless archipelago: change itself is endemic.

ENSO is responsible for many of the most impressive and unusual adaptations. Who would have imagined the marine iguana's concertina-like growth and shrinkage with its food supply, not to mention its post-El-Niño fertility resilience? Or the cormorant's unique mating pattern in response to the cold upwellings of La Niña? The cormorant's mating stands as one of just a handful of polyandry cases in all of ornithology. Consider also the giant daisy forest that topples and regrows in synchrony with the swings of ENSO, requiring a dry period for its seeds to germinate after the tall trees are down. Many of the extraordinary adaptations in the archipelago relate somehow to the climatic swings of the ENSO cycle. Visiting the islands today, for short trips with all the comforts of the modern world, one can easily underestimate the environmental rigors that shaped the unique organisms of Galápagos.[3] The cases presented in this book offer a clear sense of the unique properties of the islands and of their truly powerful "creative force" in evolution.

Because mammalian predators and competitors were not among the original terrestrial challenges of Galápagos, many organisms evolved an exuberant nonchalance. The uninhibited zeal of many Galápagos organisms is a wonder to behold,

---

[3] Sometimes even experienced fieldworkers are caught off guard or underprepared for the challenges of terrestrial Galápagos (examples are discussed in Sulloway 2005 and 2006).

from the exaggerated beak fencing of the waved albatrosses, to the showy parading and sky-pointing of boobies. It is truly as if they all "let down their guard"
to the dangers of continental regions. It is not unusual for Galápagos visitors to
have to side-step boobies, iguanas, or sea lions nonchalantly resting, walking,
or even mating in the middle of a tourist path. These encounters with wildlife
impart a feeling of being more "at one with nature" in Galápagos than almost
anywhere else on Earth. However, exuberant nonchalance also entails naiveté to
the dangers of the larger world, including terrestrial predators, competitors, and
diseases that humans have introduced. Our visits today entail a serious responsibility to avoid contributing further to these threats and, even more importantly,
to help reduce them.

*When plant and animal colonists reached multiple islands, evolution often
promoted island-by-island diversification.* Because conditions vary substantially between the islands, the artistry of evolution, if we may call it so, had a
different canvas to start with on each island. Moreover, a number of "evolution
accelerators" at play in small founding populations, such as drift, founder effect,
and inbreeding, joined with natural selection to produce rapid divergence and
adaptation among different island populations from the same source, giving
rise to adaptive radiations. The radiations—including the tortoises, finches, and
Scalesias—are responsible for much of the impressive biodiversity in Galápagos.
In many cases, the radiations took place with record-breaking speed (Valente
et al. 2015)—Darwin's finches remain the classic example, reaching 18 recognized species within 900,000 years of first divergence. But so far, the record for
number of species is held by diminutive land snails of the genus *Bulimulus*, with
more than 80 described species related by descent from one common origin
within the last 3 million years (Parent et al. 2008; Parent and Coppois 2016).

*Sometimes adaptive radiations were interactive, prompting the coevolved
adaptations or "co-divergence" of two or more species.* In Galápagos as elsewhere,
two or more species may evolve interdependent adaptations, as exemplified by
coevolved adaptations among endemic saddleback Galápagos tortoises and the
arborescent *Opuntia* of low islands, or between the ground finches, especially
cactus ground finches, and the same *Opuntia* hosts. The latter case of coevolution provided a model for preliminary tests of the hypothesis that tree daisies
and tree finches coevolved as they opened new niches for one another. When
one considers the additional documented mutualistic relations between birds
and reptiles—for example, tortoises and iguanas carefully lift themselves to a
posture that "invites" cleaning of ectoparasites by finches and mockingbirds[4]—it

---

[4] The "cleaning symbiosis" between tortoises and finches was described in 1974 by MacFarland
and Reeder; that between land iguanas and finches or mockingbirds was described in 1984 by
Eibl-Eibesfeldt.

becomes clear that quite a few coevolved adaptations were shaped over time in this isolated little world, often in surprising combinations. A study by Olesen et al. (2018) reinforced the point, finding that birds commonly serve a double mutualist role for plants, both pollinating their flowers and dispersing their seeds, for example.

But sometimes radiations did not happen at all: colonists settled on various is-lands, and evolution reshaped them in various ways, but they did not diversify into different species. These cases, too, have much to teach us about evolution on is-lands. For example, in the case of the marine iguanas, after millennia of evo-lution, populations on diverse islands differ only as subspecies; therefore, they are still capable of interbreeding. Comparative genetic analysis reveals several million years of stasis in their phylogenetic history, which means the subspe-cies differences are amazingly recent. Several worthy suggestions exist to explain this finding, but we'll need more information to know for sure why differenti-ation was delayed. In the case of Galápagos rails, we infer that descendants of the colonists evolved slightly heavier bodies and shorter wings, thus becoming reluctant flyers but not flightless. More research is needed on their flying capa-bility, but suggestions of nocturnal interisland flight help to explain rails' sudden appearance on Pinzón, their quick rebound following goat eradication, and their continued status in the islands as one species with, in great contrast to the iguanas, no subspecies. Finally, in the case of Darwin's finches, diversification is well on its way—enough that we recognize many morphospecies—but their repeated hybridization indicates that many remain "incomplete species." The finches are thus in a precarious state: human-induced changes in habitat, diet, climate, and invasive species, or all of the above, may well deflect their evolu-tion from pre-settlement directions, blurring morphological distinctions that evolved over millennia before humans arrived. And there is more ...

Although relatively new, human impact is reducing isolation and changing Galápagos ecosystems in profound ways. Human activity in the islands dates from a first visit in 1535, which literally put them on the map for the exploitation of tortoises and sea lions by buccaneers, whalers, and hunters. Lasting human set-tlement began in 1832, adding agriculture and coastal fishing to the mix and leading to endemic population reductions (e.g., of tortoises), habitat modifica-tion in the highlands (e.g., loss of *Scalesia* forests), and increasing numbers of invasive species (e.g., feral farm animals). The advent of commercial fishing and tourism in the 1970s sparked new waves of immigration and marine exploitation that fed into the social turmoil of the 1990s and 2000s. Thankfully, back in 1959 a community of scholars and conservationists laid the groundwork for a national park in Galápagos and set up several hard-working conservation institutions to help, including the Galápagos National Park Service and the Charles Darwin

Foundation. Thanks to their efforts, and to the relatively late human settlement of the archipelago, anthropogenic extinctions are still relatively few in Galápagos.[5]

As Galápagos conditions change, formerly adaptive traits become vulnerabilities. Included are both the general vulnerabilities that Galápagos organisms share with sibling species from continental areas (e.g., all flight-capable birds are vulnerable when perched on the ground), and the specific vulnerabilities from novel adaptations to Galápagos (e.g., adaptations that emerged in the absence of terrestrial predators, such as the evolved flightlessness of the cormorant or the egg-rolling behavior of waved albatrosses). Though often intriguing and even endearing to us today, many specific vulnerabilities leave organisms susceptible to anthropogenic perturbations. Evolution has no anticipatory mechanism and, in Galápagos, the past was not always helpful preparation for the changes that have taken place since people arrived in the islands. The evolution of exuberance leaves island organisms fully exposed.

Ironically, yesterday's adaptations to Galápagos are part of today's greatest conservation challenges. For example, the adaptation of a scrubby daisy to the humid highlands of Galápagos put lofty giant daisies right in the middle of the hopes of agricultural settlers on various islands. Not only were square kilometers of *Scalesia* cleared for crops and pastureland, but their soft wood was easy to use in building service structures. Moreover, the *Scalesia* life history pattern is a close match to the ENSO cycle, and *Scalesias* sprout and die in synchrony with pulses of drought and rain. When farmers imported continental tree species, many were shade tolerant and thus readily colonized the understory beneath mature daisies. When the daisies died back at the next El Niño, the introduced trees inhibited the growth of shade-intolerant daisy seedlings, and the invaders took over. The adaptation of tree finch species to *Scalesia* habitat was the next vulnerability to be exposed, as their numbers declined in turn. So today, to help the tree finches, one must also control the invasive hardwoods and bring back the daisies. It's a point that Galápagos makes especially well: an evolutionary perspective is key to effective conservation. Otherwise, we risk overlooking key relationships and the vulnerabilities that come with them.

Past environmental perturbations have endowed Galápagos organisms with different forms and degrees of resilience, in addition to those they had at colonization. The list is most impressive, ranging from specialized stress physiology (marine iguanas), to responsive mating patterns (cormorants) and quickly rebounding

---

[5] Of 109 native and endemic vertebrate species, 14 (12.8%) are now extinct (including the Pinta tortoise, with the 2012 death of Lonesome George). Six species (5.5%) went extinct prior to the arrival of humanity, and eight have since gone extinct (7.3%). The total, 12.8%, is remarkably low compared to almost anywhere else, but some of that is just plain luck. Today, several vertebrate endemics are hovering precariously near the brink, including the mangrove finch, Charles mockingbird, and some populations of vermilion flycatcher (based on tallies in Jiménez-Uzcátegui et al. 2008, updated to 2020 by the author).

reproductive capacity (penguins), to abundant and durable seed production (*Scalesias*) and seasonal migration (domed tortoises). Such resilience gives us hope, for Galápagos organisms specifically, and perhaps for life on Earth more generally. But these resiliences evolved in response to the natural perturbations of the islands, not in response to the perturbations that humans brought. The resiliences may well help with the latter, but we cannot expect them to bail us out. The remaining gap in resilience is our responsibility.

The effects of climate change are already serious in Galápagos. Marine iguanas are one example, where the unique adaptation of quiescence and shrinkage during El Niño events favors larger individuals who have the biomass to ride out the duration of food scarcity. As climate change makes El Niño events more frequent and longer lasting, the advantage of relatively bigger iguanas holds within each of the archipelago's diverse populations. Additional analysis is warranted, but it appears that natural selection therefore favored an increase in the body sizes of local populations over the interval from 1906 to 1997 (Figure 5.9). Climate change is further implicated in the case of penguins and cormorants, whose numbers are heavily affected by El Niños. As El Niño events crowd together in time, there is less chance for the La Niña phase to allow recovery, even with potential year-round mating and facultative polyandry. The recent decline of blue-footed boobies is another indicator of changing climate in the eastern Pacific. As more frequent El Niño events inhibit the cool-water sardines on which the boobies thrive, and sometimes trigger the warm-water phase of the decadal oscillation, the boobies truly will be *bobo* if they fail to leave what may have been the homeland of their specific adaptations. The Galápagos of the future would hardly be the same without these species.

Because human settlement is not even 200 years old, human social and cultural adaptations to the archipelago remain actively under way today, with key implications for sustainability. A leading example comes from fisheries management, which was unregulated until the "sea cucumber wars" and ensuing changes in governance in the late 1990s. When fishers immigrated at the height of an international boom in sea cucumber trade, government efforts to impose restrictions were met with strong resistance. Protests continued until fishers were invited to participate in decision-making, and even then, the path forward was as rocky as a Galápagos tourist trail. Only since fishers have been actively involved in writing the rules for the Fishing Chapter (*Capitulo Pesca)* of the management plan has co-management enjoyed a smoother ride. The fisheries example emphasizes that local voices must be built into the decision process for the results to last. It would be wonderful if the lessons from fishing regulation could be drawn upon now in guiding the local cultural evolution of tourism regulation, especially with the "pause" in tourism provided by the coronavirus pandemic. There is little time to waste.

Galápagos illustrates the importance of an SES approach to conservation. The SES approach ties together the social and ecological processes that influence the survival and reproduction of focal species, such as in the case of waved albatrosses. Their fertility is clearly affected by El Niño, by mosquitoes, and likely by tourism on Española. Similarly, their mortality is now a function of ecological variables plus longline fishers and the shiny catch of gill netters in the Peruvian upwelling. In another case, the little Galápagos rail all but disappeared when its skulking habitat was eaten away by invasive goats, but the rail came running—and flying—back when goats were eliminated, only to face expansive new tangles of invasive blackberry and stifling quinine. And a final case is the ebb and flow of annual brown sea cucumber densities, where off-season natural rebounds were eventually overwhelmed by overzealous human harvests, depleting the stock. In each instance, the evolutionary fitness of a native species is now a product of both natural conditions and human activity. Humans are integral elements of the SESs in which we live, in Galápagos and elsewhere. We must now manage both social and ecological aspects if the islands and their biota are to have a sustainable future. The SES framework helps with that management, and with understanding the coevolutionary dynamics between social and ecological subsystems as both respond, going forward, to changing environmental conditions.

Important scientific findings and principles continue to pour in from this land of natural experiments. Recently, we have learned from Galápagos that evolution by natural selection can take place rapidly, as demonstrated, for example, by the impact of a 1-year drought on medium ground finches. Other work on ground finches suggests that new species can also appear rapidly via hybridization, even in a single generation. We have learned key lessons about seabirds, such as Anderson's rule that "food controls the breeding success of tropical seabirds" (Tompkins et al. 2017, 3), and about plant-based carotenoids crucial to the mating success of boobies, iguanas, and other species. We have seen both sides of ENSO for penguins and cormorants: their vulnerability during El Niño events, but also their resilience with La Niñas. And this partial list doesn't even touch upon what the islands have recently taught us in other fields, such as geology, marine biology, biogeography, and pest management. It's a safe bet that some of the most important lessons of all are still to come, related to sustainability.

Advances in conservation have been huge, but there is still much to be done.[6] Invasive species are still the single biggest threat to Galápagos natives. With

---

[6] Major accomplishments were nicely summarized by Barry (2018, 2), who noted that "Over the decades, scientists and conservation managers in Galápagos have set [and reached] many seemingly impossible goals." Her list includes Project Isabela for goat eradication, the biocontrol of cottony cushion scale, the Giant Tortoise Restoration Initiative, and the ongoing restoration of Floreana Island. Hopefully, the control of *Philornis downsi* and the rebound of the mangrove finch population will soon be added to the list.

plants, the problem is all the more serious because introduced flowers and fruits are often highly seductive for native species. In fact, native birds of Galápagos interact with a much wider range of plants than do their relatives on the mainland—in what is called "interaction release" (Traveset et al. 2015), which serves to spread the newcomers far and wide. Inadvertently, tortoises also become change-makers through invasive seeds they took in and carried along, especially now that climate change is adding moisture to the lowlands (Ellis-Soto et al. 2017). On the other hand, efforts by dedicated staff of the GNPD, the Darwin Station, conservation NGOs, and, increasingly, local civilian groups add up to an amazing resilience of their own, combating invasives in the service of native species' conservation. Thanks to these heroic efforts, the area of *Scalesia pedunculata* forest is increasing today for the first time in many years (personal observation).

Conservation can only succeed in the long term with the support and participation of the local population. Although urgent, short-term protective measures, like Project Isabela, can sometimes be achieved without community support, the turmoil of the 1990s and 2000s—when protesting fishers went so far as to slaughter tortoises and to release goats on recently cleared islands—emphasized that successful long-term conservation efforts require "buy in" from local people. When conservation efforts fully include locals from the start, bringing benefits to them as well as to targeted plant and animal species, those efforts become far more effective than conservation practices of the past. I call such initiatives *sustainability efforts* to emphasize that the goal combines human welfare and the welfare of other species. How can we go about implementing more and better sustainability efforts in Galápagos? This book aims to serve as a springboard for improved ideas of the kind.

One key to sustainability lies in promoting recognition that all livelihoods in Galápagos are biodiversity-dependent. Work in tourism makes the case plainly: recognizing the link of one's livelihood to the native biota of the islands readily promotes the commitment of tourism workers to the biota's future and to sustainability thinking. But sustainability work is neither easy nor automatic: it definitely takes training and experience, as witness the impressive efforts under way today in the name of Education for Sustainability (Creamer and Cabot 2019). But emphasizing the biodiversity dependency of livelihoods has two key benefits. One, it makes clear the intrinsic incentive in the islands to take biodiversity protection seriously.[7] Two, it builds on the self-reinforcing

---

[7] As an example, consider the work of the naturalist guides of Galápagos. With livelihoods as GNPD employees, their concern for local biota is self-reinforcing. They have a clear stake in the conservation of ecological and social systems—a stake that clearly motivates their understanding and activity—and it works. One would be hard-pressed to find anywhere a group of more dedicated conservationists.

rewards of learning that Darwin appreciated years ago: the more one learns about Galápagos, the more one wants to know. Sustainable fishing is another example: it's a livelihood that incentivizes concern for the future of fish stocks and marine ecosystems, except when bloated market prices get in the way. Efforts to build recognition that all human livelihoods in Galápagos depend on sustained native biodiversity will go a long way toward mobilizing local support of sustainability. The point to be made is that everyone living in Galápagos is a stakeholder in its biodiversity protection (see Figure 10.1).

Regulated tourism is key to a sustainable future in Galápagos. As the largest and potentially most sustainable of commercial sectors in the archipelago, tourism must become a leading force for sustainability. In some ways it is already, but other measures are needed right away, such as increasing the Park entrance fee, adding a sustainability assessment (that would support the Darwin Foundation's sustainability work, and also special projects, like "payment for conservation services" in Chapter 6, artificial islands for penguins and cormorants in Chapter 8, and other innovative efforts), and limiting the total number of visitors to the archipelago, in both boat-based and land-based tours. Galápagos tourism needs to earn its keep as bona fide ecotourism, benefiting the livelihood of locals and contributing substantially to conservation and sustainability. Visitors, too,

Figure 10.1. "Let us conserve what is ours" is carved into a bench where Galápagos sea lions rest, along the esplanade of the town of Puerto Baquerizo Moreno, San Cristóbal.

need to realize the importance of their responsible decisions, so that collective conservation can become a major force for sustainability going forward.

The goal is not to return Galápagos to some hypothetical "pristine state:" rather, the goal is to manage human impact for sustainability. The time has come to curb human impact in Galápagos in order to make the archipelago indefinitely livable for its native and endemic species, and for its legal residents. Eradication efforts like those of Project Isabela may be necessary for especially harmful invasive species, but it is neither realistic nor desirable, given the high cost, to eradicate all introduced species. A more realistic goal is to reduce the rates of extinction and introduction as close as possible to natural rates, which can be estimated following methods like those in Tye (2006), and to prevent further anthropogenic habitat conversion and degradation. The goal must be to bring conditions close to indefinitely livable for all historically legitimate inhabitants of the islands, including both native biota and legal residents, their descendants and their symbionts (e.g., controlled farm animals and plants). Making Galápagos sustainable is one of the big, worthy challenges of our time, and one that would yield valuable lessons for other places.

What happens in Darwin's little world matters across the planet. Finally, we return to the question posed at the start: what makes Galápagos worth it, given all the effort required to carry out successful conservation and build steps to sustainability on tiny volcanic slopes way out in the Pacific? Some authors would justifiably say that the native organisms and habitats of Galápagos are so unusual that they intrinsically warrant our efforts. Others would surely cite the value of scientific knowledge that has come from Galápagos. A related argument points to the educational value of tourism in the islands: if one is well prepared, open-minded, and attentive, one can learn and retain more about evolution and ecology—and quite possibly cultural dynamics—in a week's thoughtful visit to Galápagos than one can learn in a month of study via media or classrooms. Moreover, lessons learned in Galápagos are often helpful for conservation efforts elsewhere. Lessons from Project Isabela, for example, and from Galápagos fisheries are widely discussed among conservationists and are used in thinking about other locations.

The evidence presented here is confirmation of *all* these arguments. The world gains these things and more from the islands, and their value will surely grow as our understanding of Galápagos evolution and sustainability improves going forward. It was no accident, after all, that Galápagos was listed among the very first World Heritage Sites by UNESCO: by 1978, the archipelago had already given the world enough insight and inspiration to warrant special recognition. But there are two further arguments that should be highlighted.

Galápagos offers us a microcosm of the whole. At first glance, Galápagos seems totally unique. It's small and isolated, its habitats are extreme, its wildlife is peculiar, and its initially tiny human population is diverse and growing rapidly. But on a larger scale, are those not all true about Earth itself? The small size and isolation of Galápagos may exacerbate some threats and challenges, its position in the heartland of El Niño may exaggerate the swings and changes of contemporary climate, and its land area for human use, just 3% of the total, is close to a record low—although human impact reaches to every corner of the archipelago. But these conditions make both problems and solutions show up faster in Galápagos than elsewhere. Galápagos is a valuable test bed for the sustainability challenges facing the whole of planet Earth. It's a worthy "model for anyone who cares about our future" and the future of all species currently sharing the planet with us (Nicholls 2014, in the quote opening this chapter). The key challenge facing Galápagos is also the key challenge for the globe: the transition to an interacting set of social-ecological systems that, on the whole, are indefinitely livable.

Finally, Galápagos needs our help. Just as the National Park Directorate and the Darwin Research Station did not stand by in the 1990s when goats overran Alcedo Volcano and threatened its tortoises, so we must not stand by today when concrete actions are needed to prevent further threats and extinctions. What can we do? First, we can support Galápagos educationally, by informing ourselves, spreading the word, and encouraging the related educational efforts of others, especially the hard-working teachers in Galápagos. Second, we can support Galápagos through purposeful visits, particularly if we prepare carefully beforehand to learn what to look for, why it matters, and how to help. It's a simple, well-known formula: change ourselves to be better visitors, and we nudge forward changes that help Galápagos. Third, we can support Galápagos from home by contributing time, talent, funds, and ideas in appropriate ways (see Appendix 2 for suggestions). Only a concerted effort by many people, including dedicated locals, visitors, donors, and researchers, will make Galápagos sustainable.

***

Ideally, this book will, in its own small way, inspire each of us to support as best we can the quest for sustainability in these endearing, enchanting islands, including the many conservation efforts under way today and those being planned for the future. Our efforts will be worthwhile for the archipelago and, truly, for the whole planet. May we be wise enough and caring enough to help Galápagos find a path to sustainability. Life on Earth may well depend upon it.

APPENDIX 1

An Evolution Primer

Here follows a brief review of first principles of evolution, intended to help as needed with terms and logic. Truly a primer, it includes little elaboration or justification and few examples—the main text provides those. This primer is like a map's key, designed to offer helpful information when you need it, but otherwise easy to ignore. It features a liberal sprinkling of quotes from Darwin (1859), first edition, with original page numbers, to illustrate the lasting value onf his insights.[1] Readers seeking a more thorough introduction are urged to consult Parker (2015) and Zimmer and Emlen (2019).[2]

The main text uses the terms and logic that are outlined below. Both the *intergenerational* nature of evolutionary change and the role played by environmental conditions in that change are emphasized.[3] As Smith (2017) pointed out, misleading shorthand explanations of evolution are often used in the literature today, which is a particular problem in the interpretive literature about Galápagos (where, presumably, many visitors go for a better understanding of evolution). Sometimes such shorthand leads us to think, for example, that organisms evolve over time because they "want" be a certain way, or "need" (or "don't need") to be that way, or because a change for them "makes sense" under the circumstances (e.g., see Lynch 2018, 110).

Giving a genuine evolutionary explanation is far more useful and only slightly more challenging. The focus here is on intergenerational change in the heritable features of populations and on the processes that cause that kind of change. In fact, we use as our definition of evolution any change in the frequency of heritable features within a population across one or more generations of time. We'll keep in mind a range of processes that can cause such frequency changes, including the "big five," mutation, lateral gene transfer, migration, drift, and natural selection, each of which is described below. But we give special attention to natural selection because of its prominent role in shaping the fascinating features of Galápagos organisms. We focus exclusively on heritable variation, because heritability is what allows changes to accumulate through time. Darwin put it succinctly, "Any variation which is not inherited is unimportant for us" (12).

One key feature of natural selection should be highlighted at the start: natural selection takes place through the interactions of individual organisms in a population with their environment. Out of those interactions come individual differences in survival and reproduction—a key measure for our purposes—associated with heritable features of the organism (size of beak, length or width of tail, etc.). With the passing of generations, those

---

[1] For the authentic Darwin experience in inexpensive paperback, see: Darwin, Charles. 1964. *On the Origin of Species: A Facsimile of the First Edition*. Cambridge, MA: Harvard University Press.

[2] For a useful online resource, see https://evolution.berkeley.edu/evolibrary/article/evo_01. Also useful for younger readers and their teachers is http://evolution.discoveringGalápagos.org.uk/

[3] This Appendix refers to genetic evolutionary change only. Cultural evolutionary change can readily occur within the span of a generation—a topic explored elsewhere in some detail (Durham 1991).

survival and reproduction differences will produce changes in the population's frequencies of the heritable features—changes we call "evolution by natural selection."[4]

Consider the fishing prowess of Galápagos cormorants to illustrate the point. The ancestors of today's cormorants lived in populations with heritable variation among individuals in body size and wing length. Cormorants lived for millennia on islands without terrestrial predators, amid one of the world's most productive ocean upwellings, dense with fish, so there was little survival and reproduction benefit to cormorant flight. Strong wings could hardly take a bird to better circumstances. Instead, cormorants born with smaller-than-normal wings—conferring less buoyancy—and larger-than-normal bodies—providing more mass in the water and less heat loss—fished more efficiently and fed their chicks more reliably than other birds. Since both wing size and body mass are heritable, their chicks also thrived with these advantages, and their "grand chicks" in turn, with consistent survival and reproductive advantage. Barring major environmental changes, the same advantage held across many generations, so that cormorants with smaller wings and larger bodies became more common in the population, an indication of evolutionary change.[5] Eventually, after many generations, their descendants were large birds with reduced wings, incapable of flight.

In this example, the population did not change toward flightlessness because organisms "needed" or "wanted" that feature, or because it "makes sense" in their habitat. The population evolved because of reproductive differences across generations among the heritable variants in the population. The reproductively successful variants increased in frequency, the hallmark of evolution by natural selection. The implications are important: today's vulnerable organisms in Galápagos and elsewhere will not evolve to a stronger state because they "need" or "want" to meet conservation challenges. They will evolve added strength, as generations go by, if heritable variants exist for strength, and the variants conferring added strength have survival and reproductive advantages. To appreciate the plight of Galápagos organisms and to take meaningful action, we must use intergenerational logic and explanation.

In one area, Darwin's otherwise thoughtful words come up short today: inheritance. In Darwin's time, "the laws governing inheritance [were] quite unknown" (1859, 13). Yet he realized that "inheritable deviations of structure" (12) were central to his theory, if changes in organisms were to endure and accumulate. Since Darwin, our understanding of heritable variation has vastly improved, with new knowledge filling many of the gaps he faced, including knowledge of the structure and function of hereditary material (DNA), of genes and chromosomes, and of entire genomes (see Mukherjee 2016 and Watson et al. 2017).[6] These advances now allow us to put genetics and evolution to work in the service of conservation, as emphasized in the main text.

---

[4] Sometimes one generation is enough to produce measurable change in the population frequency of heritable features. One year of drought proved sufficient for evolution by natural selection in the beak of medium ground finches on Daphne Island, as described in the heralded work of Peter and Rosemary Grant (see Chapter 6).

[5] For the sake of simplicity, this example ignores the impact of the El Niño-Southern Oscillation (ENSO). Chapter 9 explores the evolutionary implications of ENSO for the Galápagos cormorant and penguin populations.

[6] For our purposes, it suffices to think of a gene as a unit of genetic inheritance, made of DNA. Genes exist in longer structures of hereditary material called chromosomes (the famous "staining bodies" of early microscope work). Human beings have about 25,000 genes, normally contained within 46 different chromosomes, 2 each of 23 pairs (one of each pair is inherited from each biological parent). A genome is the complete genetic information of an organism or virus; genomics is the study of genomes.

Here follow key principles and definitions, numbered for easy reference:[7]

1. The term *evolution* refers to the proposition that all species on Earth, past and present, including humans, are related by descent from one common origin. Put differently, all species of plants and animals have been derived via generations of survival and reproduction from a single common ancestral species, which we call LUCA (Last Universal Common Ancestor).[8] Life is thus connected in "one grand system" as Darwin put it (456): "The ordinary succession by generation has never once been broken" among organisms alive today (489).

2. Two patterns are discernable in descent. One is sequential *branching*, which makes evolutionary histories bush- or tree-like, with descendant species (newer branches) diverging over time from ancestral species (older branches), so that we often speak of the "tree of life." The other pattern arises from the lateral movement of genes between branches, or *horizontal gene transfer* (HGT). Such sideways insertions make a "tangled tree" (Quammen 2018), which is especially common among single-celled organisms. HGT in multicellular organisms can cause genes to have histories of descent quite different from those of the species bearing them.

3. Descent by either pattern is not fixed or immutable: living organisms can and do change over time. They do not have lasting "essences:" over time, they *descend with modification,* as Darwin was fond of saying. He so liked this expression—effectively his definition of evolution—that the word "evolved" occurs only once in his big book, *On the Origin of Species* ([1859] 1964, 490), in the last sentence: "From so simple a beginning endless forms most beautiful and most wonderful have been, and are being, evolved."

4. Every time branching occurs, it leads to heritable changes among "sister" species, causing descendants to become increasingly different. Consequently, "From the first dawn of life, all organic [species] are found to resemble each other in descending degrees, so that they can be classed in groups under groups" (411) in a hierarchical system. The biological system of taxonomic classification, or Linnaean system (named after Carolus Linnaeus, who founded the system in the 18th century), is thus "founded on descent with modification" (420). Organisms are grouped by decreasing similarity in their features shared by descent, called *homologous features*, from a "trunk" (a common ancestor) to the tips of branches, representing individual species with new features. Near the tips, closely related species share features of recent evolutionary origin, called shared derived characters or exclusively shared innovations. Evolutionists use these descending degrees of similarity to infer the *phylogeny* of a group of organisms—that is, their evolutionary history showing branches of related species and dates of divergence.

5. The history of descent with modification among living organisms has involved only natural processes. Careful examination has produced no convincing evidence of supernatural agency in the history of life. Note that this does not deny the existence

---

[7] The sections below are adapted from the author's own earlier article (Durham 2007), with permission from the American Anthropological Association.

[8] LUCA is actually an evolutionary intermediate: there were untold living organisms before and contemporary with LUCA, but LUCA was the one species to become ancestral to all life on the planet. Genes shared among today's living organisms reveal aspects of the genetics, physiology, and habitat of LUCA (Weiss et al. 2016).

of God, a wholly separate question. The geneticist Theodosius Dobzhansky (1973, 127) proposed that evolution could be seen as God's "method of creation"—an argument that is thoroughly defensible. In fact, many world religious traditions have no quarrel with an evolutionary view of life. The same is true for Christian religions, unless one insists on a completely literal biblical interpretation (see publications of the Pew Research Center, http://www.pewforum.org/2009/02/04/religious-groups-views-on-evolution/).

6. Five natural processes cause modification over time in the heritable features of organisms, thus shaping branches of the tree of life: mutation, lateral gene transfer, migration, chance fluctuations or "drift," and natural selection. *Mutations* are changes in the chemical information of genetic material (technically, changes in the sequence of variable components called nucleotides in DNA or RNA, although we rarely need that level of detail here). Whether spontaneous or induced by physical or chemical agents, they are the original source of variation for evolutionary change in organisms and viruses. However, they generally occur at very low frequencies—the estimated average germ-line mutation rate in human beings is only 0.5×10^{-9} changes per base pair (a base pair is a chemical subunit of DNA containing two paired nucleotides) annually (Scally 2016). Something that infrequent will rarely be a main force of evolutionary change. The second process, *horizontal gene transfer*, moves genes (including mutations) between branches. It provides a second pathway for variation to come into a species, although ultimately that variation, too, stems from previous mutation in other, preceding species in the history of lateral transfers.

7. The third process, *migration*, is simply the movement of organisms from one population to another. Migration is an important source of evolutionary change under certain circumstances, especially in highly mobile organisms. It gives rise to gene flow, the movement of genes between populations. But because it simply moves variation around in space, it leads to major evolutionary trends by itself only in special circumstances. Chance fluctuation, or *drift*, is also important in particular circumstances, as when variation is accidentally reduced by a natural hazard, or when a new, separate population is started by a biased, nonrandom pool of founders, the so-called *founder effect*. The importance of drift as an evolutionary process diminishes with the size of populations. It is an important contributor to evolution in small populations, like many in Galápagos (see examples in Hedrick 2019).

8. The single most important process of evolutionary change is *natural selection*. As Darwin (1859, 6) put it, natural selection has been "the main but not exclusive means of modification" in the history of life. Natural selection refers to the preservation of heritable variation by its survival and reproductive advantage under given environmental conditions. Natural selection does not create or induce variation; variation originates in mutation (see #6), and may further be shuffled and repackaged via sexual reproduction and HGT. Natural selection simply acts as a filter, a "principle of preservation" (127), automatically favoring heritable variants that have a reproductive advantage under current conditions. Many environmental conditions (droughts, floods, food shortages, high winds, and more) and processes (competition, predation, parasitism, etc.) may cause natural selection, but the result is always the same: differential reproduction promotes the differential preservation of heritable variation in a population.

9. The term "natural selection" is an analogy. Darwin said, "I have called this principle, by which each slight variation, if useful, is preserved, by the term Natural Selection, in order to mark its relation to man's power of selection" among variants of domesticated animals (61). Darwin used this analogy to make natural selection clear and understandable for his readers, but it sometimes misleads people to think there is a purposeful agent in nature picking and choosing. Natural selection is simply the differential preservation of heritable variation in a population as a result of the survival and reproduction of its bearers.

10. An environmental cause of differences in survival and reproduction among individuals in a population is called a *selection pressure*. These pressures are not accidental or random, as sometimes portrayed in popular media. When selection pressures continue for extended periods of time, the cumulative power of natural selection shows itself, incrementally building intricate designs, like crypsis (camouflage), complex organs (like the vertebrate eye), impressive locomotory systems, and more. Natural selection always works by the selective retention of naturally occurring variation that has origins in random mutation. The key point is selection's ability to add up small beneficial differences over time, even though they are introduced at random: "It is the steady accumulation, through natural selection, of such differences, when beneficial to the individual, that gives rise to all the more important modifications of structure" (170). Thus, from the random mill of mutations, only the beneficial ones, with some tag-along neutral ones, are retained; the harmful ones mean that their carriers do not survive and reproduce as well, and thus harmful mutations are lost over time. Selection's ability to accumulate "successive slight variations" (457) fuels most directional trends in evolution.

11. Although not at all random, natural selection *is* opportunistic. In Francois Jacob's (1977) famous analogy, natural selection works only on *existing* variation in a population, rather like a tinkerer who works only with parts on hand, in contrast to a design engineer. The amazing—if imperfect—designs of living organisms result from the cumulative culling action of natural selection, often over many generations, on naturally occurring variation.

12. Two main products of evolution by natural selection are: *adaptation*—evolved features of morphology, physiology, or behavior that promote the survival and reproduction of the individual organism in its environment—and *diversification* or *speciation*, the origin of new species. These products can be called the vertical and horizontal dimensions of evolutionary change, or "anagenesis"—the shape of a given branch—and "cladogenesis"—the branching pattern—respectively (Mayr 1982, 400). The "tree of life" is a complex product of both, with the added element of HGT. It sounds simple, but complexity lies in the details: how and when branches form, how many branches, how fast they change, etc. *Clade*, another key term, refers to any complete branch, including all descendant species on its tips. Also called a "monophyletic group," a clade includes a given ancestor at any point on the tree plus all its descendant species.

13. Darwin was the first to recognize that adaptations *are individually beneficial*, or as he put it, "good for the individual possessor" (459). To Darwin, this conclusion followed from the view that natural selection acts solely "by the preservation of individuals with any favorable [heritable] variation" (194). Thus, natural selection will rarely "produce in a being anything injurious to itself. . . . If a [fair measure be made] between the good and evil caused by each part [of an organism], each will be

found on the whole advantageous" (201).[9] If conditions change, such that "any part becomes injurious, it will be modified" or the species will go extinct (201). Thus, the features of living organisms have evolved for their adaptive value *to the individual, not to the species* as was once widely believed. Living organisms rarely, if ever, show a "species preservation instinct"; any instinct they have is for their own individual survival and reproduction.[10]

14. What is *fitness*? In evolution, fitness is a measure of "adaptedness," the state of being adapted to particular environmental circumstances. Darwin thought of it as "success in leaving progeny" (62). An organism was "fit" if it had a lot of offspring: the more offspring, the higher its fitness. But that definition was muddied when sociologist Herbert Spencer persuaded Darwin to call natural selection "the survival of the fittest." Since "success in leaving progeny" requires the survival of the parent organism and its progeny, "survival of the fittest" can be viewed as the "survival of the survivors," which is confusing and nearly tautological. Today, we define evolutionary fitness as the effectiveness of design for survival and reproduction (after Williams 1966, 158). Hence "survival of the fittest" means the differential persistence over time of effective designs for survival and reproduction of a given organism in a given environment, which is not tautological. Accordingly, we expect Galápagos to be populated today with organisms sporting time-tested designs for individual survival and reproduction in their habitats of that special place. The challenge is to figure out how and when (in what circumstances) the observed features generate their fitness benefits.

15. This improved conception of fitness allows us to offer a more concise, testable definition for adaptation. An *adaptation* is any heritable feature (morphological,

---

[9] As the saying goes, "never say never." I've changed Darwin's wording here from "will never produce" to "will rarely produce" because there are special cases where natural selection does produce injurious conditions. Without a deeper understanding of heredity, Darwin could not have anticipated these cases. They arise in diploid organisms (i.e., organisms with typically two copies of each gene) where deleterious conditions can be produced—and maintained at considerable frequency in populations—as byproducts of natural selection. In these organisms, the two copies of a given gene can match (i.e., they can be the same genetic variant)—in which case we call individuals "homozygotes"—or they may be different variants, which we call "heterozygotes." In certain circumstances, heterozygotes can have a survival and reproductive advantage over homozygotes; one of the most carefully documented of all such cases is in humans, in environments where malaria and its vectoring mosquitoes are common. In such places, heterozygotes (AS) for the blood disorder sickle cell anemia have a degree of protection from malaria that homozygotes for the normal adult gene (AA) do not have, and the heterozygotes do not suffer from the anemia as do SS homozygotes. (The protection confirms that the AS condition is highly adaptive: the malarial parasite actually triggers sickling in the blood of normally nonsickling AS persons, killing the infected cells and the parasite along with them.) As a result, there's an enduring survival and reproductive advantage to the AS type. Over time, natural selection will increase the frequency of ASs in the population, and from their reproduction (AS with AS) will come, by the regularities of heredity, recurring cases of sickle cell anemia (SS). Without medical intervention, sickle cell anemia is certainly injurious: SS individuals typically die at young ages from sickling complications. Thus, SSs are an evolutionary byproduct of the advantage enjoyed by AS heterozygotes in these environments. In sum, natural selection does sometimes "produce in a being features injurious to itself" as a byproduct. For full discussion of sickle cell disease, in theory and examples, see Durham (1991, Chapter 3).

[10] As Darwin (1859) was first to realize (see Chapter VI of *On the Origin of Species*), the behavior of castes (nonreproductive "workers," "soldiers," "nurses," and more) in the eusocial insects are an important "exception that tests the rule," only to confirm it via the unusual genetics of their sexual reproduction (see Zimmer and Emlen 2019).

physiological, and/or behavioral) of an organism that confers fitness, which is considered the "function" of the feature, and that was shaped by natural selection for that function. One should thus ask two questions of each potential adaptation: (1) What is its function for the organism? (How does it enhance an individual's design for survival and reproduction?) (2) Does the feature have a history of being naturally selected for that function? Sometimes features have a function but do not have a history of natural selection for that function. For example, people's noses today often hold eyeglasses in place, a good design in the modern world. But since eyeglasses have not been around long enough to influence the evolution of noses, that function has not been naturally selected for, and so glasses-holding by noses is *not* an adaptation. It is adaptive "after the fact," *ex aptus*, which evolutionists often call an *exaptation*.[11]

16. Darwin recognized that while all adaptations have their function, not all functions render advantage with respect to habitat. In sexually reproducing organisms, some functions are advantageous in terms of access to mates and reproductive opportunity. Often it is the males who evolve traits enhancing their reproduction via improved access to females, which we call *sexual selection*. This can happen two ways. First, there is *intrasexual selection*, wherein competition over access to mates selects for "characters useful to the males alone in their struggles with other males" (127), favoring behaviors and 'equipment' for dominance over competitors (e.g., size, armament, and/or male–male aggression). Second is *intersexual selection*, competition to attract mates. Driven by the mate choice of the other sex(es), intersexual selection favors traits (color, shape, ornamentation, and/or behavior) that are attractive to the choosing sex. It is all about beauty in the proverbial "eyes of the beholder" (potential mates), not domination (see Prum 2017). Bright, eye-catching colors are often a clue to a history of intersexual selection in a species. By either mode, sexual selection renders the two (or more) sexes different from one another, which is called *sexual dimorphism* or sexual polymorphism.

17. Evolutionists now speak of *inclusive fitness,* an important adjustment to Darwin's focus on individuals. It includes in an individual's fitness the expected impact of that individual's actions on survival and reproduction of its close genetic relatives.[12] Natural selection can be seen as the preservation of genes through time via their net effect on the survival and reproduction of the organisms who carry them. Since close relatives share many identical genes through inheritance, the representation of one's genes in the next generation is influenced by not just one's own survival and reproduction but also by one's influence on the survival and reproduction of close relatives, which can increase representation of those genes in the population. This preservation of variants through their impact on survival and reproduction of close genetic relatives is called *kin selection*. Kin selection can propel the evolution of characteristics that favor the survival and reproduction of an individual's close relatives, and even a degree of altruism (reproductive self-sacrifice) toward those relatives.

[11] Such a feature is *ex aptus*—taken advantage of in the service of survival and reproduction—not *ad aptus*—shaped for that function (see Gould and Vrba 1982).
[12] The insight was actually foreshadowed by Darwin himself (1859, 202), in his discussion of a bee's "power of stinging" and other hive-bee behaviors as "useful to the community [thus fulfilling] all the requirements of natural selection."

18. Diverse forms of natural selection can thus be distinguished: *environmental selection* promotes adaptations to habitat, *sexual selection* promotes adaptations to mating and mate choice, and *kin selection* promotes adaptations to living in groups that include close genetic relatives.

19. The second product of Darwin's theory, *diversification,* pertains to the branching pattern of the tree of life, also called *speciation.* Generally, the term "species" refers to any single evolving branch on the tree. The evolution of a branch/species may not be fully independent, owing to processes like hybridization (mating between individuals of different branches) and lateral gene transfer. But diversification requires a degree of isolation—enough for genetic differences to accumulate over time among branches, eventually evolving lasting mechanisms of reproductive isolation. The "high importance of barriers," wrote Darwin, explains why oceanic islands, as one example, "should have few inhabitants, but of these a great number should be endemic" (isolation allows their genetic differentiation) and "why all the inhabitants of an archipelago . . . should be closely related to each other, and likewise be related, but less closely, to those of the nearest continent or other source whence immigrants were probably derived" (408–409). It's a good bet that Darwin was thinking about Galápagos when writing that passage.

20. The word *species* is notoriously difficult to define. It is helpful to distinguish *completed species* from *morphospecies.* Completed species are reproductively isolated from one another, having evolved enough differences to be reproductively incompatible, thus to remain separate species.[13] Morphospecies look to us like different species, and so we give them names like "small," "medium," and "large" ground finch. But confusion arises because many morphospecies are "species before speciation is complete"—not yet reproductively isolated and thus able to mate and form viable hybrid offspring. After sibling species split into separate populations, it can take 10 million years or more for the sibling species to become completed species (Grant and Grant 2008). Another useful term is *subspecies,* which refers to populations of the same species living in different areas and having a recognizably different feature or two, like size or color (but not being so different as to be considered separate morphospecies). The numerous subspecies of marine iguanas in Galápagos are a good example.

21. The general pattern to species formation can be simplified to three steps or phases.[14]

  (a) *Separation into reproductively isolated populations.* To begin, multiple individuals of an ancestral species, or at least one fertilized female, establish themselves in a new environment for that species, like a new Galápagos island. They may separate by dispersal, vicariance, or hybridization. *Dispersal* refers to walking, swimming, drifting, or flying to a new habitat. Dispersal most likely

[13] Defining speciation by reproductive isolation is known as the "biological species concept." The biological species concept was made popular by evolutionist Ernst Mayr, but the notion goes back to at least Georges-Louis Leclerc, Comte de Buffon, in 1753 (Mayr 1982). Indeed, Darwin also used the fact that all pigeon breeds can mate and produce fertile young as evidence that they were varieties of one species (1859, 26). However, Darwin held back from generalizing: "In some [cases], sterility [is] an unfailing test" he wrote, "with others it is not worth a farthing" (from Nicholls 2014, 82).

[14] Grant and Grant (2008) is recommended to readers seeking a full description and discussion of this process.

involves small numbers of organisms, which can lead to rapid evolutionary change via founder effect, drift, and inbreeding.[15] These latter processes are sometimes called *evolution accelerators*. The second pathway to separation, *vicariance,* entails splitting of an initial population into two or more geographically separated sister populations. Vicariance usually results in relatively large numbers in each population—a key difference from dispersal. In Galápagos, vicariance could have come from an island's subsiding until its various volcanos became separated by the sea, or from tectonic movement that broke an island into pieces. Evolutionary change following vicariance is likely to be slower, lacking most or all the evolution accelerators at work in small populations.

The third pathway to separation involves *hybridization*, the cross-breeding of genetically different parents. Two types of hybridization are distinguished. One involves the accidental duplication of one parental genome, with self-fertilization of its gametes creating *polyploid* progeny (having multiple chromosome copies). If viable, such progeny are automatically reproductively isolated from others in the population: with duplicated chromosome numbers, their germ cells (e.g., pollen and ova) are incompatible with normal monoploid (single-copy) germ cells. This process is relatively common in plants, giving rise in a single generation to a new, reproductively isolated population. Another kind of hybridization occurs between similar morphospecies, including animals, and produces *monoploid* hybrids with normal chromosome numbers. When such interspecies hybrids are viable (as from matings between incomplete species), reproductive isolation may occur by some novel means, not chromosome mismatch. In Galápagos, for example, two species of Darwin's finches cross-bred and produced novel traits in the hybrid offspring, including a bigger overall size, a bigger beak, and a novel song learned from the male parent. These traits caused the hybrids to mate exclusively and successfully among themselves, despite repeated inbreeding (Lamichhaney et al. 2018).

(b) *Genetic differentiation*. With the initial population split into two or more reproductively isolated pieces (often designated A, B, C, etc.), the "sister populations" will inevitably live under different conditions (whether geophysical differences, differences in accompanying biota, differences in food resources, or other things). Evolutionary processes will operate differently in the separated populations, and thus genetic differences will accumulate between them. Assuming low rates of migration (or *gene flow*) between the populations, differences will grow with time, and the sister populations will diverge genetically and eventually in morphology, physiology, and/or behavior.

(c) *Species formation*. Given enough isolation and time, genetic differences between sister populations will grow until large enough to prevent successful mating between the two. Once reproductive isolation is reached, the populations are *completed species*. When the differences are not yet large enough to prevent mating between the sister populations, they can still hybridize and are thus *incomplete species* (also called *incipient species*, or species before speciation is complete; Grant and Grant 2008). Many of the

---

[15] Inbreeding, the mating of close genetic relatives, increases the exposure to natural selection of otherwise hidden recessive genes, and is thus considered an accelerator of evolutionary change in small populations.

morphospecies identified in Galápagos—including many of Darwin's finches—are incomplete or incipient species, which allows them to hybridize.

22. Many more species have been formed than survive today, having been lost through *extinction*. As Raup (1994, 6758) said, "The number of species extinctions in the history of life is almost the same as the number of originations; present day biodiversity is the result of a trivial surplus of originations, cumulated over millions of years." A similar relationship likely holds in Galápagos, although Steadman et al. (1991) found low rates of prehuman extinction in the fossil record of the islands. In general, small islands are likely to witness high extinction rates, a concept integral to the theory of island biogeogrxaphy (see Chapter 3). Human activity on islands greatly increases the rate of extinction, and it is a major concern in Galápagos today.

23. Students today sometimes ask if an evolutionary view of Earth's biodiversity doesn't "diminish the wonder and glory of life," making it seem "too mechanical or routine." I ask them to consider other forms of wonder and glory, like the *unbroken line of descent* that connects each of us to the beginning of life, and to think about the related *kinship of life*, which means that each of us is related to all other species, past and present, like distant cousins of varying degrees. For the future of our planet, I say, would it not help to think more often about ourselves as kin to all life? We can surely do more to help the survival and reproduction of our own wide-ranging relatives, in Galápagos and elsewhere. Finally, I point out each of us human beings, like all other organisms, is among *the pretty good designs* that evolution has produced for the environments that shaped them. I emphasize that these implications of evolution, and more like them, deserve to be recognized and celebrated. I usually end with Darwin's words, "When I view all beings not as special creations, but as the lineal descendants of some few beings which lived long [ago] . . . they seem to me to become ennobled" (489). If you stop to look at the big picture, you will see what Darwin meant: "There is grandeur in this view of life" (490).

APPENDIX 2

Supporting Sustainability in Galápagos

You Can Make a Difference in Many Ways

Here are some things you can do to support Galápagos on its path to sustainability. This list is intended as a springboard to prompt your own thinking about ways to help. At the outset, I recommend reviewing the book's section on human history in the archipelago (Chapter 4), to refresh your appreciation of the complex social reality of Galápagos today and how it came to be that way. Our support must always be appropriate to the circumstances, and culturally sensitive. If you are considering a visit, please check early on with the Ecuadorian Embassy near you regarding travel conditions, health information, and visa requirements for visiting Ecuador and Galápagos. The COVID-19 pandemic has prompted changes in tourism to the islands, but there remain many safe ways you can help aside from visiting, and still more safe and healthy options will open up with time.

1. *Keep learning.* Don't stop here: let this book be just an introduction to the archipelago's amazing organisms, history, and conservation challenges. Keep going: find other books and articles to read and enjoy.[1] Watch some of the splendid videos about Galápagos.[2] Study available Web resources.[3] Think over how Galápagos has changed your views of life on Earth and explore new directions in that thinking. There is always much more to learn from Darwin's "little world."
2. *Spread the word.* Read a Galápagos book in your book club. Hang a picture in the hallway. Share Galápagos resources with friends and family to spread appreciation for the islands. Help others feel that special connection to life that blooms in Galápagos and share some of the ways it motivates deeper understanding of ourselves, other organisms, and conservation. Show skeptics the value in the insights you have gained from Galápagos. Explain the examples you find especially convincing or valuable, and help them appreciate the predicament organisms face in the archipelago today.

---

[1] Especially recommended are: De Roy, Tui. 2020. *A Lifetime in Galápagos.* Princeton University Press; De Roy, Tui, ed. 2016. *Galápagos: Preserving Darwin's Legacy.* Bloomsbury; Nicholls, Henry. 2014. *The Galápagos: A Natural History.* Basic Books; Watkins, Graham, and Pete Oxford. 2009. *Galápagos: Both Sides of the Coin.* Imagine! Publishing; Stewart, Paul D. ed. 2006. *Galápagos: The Islands That Changed the World.* New Haven, CT: Yale University Press.

[2] Among the splendid videos are *Galápagos 3D,* with David Attenborough. Sky3D, 2013; and *Galápagos: Islands that Changed the World,* narrated by Tilda Swinton. BBC Videos, 2007. For a fascinating documentary of human settlers in the 1930's and their mysterious disappearances from Floreana, see Dana Goldfine and Daniel Geller's *The Galápagos Affair: Satan Came to Eden.* New York: Zeitgeist Films.

[3] A good starting point is https://www.discoveringGalápagos.org.uk/

3. *Support Galápagos conservation.* Consider donating time, talent, or resources to help conservation efforts underway in Galápagos. Consider a financial contribution to waved albatross conservation, for example, or tortoise repatriation, *Scalesia* restoration, blue-footed booby monitoring, penguin condo construction, and the like. Or contribute to campaigns to control or eliminate introduced species, or to provide protective refugia for the fragile denizens of the archipelago. Can you volunteer for a month or more at one of the hard-working NGOs in Galápagos or at the Galápagos National Park Directorate?[4] Are you skilled at some specific technology that might advance sustainability goals in the islands, of great net benefit to the Park or the Charles Darwin Foundation? (e.g., remote sensing, telemetry, bird call recording, electric vehicles, desalination) Can you help schools with remote learning supplements or with updates in computer technology? If so, share your ideas first with the Ecuadorian Embassy nearest you, or with an NGO whose work you respect. Incorporate their feedback and suggestions for your idea: does it still make sense?

4. *Encourage environmental education.* A concerted effort is underway today in Galápagos to increase local understanding of ecology and evolution, to foster local biodiversity appreciation and conservation ethics, and to build a better, more sustainable life for residents through education. Consider ways to support these efforts, such as a donation to help Galápagos schoolchildren get out to visit other islands, or to help build a first-rate children's library. Consider funding a scholarship at the Tomás de Berlanga School, a private bilingual school in Puerto Ayora, Santa Cruz. Or volunteer to work with local NGOs and teachers on environmental education, sharing expertise that you may have in areas like taxonomy, botany, ornithology, or climate change. Local environmental education is one of the highest priorities for a sustainable Galápagos.

5. *Take steps to make your own lifestyle more sustainable and explain the steps to others.* Consider what it would take for you, me, all of us, to live on Earth as if it were one big Galápagos, and for us to appreciate and protect the interconnections of species in this larger "little world." Think of ways to shrink your own ecological "footprint" in how you live, work, and travel. Share what you are doing with family, friends, and neighbors, and offer to help them reduce their own footprint. Help children understand why habitat conversion, species loss, and climate change are important topics today. Help them understand why you care about some little islands way out in the Pacific Ocean.

6. *Work on solutions locally.* Let Galápagos inspire you to help work on environmental problems where you live. Conservation is sorely needed all around us. Learn about threatened or endangered species living near you, and help raise awareness of their plight. Join an organization promoting local biodiversity conservation, and share your insights about Galápagos with them. Volunteer for habitat restoration or invasive species removal, akin to efforts in Galápagos. Join or start a book club, and include a few Galápagos titles: these topics and activities are all livelier and more fun when they are shared and discussed.

7. *Think big, but be realistic.* Remember the #1 goal is sustainability: helping Galápagos—indeed the planet—on the path to becoming indefinitely livable for current and future inhabitants, both native species and humanity. The goal is not to

---

[4] Among many worthy examples are the Charles Darwin Foundation (CDF), FUNDAR (Foundation for Responsible Alternative Development in Galápagos), World Wildlife Fund (WWF), and Galápagos Conservancy.

return the islands to some hypothetical pristine state, nor to prevent extinction and immigration of every single species, because these processes are inherent to life on islands. Rather, it is the subtler goal of reducing our direct and indirect impact on habitats and organisms in Galápagos to where rates of species immigration and extinction are close to natural levels (far below their current values), and the social and environmental well-being of human inhabitants does not erode over time. As you produce good, practical ideas for sustainability, run them by your friends for feedback and discussion. If you are convinced of their value, write out brief descriptions and send them to NGOs like the Charles Darwin Foundation (Ecuador), Galápagos Conservancy (U.S.) and Galápagos Conservation Trust (U.K.), who can consider them in light of their experience. They will have a good sense for how best to bring worthy ideas to the attention of pertinent authorities in Galápagos.

8. *Visit the islands if you can, and do so responsibly, with the goal of making a positive contribution to Galápagos sustainability.* Not only will your trip be memorable, but also it will help Galápagos if you take these simple steps:

   (a) *Make it an educational odyssey.* Read a lot beforehand and continue to read during the trip, so that you know what to look for each day. Consider a hybrid format, taking full advantage of boat-based tours to remote visitor sites, yet including some nights on land to experience the local communities and to take some responsibly run day trips by local vendors. What strikes you as really different in Galápagos, and what seems a lot like home? How do you see Galápagos as emblematic of the wider challenges facing life on Earth? What carryover value do you see from Galápagos for your daily life?

   (b) *When you arrive, don't hurry.* It will be exciting, with much to see and do, but take it slow: give yourself time to settle into the experience, look around, and notice details. Focus on each new species you encounter. What do you notice about the organism? About its environmental context? How does it interact with other organisms? If your guide moves along too fast, ask them to slow down. Ask questions. If you are with children, gently encourage them to ask questions, too.

   (c) *Feel your connection to life.* One of the most transformative experiences of Galápagos is sensing the unity of life and your place within it. Enjoy the fact that most animals will not run from you, accepting you casually as just another species. Sit nearby and commune for a moment: let yourself feel close and connected. Look a tortoise straight in the eye and notice all the details of its face. Watch an albatross feeding its hatchling close by, or a booby doing the same, and consider all that goes into that effort. Think of yourself as another branch on the very same tree. Enjoy being part of it all.

   (d) *Don't ignore the plants.* While obviously not as lively and charismatic as animals, the plants of Galápagos are equally fascinating if you give them a chance. You can start by looking for *Scalesias* on every island you visit (ask your guide for help). When you come to an island that has giant species (e.g., Santa Cruz), go see them if you can. Look for their flowers and look for tree finches amid their branches. Another good starting place with plants is cacti, where the genera are easy to distinguish (again, your guide can help). Look for signs of cactus predation: how does predation help explain the differences among cacti on different islands? And speaking of plants, look for lantanas. Can you find both the endemic and invasive lantana species? Which do you see more of, and why? What hardwood trees do you see? Are they endemic?

(e) *Keep a journal or other writing.* Take time each day to write some notes and reflections for yourself. Keep a notebook in your pocket or daypack for observations and questions that arise. Jot down the names of species you see; these notes are really helpful later when you're adding IDs to photos and learning more about what you saw. And be sure to avoid Darwin's mistake: always note the island of your observations and your photographs. Consider carrying a map, marking your route, the time of day you are there, and anything unusual about the location or context.

(f) *Take pictures and/or video, but don't just do that.* One can enjoy Galápagos fully with and without a camera, and both have advantages. Each of us must find our own balance of "being there" and "recording it" (sometimes couples find a pairwise balance). On a first visit, I recommend a happy medium, which is greatly facilitated if you already know your camera very well. Practice ahead of time. Bring its manual, and don't wait until your flights to Galápagos to learn what each button or setting does—especially to turn off the flash, which is not permitted at visitor sites. Consider using iNaturalist, an easy-to-use app for cell phones that can help you to identify and appreciate what you see. When you see something exiting, ask your guide if the sighting could be of use to science. Citizen science is becoming more important to the conservation of places like Galápagos.

(g) *Support the human communities of Galápagos.* Spend at least a few days and nights in one or more of the towns, checking out local restaurants, cafes, bookstores, and handicraft stores. Visit some noteworthy sites of human activity, like Darwin's initial landing site on San Cristóbal, the "wall of tears" on Isabela, and the "pirate caves" on Floreana (see Moore and Cameron 2019 for good ideas). Or experience a day trip of artisanal fishing with local fishermen from the larger towns, working hard in tourism to share their knowledge and lore of the sea. You can snorkel and even have fresh sushi prepared from the recent catch. While in Puerto Ayora, visit the Charles Darwin Research Station, both its indoor and its outdoor exhibits. Visit the town's fish market at peak operation, often mid- to late morning. Buy some "conservation coffee" produced in the islands and take it back for friends at home. Ask your naturalist guide about locally produced chocolates and where you can go to taste them. Make an effort to converse with locals, many of whom speak English and other languages. Ask about their concerns about the future of Galápagos. Consider donating to a local organization or project that moves you.

(h) *Share Galápagos back home.* Once back, tell others about your experiences and share your pictures, perhaps in a slide show of highlights. Host a coffee hour for friends with a "conservation coffee" produced in Galápagos and explain its story. Add some Galápagos chocolate products for a special treat. Or lead a discussion in your book club about Galápagos. Explain why Galápagos organisms have such peculiar, unique properties. Write something for social media, a newspaper, or a blog. Share some of the first-rate books for younger readers with children you know.[5] Help them understand why Galápagos organisms are

---

[5] Quality children's books include: Arnold, Marsha Diane. 2018. *Galápagos Girl (Galapagueña).* Children's Book Press (Lee & Low Books); Collard, Sneed B. III. 2018. *One Iguana, Two Iguanas: A Story of Accident, Natural Selection, and Evolution.* Tilbury House Publishers; Heller, Ruth. 2000. *"Galápagos" Means "Tortoises."* Sierra Club Books for Children. And there is a lovely children's book about Darwin: Radeva, Sabina. 2019. *Charles Darwin's On the Origin of Species.* Crown Books for Young Readers.

so unusual, and why they are threatened today. Offer to speak and show pictures to a school or youth group.

Think of ways to help going forward. Consider what you can do personally to help Galápagos move toward sustainability. How can you help from a distance? Some of the topics 1 to 7 above may be helpful. Ask the closest Ecuadorean embassy or consulate how you can help. There may sometimes be opportunities to help school or public libraries in Galápagos, for example, or to help with scholarship opportunities for students from the islands who are seeking advanced study. Help spread the word about valiant efforts underway by the National Park Directorate and the Darwin Research Station. Finally, help persuade other thoughtful, responsible visitors to visit the islands, and share your tips for preparation. And, in fact, I say . . .

9. *Go again*: Repeat step 8 if you can. I'm not saying, "Go indiscriminately," but I *am* saying "Go, when by going, you can help." There are so many things to see and take in during a first visit that you may well have a deeper, more satisfying experience on a second visit, and you may also come away with more and better ideas for helping.

10. The bottom line is: Do what you can to help get word out and contribute. Galápagos needs our support!

"Galápagos changes people's lives."

Graham Watkins, former Executive Director,
Charles Darwin Foundation
(cited in Nash 2009, 107)

References

"350.org." Accessed September 28, 2018. https://350.org/.

Abe, Takuzo, Keiko Sekiguchi, Hiroji Onishi, Kota Muramatsu, and Takehiko Kamito. 2012. "Observations on a School of Ocean Sunfish and Evidence for a Symbiotic Cleaning Association with Albatrosses." *Marine Biology* 159, no. 5: 1173–76. https://doi.org/10.1007/s00227-011-1873-6.

Abzhanov, Arhat, Winston P. Kuo, Christine Hartmann, B. Rosemary Grant, Peter R. Grant, and Clifford J. Tabin. 2006. "The Calmodulin Pathway and Evolution of Elongated Beak Morphology in Darwin's Finches." *Nature* 442, no. 7102: 563–67. https://doi.org/10.1038/nature04843.

Acuña-Marrero, David, Adam N. H. Smith, Pelayo Salinas-de-León, Euan S. Harvey, Matthew D. M. Pawley, and Marti J. Anderson. 2018. "Spatial Patterns of Distribution and Relative Abundance of Coastal Shark Species in the Galápagos Marine Reserve." *Marine Ecology Progress Series* 593:73–95.

Adsersen, Henning. 1989. "The Rare Plants of the Galápagos Islands and Their Conservation." *Biological Conservation* 47, no. 1: 49–77. https://doi.org/10.1016/0006-3207(89)90019-0.

Alfaro-Shigueto, Joanna, Jeffrey C. Mangel, Mariela Pajuelo, Peter H. Dutton, Jeffrey A. Seminoff, and Brendan J. Godley. 2010. "Where Small Can Have a Large Impact: Structure and Characterization of Small-scale Fisheries in Peru." *Fisheries Research* 106, no. 1: 8–17. https://doi.org/10.1016/j.fishres.2010.06.004.

Altamirano, Manfred, Veronica Toral-Granda, and Eliecer Cruz. 2004. "The Application of the Adaptive Principle to the Management and Conservation of *Isostichopus fuscus* in the Galápagos Marine Reserve." In *Advances in Sea Cucumber Aquaculture and Management,* edited by Allessandro Lovatelli, Chantal Conand, Steve Purcell, Sven Uthicke, Jean-François Hamel, and Annie Mercier, 247–57. FAO Fisheries Technical Paper, No. 463. Rome, Italy: FAO.

Anchundia, David, Kathryn P. Huyvaert, and David J. Anderson. 2014. "Chronic Lack of Breeding by Galápagos Blue-Footed Boobies and Associated Population Decline." *Avian Conservation and Ecology* 9, no. 1: 6. https://doi.org/10.5751/ACE-00650-090106.

Anderson, David J. 1989a. "Differential Responses of Boobies and Other Seabirds in the Galápagos to the 1986–87 El Niño-Southern Oscillation Event." *Marine Ecology Progress Series* 52, no. 3: 209–16.

Anderson, David J. 1989b. "The Role of Hatching Asynchrony in Siblicidal Brood Reduction of Two Booby Species." *Behavioral Ecology and Sociobiology* 25, no. 5: 363–68. https://doi.org/10.1007/BF00302994.

Anderson, David J. 1990a. "Evolution of Obligate Siblicide in Boobies. 1. A Test of the Insurance-Egg Hypothesis." *The American Naturalist* 135, no. 3: 334–50. https://doi.org/10.1086/285049.

Anderson, David J. 1990b. "Evolution of Obligate Siblicide in Boobies. 2: Food Limitation and Parent–Offspring Conflict." *Evolution* 44, no. 8: 2069–82. https://doi.org/10.1111/j.1558-5646.1990.tb04312.x.

Anderson, David. 2006. "Basic Science May Save the Waved Albatross." *Galápagos News* (Fall): 6–7.

Anderson, David. 2008. "Waved Albatross." In *Albatross: Their World, Their Ways*, edited by Tui De Roy, Mark Jones, and Julian Fitter, 160–61. Buffalo, New York: Firefly Books.

Anderson, David. 2016. "Nazca Booby Behavior: Some Evolutionary Surprises." In *Galápagos: Preserving Darwin's Legacy*, edited by Tui de Roy, 138–45. London: Bloomsbury.

Anderson, David. 2018. "Status of the Blue-Foots." *Galápagos News* (Spring-Summer): 14.

Anderson, David J. and Sharon Fortner. 1988. "Waved Albatross Egg Neglect and Associated Mosquito Ectoparasitism." *The Condor* 90, no. 3: 727–29. https://doi.org/10.2307/1368369.

Anderson, David J., Kathryn P. Huyvaert, Victor Apanius, Howard Townsend, Cynthia L. Gillikin, Lori D. Hill, Frans Juola, et al. 2002. "Population Size and Trends of the Waved Albatross (*Phoebastria irrorata*)." *Marine Ornithology* 30:63–69.

Anderson, David J., Kathryn P. Huyvaert, Dana R. Wood, Cynthia L. Gillikin, Barrie J. Frost, and Henrik Mouritsen. 2003. "At-Sea Distribution of Waved Albatrosses and the Galápagos Marine Reserve." *Biological Conservation* 110, no. 3: 367–73. https://doi.org/10.1016/S0006-3207(02)00238-0.

Anderson, David, Elaine Porter, and Elise Ferree. 2004. "Non-breeding Nazca Boobies (*Sula granti*) Show Social and Sexual Interest in Chicks: Behavioural and Ecological Aspects." *Behaviour* 141, no. 8: 959–77. https://doi.org/10.1163/1568539042360134.

Anderson, David J., and Robert E. Ricklefs. 1995. "Evidence of Kin-Selected Tolerance by Nestlings in a Siblicidal Bird." *Behavioral Ecology and Sociobiology* 37, no. 3: 163–68. https://doi.org/10.1007/BF00176713.

Anderson, Sean C., Joanna Mills Flemming, Reg Watson, and Heike K. Lotze. 2011. "Serial Exploitation of Global Sea Cucumber Fisheries." *Fish and Fisheries* 12, no. 3: 317–39. https://doi.org/10.1111/j.1467-2979.2010.00397.x.

Ardoin, Nicole M., Mele Wheaton, Carter A. Hunt, Janel S. Schuh, and William H. Durham. 2016. "Post-trip Philanthropic Intentions of Nature-Based Tourists in Galápagos." *Journal of Ecotourism* 15, no. 1: 21–35. https://doi.org/10.1080/14724049.2016.1142555.

Atkinson, Rachel, Jorge Luis Rentería, and Walter Simbana. 2008. "The Consequences of Herbivore Eradication on Santiago: Are We in Time to Prevent Ecosystem Degradation Again?" In *Galápagos Report 2007–2008*, 21–24. Puerto Ayora, Galápagos: Charles Darwin Foundation, Galápagos National Park, and National Galápagos Institute.

Austin, Jeremy J., E. Nicholas Arnold, and Roger Bour. 2003. "Was There a Second Adaptive Radiation of Giant Tortoises in the Indian Ocean? Using Mitochondrial DNA to Investigate Speciation and Biogeography of *Aldabrachelys* (Reptilia, Testudinidae)." *Molecular Ecology* 12, no. 6: 1415–24. https://doi.org/10.1046/j.1365-294X.2003.01842.x.

Awkerman, Jill A., Sebastian Cruz, Carolina Proaño, Kathryn P. Huyvaert, Gustavo Jiménez Uzcátegui, Andres Baquero, Martin Wikelski, and David J. Anderson. 2014. "Small Range and Distinct Distribution in a Satellite Breeding Colony of the Critically Endangered Waved Albatross." *Journal of Ornithology* 155, no. 2: 367–78. https://doi.org/10.1007/s10336-013-1013-9.

Awkerman, Jill A., Akira Fukuda, Hiroyoshi Higuchi, and David J. Anderson. 2005a. "Foraging Activity and Submesoscale Habitat Use of Waved Albatrosses *Phoebastria irrorata* during Chick-Brooding Period." *Marine Ecology Progress Series* 291: 289–300. https://doi.org/10.3354/meps291289.

Awkerman, Jill A., Kathryn P. Huyvaert, and David J. Anderson. 2005b. "Mobile Incubation in Waved Albatross (*Phoebastria irrorata*): Associated Hatching Failure and Artificial Mitigation." *Avian Conservation and Ecology* 1, no. 1: 2.

Awkerman, Jill A., Kathryn P. Huyvaert, Jeffrey Mangel, Joanna Alfaro-Shigueto, and David J. Anderson. 2006. "Incidental and Intentional Catch Threatens Galápagos Waved Albatross." *Biological Conservation* 133, no. 4: 483–89. https://doi.org/10.1016/j.biocon.2006.07.010.

Awkerman, Jill, Mark Westbrock, Kathryn Huyvaert, and David Anderson. 2007. "Female-Biased Sex Ratio Arises after Parental Care in the Sexually Dimorphic Waved Albatross (*Phoebastria irrorata*)." *The Auk* 124, no.4: 1336–46. https://doi.org/10.1642/0004-8038(2007)124[1336:FSRAAP]2.0.CO;2.

Baião, Patricia C., and Patricia G. Parker. 2008. "Maintenance of Plumage Polymorphism in Red-Footed Boobies in the Galápagos Archipelago: Observations of Mate Choice and Habitat Association." *The Condor* 110, no. 3: 544–48. https://doi.org/10.1525/cond.2008.8486.

Baião, Patricia C., E. A. Schreiber, and Patricia G. Parker. 2007. "The Genetic Basis of the Plumage Polymorphism in Red-Footed Boobies (*Sula sula*): A Melanocortin-1 Receptor (MC1R) Analysis." *The Journal of Heredity* 98, no. 4: 287–92. https://doi.org/10.1093/jhered/esm030.

Baker-Médard, Merrill, and Kristina Natalia Ohl. 2019. "Sea Cucumber Management Strategies: Challenges and Opportunities in a Developing Country Context." *Environmental Conservation* 46, no. 4: 267–77. https://doi.org/10.1017/S0376892919000183.

Barham, Peter J., Les G. Underhill, Robert J. M. Crawford, Res Altwegg, T. Mario Leshoro, Duncan A. Bolton, Bruce M. Dyer, and Leshia Upfold. 2008. "The Efficacy of Hand-Rearing Penguin Chicks: Evidence from African Penguins (*Spheniscus demersus*) Orphaned in the Treasure Oil Spill in 2000." *Bird Conservation International* 18, no. 2: 144–52. https://doi.org/10.1017/S0959270908000142.

Barnett, Bruce D., and Robert L. Rudd. 1983. "Feral Dogs of the Galápagos Islands: Impact and Control." *International Journal for the Study of Animal Problems* 4, no. 1: 44–58.

Barragan-Paladines, Maria Jose, and Ratana Chuenpagdee. 2017. "A Step Zero Analysis of the Galapagos Marine Reserve." *Coastal Management* 45, no. 5: 339–59. http://dx.doi.org/10.1080/08920753.2017.1345606.

Barry, Johannah. 2018. "From the President." *Galápagos News* (Spring-Summer): 2.

Bartholomew, George A. 1966. "A Field Study of Temperature Relations in the Galápagos Marine Iguana." *Copeia* 1966, no. 2: 241–50. https://doi.org/10.2307/1441131.

Bassett, Carol A. 2009. *Galápagos at the Crossroads: Pirates, Biologists, Tourists, and Creationists Battle for Darwin's Cradle of Evolution*. Washington, D.C.: National Geographic Books.

Beebe, William. (1924) 2012. *Galápagos: World's End*. North Chelmsford, MA: Courier Corporation.

Beheregaray, Luciano B., Claudio Ciofi, Dennis Geist, James P. Gibbs, Adalgisa Caccone, and Jeffrey R. Powell. 2003. "Genes Record a Prehistoric Volcano Eruption in the Galápagos." *Science* 302, no. 5642: 75.

Beltram, Fiona L., Robert W. Lamb, Franz Smith, and Jon D. Witman. 2019. "Rapid Proliferation and Impacts of Cyanobacterial Mats on Galápagos Rocky Reefs during the 2014–2017 El Niño Southern Oscillation." *Journal of Experimental Marine Biology and Ecology* 514:18–26. https://doi.org/10.1016/j.jembe.2019.03.007.

Benavides, Edgar, Rebecca Baum, Heidi M. Snell, Howard L. Snell, and Jack W. Sites, Jr. 2009. "Island Biogeography of Galápagos Lava Lizards (Tropiduridae: Microlophus): Species Diversity and Colonization of the Archipelago." *Evolution* 63, no. 6: 1606–26.

Bendik, N. F., and A. G. Gluesenkamp. 2013. "Body Length Shrinkage in an Endangered Amphibian Is Associated with Drought." *Journal of Zoology* 290, no. 1: 35–41. https://doi.org/10.1111/jzo.12009.

Benitez, S. 2001. *Visitor Use Fees and Concession Systems in Protected Areas: Galápagos National Park Case Study.* Arlington, Virginia: Nature Conservancy.

Benitez-Capistros, F., G. Camperio, J. Hugé, F. Dahdouh-Guebas, and N. Koedam. 2018. "Emergent Conservation Conflicts in the Galápagos Islands: Human–Giant Tortoise Interactions in the Rural Area of Santa Cruz Island." *PLOS One* 13, no. 9: e0202268. https://doi.org/10.1371/journal.pone.0202268

Bensted-Smith, R. 2002. *A Biodiversity Vision for the Galápagos Islands.* Puerto Ayora, Galápagos: Charles Darwin Foundation and World Wildlife Fund.

Benz, Richard. 2000. *Ecology and Evolution: Islands of Change.* Arlington: National Science Teachers Association Press.

Berger, Mark J., and Gill Bejerano. 2017. "Comment on 'A Genetic Signature of the Evolution of Loss of Flight in the Galápagos Cormorant.'" *BioRxiv* 181826. https://doi.org/10.1101/181826.

Berger, Silke, Martin Wikelski, L. Michael Romero, Elisabeth K. V. Kalko, and Thomas Rödl. 2007. "Behavioral and Physiological Adjustments to New Predators in an Endemic Island Species, the Galápagos Marine Iguana." *Hormones and Behavior* 52, no. 5: 653–63. https://doi.org/10.1016/j.yhbeh.2007.08.004.

Berkes, Fikret, and Carl Folke, eds. 1998. *Linking Social and Ecological Systems: Management Practices and Social Mechanisms for Building Resilience.* Cambridge: Cambridge University Press.

Berkovitz, Barry and Robert Peter Shellis. 2017. *The Teeth of Non-mammalian Vertebrates.* London: Elsevier.

Bertelli, Sara, and Norberto P. Giannini. 2005. "A Phylogeny of Extant Penguins (Aves: Sphenisciformes) Combining Morphology and Mitochondrial Sequences." *Cladistics: The International Journal of the Willi Hennig Society* 21, no. 3: 209–39. https://doi.org/10.1111/j.1096-0031.2005.00065.x.

Besserer, Johann. 2019. "The Political Ecology of Conservation and Development in Galápagos—Sustainability between the Global and the Local." PhD diss., University of Miami.

BirdLife International. 2020. "Marine IBA E-Atlas." https://maps.birdlife.org/marineIBAs/default.html

Blake, Stephen, Anne Guézou, Sharon L. Deem, Charles B. Yackulic, and Fredy Cabrera. 2015. "The Dominance of Introduced Plant Species in the Diets of Migratory Galápagos Tortoises Increases with Elevation on a Human-Occupied Island." *Biotropica* 47, no. 2: 246–58. https://doi.org/10.1111/btp.12195.

Blake, Stephen, Martin Wikelski, Fredy Cabrera, Anne Guezou, Miriam Silva, E. Sadeghayobi, Charles B. Yackulic, and Patricia Jaramillo. 2012. "Seed Dispersal by Galápagos Tortoises." *Journal of Biogeography* 39, no. 11: 1961–72. https://doi.org/10.1111/j.1365-2699.2011.02672.x.

Blake, Stephen, Charles B. Yackulic, Fredy Cabrera, Washington Tapia, James P. Gibbs, Franz Kümmeth, and Martin Wikelski. 2013. "Vegetation Dynamics Drive Segregation by Body Size in Galápagos Tortoises Migrating across Altitudinal Gradients." *The Journal of Animal Ecology* 82, no. 2: 310–21. https://doi.org/10.1111/1365-2656.12020.

Blaschke, Jeremy D., and Roger W. Sanders. 2009. "Preliminary Insights into the Phylogeny and Speciation of *Scalesia* (Asteraceae), Galápagos Islands." *Journal of the Botanical Research Institute of Texas* 3, no. 1: 177–91.

Bocci, Paolo. 2017. "Tangles of Care: Killing Goats to Save Tortoises on the Galápagos Islands." *Cultural Anthropology* 32, no. 3: 424–49. https://doi.org/10.14506/ca32.3.08.

Boersma, P. Dee. 1998. "Population Trends of the Galápagos Penguin: Impacts of El Niño and La Niña." *The Condor* 100, no. 2: 245–53. https://doi.org/10.2307/1370265.

Boersma, P. Dee, C. D. Cappello, and G. Merlen. 2017. "First Observations of Post-fledging Care in Galápagos Penguins (*Spheniscus mendiculus*)." *The Wilson Journal of Ornithology* 129, no. 1: 186–91. https://doi.org/10.1676/1559-4491-129.1.186.

Boersma, P. Dee, A. Steinfurth, G. Merlen, G. Jiménez-Uzcátegui, F. H. Vargas, and P. G. Parker. 2013. "Galápagos Penguin (*Spheniscus mendiculus*)." In *Penguins: Natural History and Conservation,* edited by Pablo Garcia Borboroglu and P. Dee Boersma, 284–302. Seattle: University of Washington.

Boersma, P. Dee, Hernan Vargas, and Godfrey Merlen. 2005. "Living Laboratory in Peril." *Science* 308, no. 5724: 925. https://doi.org/10.1126/science.1114395.

Bollmer, Jennifer L., F. Hernán Vargas, and Patricia G. Parker. 2007. "Low MHC variation in the endangered Galapagos penguin (*Spheniscus mendiculus*)." *Immunogenetics* 59, no. 7: 593–602.

Bordbar, Sara, Farooq Anwar, and Nazamid Saari. 2011. "High-Value Components and Bioactives from Sea Cucumbers for Functional Foods—A Review." *Marine Drugs* 9, no. 10: 1761–1805. https://doi.org/10.3390/md9101761.

Bremner, Jason, and Jaime Perez. 2002. "A Case Study of Human Migration and the Sea Cucumber Crisis in the Galápagos Islands." *AMBIO: A Journal of the Human Environment* 31, no. 4: 306–10. https://doi.org/10.1579/0044-7447-31.4.306.

Brightsmith, Donald, Jenifer Hilburn, Alvaro del Campo, Janice Boyd, Margot Frisius, Richard Frisius, Dennis Janik, and Federico Guillen. 2005. "The Use of Hand-Raised Psittacines for Reintroduction: A Case Study of Scarlet Macaws (*Ara macao*) in Peru and Costa Rica." *Biological Conservation* 121, no. 3: 465–72. https://doi.org/10.1016/j.biocon.2004.05.016.

Brooke, James. 1993. "Ban on Harvesting Sea Cucumber Pits Scientists against Fishermen." *New York Times,* November 2.

Brown, K. 2014. "Global Environmental Change I: A Social Turn for Resilience?" *Progress in Human Geography* 38, no. 1: 107–17. https://doi.org/10.1177/0309132513498837.

Bucaram, Santiago J., Alex Hearn, Ana M. Trujillo, Willington Rentería, Rodrigo H. Bustamante, Guillermo Morán, Gunther Reck, and José L. García. 2018. "Assessing Fishing Effects Inside and Outside an MPA: The Impact of the Galápagos Marine Reserve on the Industrial Pelagic Tuna Fisheries during the First Decade of Operation." *Marine Policy* 87 (January): 212–25. https://doi.org/10.1016/j.marpol.2017.10.002.

Buddenhagen, Christopher E., and Alan Tye. 2015. "Lessons from Successful Plant Eradications in Galápagos: Commitment Is Crucial." *Biological Invasions* 17, no. 10: 2893–2912. https://doi.org/10.1007/s10530-015-0919-y.

Buglass, Salomé, Harry Reyes, Jorge Ramirez-González, Tyler D. Eddy, Pelayo Salinas-de-León, and José Marin Jarrin. 2018. "Evaluating the Effectiveness of Coastal No-Take Zones of the Galápagos Marine Reserve for the Red Spiny Lobster, *Panulirus penicillatus."* *Marine Policy* 88 (February): 204–12. https://doi.org/10.1016/j.marpol.2017.11.028.

Burbano, Diana V., Thomas C. Meredith, and Monica E. Mulrennan. 2020. "Exclusionary Decision-making Processes in Marine Governance: The Rezoning Plan for the Protected Areas of the 'Iconic' Galápagos Islands, Ecuador." *Ocean & Coastal Management* 185: 105066. https://doi.org/10.1016/j.ocecoaman.2019.105066.

Burga, Alejandro, Weiguang Wang, Eyal Ben-David, Paul C. Wolf, Andrew M. Ramey, Claudio Verdugo, Karen Lyons, Patricia G. Parker, and Leonid Kruglyak. 2017. "A Genetic Signature of the Evolution of Loss of Flight in the Galápagos Cormorant." *Science* 356, no. 6341: eaal3345. https://doi.org/10.1126/science.aal3345.

Burga, Alejandro, Weiguang Wang, Paul C. Wolf, Andrew M. Ramey, Claudio Verdugo, Karen Lyons, Patricia G. Parker, and Leonid Kruglyak. 2016. "Loss of Flight in the Galápagos Cormorant Mirrors Human Skeletal Ciliopathies." *BioRxiv* 061432. https://doi.org/10.1101/061432.

Burns, Kevin J., Shannon J. Hackett, and Nedra K. Klein. 2002. "Phylogenetic Relationships and Morphological Diversity in Darwin's Finches and Their Relatives." *Evolution* 56, no. 6: 1240–1252.

Bush, Mark B., Alejandra Restrepo, and Aaron F. Collins. 2014. "Galápagos History, Restoration, and a Shifted Baseline." *Restoration Ecology* 22, no. 3: 296–98. https://doi.org/10.1111/rec.12080.

Cabrera, Mercedes. 2016. *"Pescadores Protestan En San Cristóbal."* (Fishermen Protest in San Cristóbal). *El Universo*, August 11. https://www.eluniverso.com/noticias/2016/08/11/nota/5736654/pescadores-protestan-san-cristobal.

Cai, Wenju, Guojian Wang, Boris Dewitte, Lixin Wu, Agus Santoso, Ken Takahashi, Yun Yang, Aude Carréric, and Michael J. McPhaden. 2018. "Increased Variability of Eastern Pacific El Niño under Greenhouse Warming." *Nature* 564, no. 7735: 201–6. https://doi.org/10.1038/s41586-018-0776-9.

Cai, Wenju, Agus Santoso, Guojian Wang, Sang-Wook Yeh, Soon-Il An, Kim M. Cobb, Mat Collins, et al. 2015a. "ENSO and Greenhouse Warming." *Nature Climate Change* 5, no. 9: 849–59. https://doi.org/10.1038/nclimate2743.

Cai, Wenju, Guojian Wang, Agus Santoso, Michael J. McPhaden, Lixin Wu, Fei-Fei Jin, Axel Timmermann, et al. 2015b. "Increased Frequency of Extreme La Niña Events under Greenhouse Warming." *Nature Climate Change* 5, no. 2: 132–37. https://doi.org/10.1038/nclimate2492.

Cairns, Rose. 2011. "A Critical Analysis of the Discourses of Conservation and Science on the Galápagos Islands." PhD diss., University of Leeds.

Campbell, Karl J., Joe Beek, Charles T. Eason, Alistair S. Glen, John Godwin, Fred Gould, Nick D. Holmes, et al. 2015. "The Next Generation of Rodent Eradications: Innovative Technologies and Tools to Improve Species Specificity and Increase Their Feasibility on Islands." *Biological Conservation* 185, no. 2015: 47–58. https://doi.org/10.1016/j.biocon.2014.10.016.

Caria, Sara, and Rafael Domínguez. 2016. "Ecuador's *Buen Vivir*" (Ecuador's "Good Living"). *Latin American Perspectives* 43, no. 1: 18–33. https://doi.org/10.1177/0094582X15611126.

Carlquist, Sherwin. 1967. "The Biota of Long-Distance Dispersal. V. Plant Dispersal to Pacific Islands." *Bulletin of the Torrey Botanical Club* 94, no. 3: 129–62. https://doi.org/10.2307/2484044.

Carlquist, Sherwin. 1974. *Island Biology.* New York: Columbia University Press.

Carmi, Ore, Christopher C. Witt, Alvaro Jaramillo, and John P. Dumbacher. 2016. "Phylogeography of the Vermilion Flycatcher Species Complex: Multiple Speciation Events, Shifts in Migratory Behavior, and an Apparent Extinction of a Galápagos-Endemic Bird Species." *Molecular Phylogenetics and Evolution* 102 (May): 152–73. https://doi.org/10.1016/j.ympev.2016.05.029.

Carpenter, C. C. 1966. "The Marine Iguana of the Galápagos Islands: Its Behavior and Ecology." *Proceedings of the California Academy of Sciences* 34, no. 4: 329–76.

Carrion, Victor, C. Josh Donlan, Karl J. Campbell, Christian Lavoie, and Felipe Cruz. 2011. "Archipelago-wide Island Restoration in the Galápagos Islands: Reducing Costs of Invasive Mammal Eradication Programs and Reinvasion Risk." *PLOS One* 6, no. 5: e18835. https://doi.org/10.1371/journal.pone.0018835.

Carroll, Scott P., and Charles W. Fox. 2008. *Conservation Biology: Evolution in Action*. Oxford University Press.

Castañeda-Rico, Susette, Sarah A. Johnson, Scott A. Clement, Robert C. Dowler, Jesús E. Maldonado, and Cody W. Edwards. 2019. "Insights into the Evolutionary and Demographic History of the Extant Endemic Rodents of the Galápagos Islands." *Therya* 10, no. 3: 213–28.

Castrejón, Mauricio. 2012. "The Reform of the PARMA Licensing System: The First Step in Eliminating the Race for Fish in the Galápagos Marine Reserve." In *Galápagos Report 2011–2012*, 136–42. Puerto Ayora, Galápagos: Galápagos National Park Service, Governing Council for the Special Regime of Galápagos, Charles Darwin Foundation, and Galápagos Conservancy.

Castrejón, Mauricio, and Anthony Charles. 2013. "Improving Fisheries Co-management Through Ecosystem-Based Spatial Management: The Galápagos Marine Reserve." *Marine Policy* 38 (March): 235–45. https://doi.org/10.1016/j.marpol.2012.05.040.

Castrejón, Mauricio, and Omar Defeo. 2015. "Co-governance of Small-Scale Shellfisheries in Latin America: Institutional Adaptability to External Drivers of Change." In *Interactive Governance for Small-Scale Fisheries*, edited by Svein Jentoft and Ratana Chuenpagdee, 605–25. MARE Publication Series. Cham, Switzerland: Springer. https://doi.org/10.1007/978-3-319-17034-3_31.

Castrejón, Mauricio, Omar Defeo, Günther Reck, and Anthony Charles. 2014. "Fishery Science in Galápagos: From a Resource-Focused to a Social–Ecological Systems Approach." In *The Galápagos Marine Reserve*, edited by Judith Denkinger and Luis Vinueza, 159–85. Cham, Switzerland: Springer. https://doi.org/10.1007/978-3-319-02769-2.

Castro, Isabel, and Antonia Phillips. 1996. *A Guide to the Birds of the Galápagos Islands*. Princeton, NJ: Princeton University Press.

Castro, Xavier. 2005. *Analysis of the Current Socio Economic Situation of the "Galápagos Artisanal Fishing Community."* Tokyo: JICA/Parque Nacional Galápagos.

Causton, Charlotte. 2007. "Risks Associated with Current and Proposed Air Routes to the Galápagos Islands." In *Galápagos Report 2006–2007*, 55–59. Puerto Ayora, Galápagos: Charles Darwin Foundation, Galápagos National Park, and National Galápagos Institute.

Causton, Charlotte. 2016. "Success in Biological Control: The Scale and the Ladybird." In *Galápagos: Preserving Darwin's Legacy*, edited by Tui De Roy, 184–90. London: Bloomsbury.

Causton, Charlotte E., Heinke Jäger, María Verónica Toral Granda, Marilyn Cruz, Manuel Mejía, Erika Guerrero, and Christian Sevilla. 2017. "Total Number and Current Status of Species Introduced and Intercepted in the Galápagos Islands." In *Galápagos Report 2015–2016*, 181–83. Puerto Ayora, Galápagos: Galápagos National Park Directorate, Governing Council for the Special Regime of Galápagos, Charles Darwin Foundation, and Galápagos Conservancy.

Cayot, Linda J. 2008. "The Restoration of Giant Tortoise and Land Iguana Populations in Galápagos." *Galápagos Research* 65:39–43.

Cayot, Linda J. 2016. "When It Rains." *Galápagos News* (Spring-Summer): 6.

Celata, Filippo, and Venere Stefania Sanna. 2012. "The Post-political Ecology of Protected Areas: Nature, Social Justice and Political Conflicts in the Galápagos Islands." *Local Environment* 17, no. 9: 977–90. https://doi.org/10.1080/13549839.2012.688731.

Chambers, Paul. 2006. *A Sheltered Life: The Unexpected History of the Giant Tortoise.* New York: Oxford University Press.

Charles Darwin Foundation. 2019. "Alert on the Current Conservation Status of the Galápagos Natural Assets: Institutional Position of the Charles Darwin Foundation for the Galápagos Islands." Puerto Ayora, Galápagos: Charles Darwin Foundation.

Chaves, Jaime A., Pedro J. Martinez-Torres, Emiliano A. Depino, Sebastian Espinoza-Ulloa, Jefferson García-Loor, Annabel C. Beichman, and Martin Stervander. 2020. "Evolutionary history of the Galápagos Rail revealed by ancient mitogenomes and modern samples." *Diversity* 12, no. 11: 425. https://doi.org/10.3390/d12110425.

Chavez, Francisco P., John Ryan, Salvador E. Lluch-Cota, and Miguel Niquen C. 2003. "From Anchovies to Sardines and Back: Multidecadal Change in the Pacific Ocean." *Science* 299, no. 5604: 217–21. https://doi.org/10.1126/science.1075880.

Chiari, Ylenia, Scott Glaberman, Pedro Tarroso, Adalgisa Caccone, and Julien Claude. 2016. "Ecological and Evolutionary Influences on Body Size and Shape in the Galápagos Marine Iguana (*Amblyrhynchus cristatus*)." *Oecologia* 181, no. 3: 885–94. https://doi.org/10.1007/s00442-016-3618-1.

Chiari, Ylenia, Arie van der Meijden, Adalgisa Caccone, Julien Claude, and Benjamin Gilles. 2017. "Self-Righting Potential and the Evolution of Shell Shape in Galápagos Tortoises." *Scientific Reports* 7, no. 1: 15828. https://doi.org/10.1038/s41598-017-15787-7.

Cimadom, Arno, Charlotte Causton, Dong H. Cha, David Damiens, Birgit Fessl, Rebecca Hood-Nowotny, Piedad Lincango, et al. 2016. "Darwin's Finches Treat Their Feathers with a Natural Repellent." *Scientific Reports* 6, no. 1: 34559. https://doi.org/10.1038/srep34559.

Cleary, Z., D. M. Schwartz, E. Mittelstaedt, and K. Harpp. 2020. "Dynamic Magma Storage at Near-Ridge Hotspots: Evidence from New Galápagos Gravity Data." *Geochemistry, Geophysics, Geosystems* 21, no. 3: e2019GC008722.

Clifford, L. D., and D. J. Anderson. 2001. "Food Limitation Explains Most Clutch Size Variation in the Nazca Booby." *Journal of Animal Ecology* 70, no. 4: 539–45. https://doi.org/10.1046/j.1365-2656.2001.00521.x.

Coffey, Emily E. D., Cynthia A. Froyd, and Katherine J. Willis. 2012. "Lake or Bog? Reconstructing Baseline Ecological Conditions for the Protected Galápagos Sphagnum Peatbogs." *Quaternary Science Reviews* 52 (October): 60–74. https://doi.org/10.1016/j.quascirev.2012.08.002.

Colding, Johan, and Stephan Barthel. 2019. "Exploring the Social-Ecological Systems Discourse 20 Years Later." *Ecology and Society* 24, no. 1. https://doi.org/10.5751/ES-10598-240102.

Coleridge, Samuel T. (1857) 2000. "The Rime of the Ancient Mariner." In *The Major Works*, by S. T. Coleridge. New York: Oxford University Press.

Colossus Productions. 2013. *Galápagos 3D with David Attenborough*. Go Entertainment Group.

Conroy, Jessica L., Jonathan T. Overpeck, Julia E. Cole, Timothy M. Shanahan, and Miriam Steinitz-Kannan. 2008. "Holocene Changes in Eastern Tropical Pacific Climate Inferred from a Galápagos Lake Sediment Record." *Quaternary Science Reviews* 27, no. 11/12: 1166–80. https://doi.org/10.1016/j.quascirev.2008.02.015.

Consejo de Gobierno del Régimen Especial de Galápagos. 2020. *Plan de Desarrollo Sustentable y Ordenamiento Territorial del Régimen Especial de Galápagos: Plan Galápagos 2030 (Borrador)*. (Plan for Sustainable Development and Territorial Organization of the Special Government of Galápagos: Plan Galápagos 3030 (Draft)). Puerto Baquerizo Moreno, Galápagos: Consejo de Gobierno del Régimen Especial de Galápagos.

Conservación y Desarrollo. 2017. "Smart Voyager." http://www.ccd.ec/turismo.

Cook, Timothée R., Akiko Kato, Hideji Tanaka, Yan Ropert-Coudert, and Charles-André Bost. 2010. "Buoyancy under Control: Underwater Locomotor Performance in a Deep Diving Seabird Suggests Respiratory Strategies for Reducing Foraging Effort." *PLOS One* 5, no. 3: e9839. https://doi.org/10.1371/journal.pone.0009839.

Corredor Acosta, Andrea, Alberto Acosta, Phillipe Gaspar, and Beatriz Calmettes. 2011. "Variation in the Surface Currents in the Panama Bight during El Niño and La Niña Events from 1993 to 2007." *Bulletin of Marine and Coastal Research* 40:33–56. https://doi.org/10.25268/bimc.invemar.2011.40.0.127.

Cotner, Sehoya, Hannah Graczyk, José Luis Rodríguez Garcia, and Randy Moore. 2016. "In Galápagos . . . and Uncomfortable with Evolution." *Journal of Biological Education* 50, no. 2: 115–19. https://doi.org/10.1080/00219266.2016.1175758.

Courtillot, Vincent, Anne Davaille, Jean Besse, and Joann Stock. 2003. "Three Distinct Types of Hotspots in the Earth's Mantle." *Earth and Planetary Science Letters* 205: 295–308. https://doi.org/10.1016/S0012-821X(02)01048-8.

Cox, Michael, Gwen Arnold, and Sergio Villamayor Tomás. 2010. "A Review of Design Principles for Community-Based Natural Resource Management." *Ecology and Society* 15, no. 4: 38.

Crawford, R., U. Ellenberg, E. Frere, C. Hagen, K. Baird, P. Brewin, S. Crofts, et al. 2017. "Tangled and Drowned: A Global Review of Penguin Bycatch in Fisheries." *Endangered Species Research* 34 (November): 373–96. https://doi.org/10.3354/esr00869.

Creamer, Monserrat, and Nick Cabot. 2019. *The Education for Sustainability in Galápagos Program: Origin, Purpose, Structure, Philosophy and Initial Progress at the Program's Midpoint*. Puerto Ayora, Galápagos: Galápagos Conservancy and Fundación Scalesia.

Crona, B. I., T. Van Holt, M. Petersson, T. M. Daw, and E. Buchary. 2015. "Using Social–Ecological Syndromes to Understand Impacts of International Seafood Trade on Small-Scale Fisheries." *Global Environmental Change* 35 (November): 162–75. https://doi.org/10.1016/j.gloenvcha.2015.07.006.

Cruz, Felipe, Victor Carrion, Karl J. Campbell, Christian Lavoie, and C. Josh Donlan. 2009. "Bio-economics of Large-Scale Eradication of Feral Goats from Santiago Island, Galápagos." *Journal of Wildlife Management* 73, no. 2: 191–200. https://doi.org/10.2193/2007-551.

Cruz, Austin R., Samantha T. Selby, and William H. Durham. 2017a. "Place-Based Education for Environmental Behavior: A 'Funds of Knowledge' and Social Capital

Approach." *Environmental Education Research* 24, no. 5: 1–21. https://doi.org/10.1080/13504622.2017.1311842.

Cruz, Marilyn, Mónica Ramos, Viviana Duque, Mariela Cedeño, Martín Espinosa, Alberto Vélez, Ronal Azuero, Manuel Mejía, David Arana, and Rommel Iturbide. 2017b. "Biosecurity in Galápagos is Vital for Protecting Human Health, the Local Economy and Biodiversity." In *Galápagos Report 2015–2016*, 27–32. Puerto Ayora, Galápagos: Galápagos National Park Directorate, Governing Council for the Special Regime of Galápagos, Charles Darwin Foundation, and Galápagos Conservancy.

Culik, B., J. Hennicke, and T. Martin. 2000. "Humboldt Penguins Outmaneuvering El Niño." *The Journal of Experimental Biology* 203, no. 15: 2311–22.

Curran, L. M., and M. Leighton. 2000. "Vertebrate Responses to Spatiotemporal Variation in Seed Production of Mast-Fruiting Dipterocarpaceae." *Ecological Monographs* 70, no. 4: 101–28. https://doi.org/10.1890/0012-9615(2000)070[0101:VRTSVI]2.0.CO;2.

Daigre, Maximiliano, Paulina Arce, and Alejandro Simeone. 2012. "Fledgling Peruvian Pelicans (*Pelecanus thagus*) Attack and Consume Younger Unrelated Conspecifics." *The Wilson Journal of Ornithology* 124, no. 3: 603–7. https://doi.org/10.1676/12-011.1.

Daoust, P. Y., G. V. Dobbin, R. C. Ridlington Abbot, and S. D. Dawson. 2008. "Descriptive Anatomy of the Subcutaneous Air Diverticula in the Northern Gannet, *Morus bassanus*." *Seabird* 21: 64–76.

Darwin, Charles. (1836) 1963. *Darwin's Ornithological Notes*, edited by Nora Barlow. Vol. 2. Historical Series. London: Bulletin of the British Museum (Natural History).

Darwin, C. R. 1839a. *Narrative of the Surveying Voyages of His Majesty's Ships Adventure and Beagle Between the Years 1826 and 1836, Describing Their Examination of the Southern Shores of South America, and the Beagle's Circumnavigation of the Globe. Journal and Remarks. 1832–1836.* London: Henry Colburn.

Darwin, C. R. 1839b. *Birds Part 3 No. 4 of The Zoology of the Voyage of H.M.S. Beagle by John Gould.* London: Smith Elder and Co.

Darwin, C. R. 1845. *Journal of Researches into the Natural History and Geology of the Countries Visited During the Voyage of H.M.S. Beagle Round the World, Under the Command of Capt. Fitz Roy, R. N.* 2nd ed. London: John Murray.

Darwin, C. R. 1846. "Letter to J. D. Hooker (Letter No. 986)." *Darwin Correspondence Project.* http://www.darwinproject.ac.uk/DCP-LETT-986.

Darwin, C. R. 1859. *On the Origin of Species by Means of Natural Selection, or the Preservation of Favoured Races in the Struggle for Life.* London: John Murray.

Dawson, E. Y. 1966. "Cacti in the Galápagos Islands, with Special Reference to Their Relations with Tortoises." In *The Galápagos: Proceedings of the Symposia of the Galápagos International Scientific Project*, edited by R. I. Bowman, 209–14. Berkeley: University of California Press.

Dawson, William R., George A. Bartholomew, and Albert F. Bennett. 1977. "A Reappraisal of the Aquatic Specializations of the Galápagos Marine Iguana (*Amblyrhynchus cristatus*)." *Evolution* 31, no. 4: 891–97. https://doi.org/10.2307/2407452.

De Busschere, Charlott, Léon Baert, Steven Marcel Van Belleghem, Wouter Dekoninck, and Frederik Hendrickx. 2012. "Parallel Phenotypic Evolution in a Wolf Spider Radiation on Galápagos." *Biological Journal of the Linnean Society* 106, no. 1: 123–36. https://doi.org/10.1111/j.1095-8312.2011.01848.x.

De León, L. F., J. Podos, T. Gardezi, A. Herrel, and A. P. Hendry. 2014. "Darwin's Finches and Their Diet Niches: The Sympatric Coexistence of Imperfect Generalists." *Journal of Evolutionary Biology* 27, no. 6: 1093–1104. https://doi.org/10.1111/jeb.12383.

De León, Luis F., Diana M. T. Sharpe, Kiyoko M. Gotanda, Joost A. M. Raeymaekers, Jaime A. Chaves, Andrew P. Hendry, and Jeffrey Podos. 2018. "Urbanization Erodes Niche Segregation in Darwin's Finches." *Evolutionary Applications* 12, no. 7: 1329–43. https://doi.org/10.1111/eva.12721.

De León, Luis Fernando, Joost A. M. Raeymaekers, Eldredge Bermingham, Jeffrey Podos, Anthony Herrel, and Andrew P. Hendry. 2011. "Exploring Possible Human Influences on the Evolution of Darwin's Finches." *Evolution* 65, no. 8: 2258–72. https://doi.org/10.1111/j.1558-5646.2011.01297.x.

De Queiroz, Alan. 2014. *The Monkey's Voyage: How Improbable Journeys Shaped the History of Life*. New York: Basic Books.

De Roy, Tui. 2008. "Under the Tropical Sun: The Galápagos Albatross." In *Albatross: Their World, Their Ways*, edited by Tui De Roy, Mark Jones, and Julian Fitter, 124–35. Buffalo, New York: Firefly Books.

De Roy, Tui, ed. 2016. *Galápagos: Preserving Darwin's Legacy*. London: Bloomsbury.

De Vries, Tjitte. 1975. "The Breeding Biology of the Galápagos Hawk, *Buteo galapagoensis*." *Le Gerfaut* 65: 29–57.

De Vries, Tjitte. 2015. *The Galápagos Hawk*. Monografías Zoológicas, Serie Neotropical, Vol. 1. Almenara (Castellón), Spain: Tundra Ediciones.

Dechmann, Dina K. N., Scott LaPoint, Christian Dullin, Moritz Hertel, Jan R. E. Taylor, Karol Zub, and Martin Wikelski. 2017. "Profound Seasonal Shrinking and Regrowth of the Ossified Braincase in Phylogenetically Distant Mammals with Similar Life Histories." *Scientific Reports* 7 (February): 42443. https://doi.org/10.1038/srep42443.

Defeo, Omar, and Juan Carlos Castilla. 2012. "Governance and Governability of Coastal Shellfisheries in Latin America and the Caribbean: Multi-scale Emerging Models and Effects of Globalization and Climate Change." *Current Opinion in Environmental Sustainability* 4, no. 3: 344–50. https://doi.org/10.1016/j.cosust.2012.05.002.

Defeo, Omar, Mauricio Castrejón, Roberto Pérez-Castañeda, Juan C. Castilla, Nicolás L. Gutiérrez, Timothy E. Essington, and Carl Folke. 2016. "Co-management in Latin American Small-Scale Shellfisheries: Assessment from Long-Term Case Studies." *Fish and Fisheries* 17, no. 1: 176–92. https://doi.org/10.1111/faf.12101.

Denkinger, Judith, and Luis Vinueza. 2014. *The Galápagos Marine Reserve: A Dynamic Social-Ecological System*. Cham, Switzerland: Springer. https://doi.org/10.1007/978-3-319-02769-2.

Dillard, Annie. 1982. *Teaching a Stone to Talk: Expeditions and Encounters*. New York: Harper & Row.

Dirección del Parque Nacional Galápagos. Plan de Manejo de las Áreas Protegidas de Galápagos para el Buen Vivir (Management Plan of the Protected Areas of Galápagos for Good Living). 2014. Puerto Ayora, Galápagos, Ecuador.

Dirección del Parque Nacional Galápagos, Comisión Técnica Pesquera and Sector Pesquero Artesanal de Galápagos. "Calendario Pesquero 2016–2021: Estudio Técnico" (Fishing Calendar 2016–2021: Technical Study). Puerto Ayora, Galápagos, Ecuador.

Dirección del Parque Nacional Galápagos y Observatorio de Turismo de Galápagos. 2018. *Informe Anual de Visitantes a las Áreas Protegidas de Galápagos del Año 2017 (Report of Annual Visitors to the Protected Areas of Galapagos for the Year 2017)*. Galápagos, Ecuador: Dirección del Parque Nacional Galápagos.

Dobzhansky, T. 1973. "Nothing in Biology Makes Sense Except in the Light of Evolution." *The American Biology Teacher* 35 (January): 125–29. https://doi.org/10.2307/4444260.

Donlan, C. Josh, Karl Campbell, Wilson Cabrera, Christian Lavoie, Victor Carrion, and Felipe Cruz. 2007. "Recovery of the Galápagos Rail (*Laterallus spilonotus*) following the Removal of Invasive Mammals." *Biological Conservation* 138, no. 3/4: 520–24. https://doi.org/10.1016/j.biocon.2007.05.013.

D'Ozouville, Noémi, Giuseppe Di Carlo, Fernando Ortiz, Free De Koning, Scott Henderson, and Emily Pidgeon. 2010. "Galápagos in the Face of Climate Change: Considerations for Biodiversity and Associated Human Well-Being." In *Galápagos Report 2009–2010*, 170–75. Puerto Ayora, Galápagos: Charles Darwin Foundation, Galápagos National Park, and Governing Council of Galápagos.

Duffie, Caroline V., Travis C. Glenn, F. Hernan Vargas, and Patricia G. Parker. 2009. "Genetic Structure within and between Island Populations of the Flightless Cormorant (*Phalacrocorax harrisi*)." *Molecular Ecology* 18, no. 10: 2103–11. https://doi.org/10.1111/j.1365-294X.2009.04179.x.

Duffy, David Cameron. 2020. "Charles Darwin and the Case of the Missing Cormorants." *Galápagos Research* 69: 45–48.

Dunson, William. 1969. "Electrolyte Excretion by the Salt Gland of the Galápagos Marine Iguana." *The American Journal of Physiology* 216, no. 4: 995–1002. https://doi.org/10.1152/ajplegacy.1969.216.4.995.

Durham, William H. 1991. *Coevolution: Genes, Culture, and Human Diversity*. Stanford, CA: Stanford University Press.

Durham, William H. 2007. "The Elephant in the Room: Evolution in Anthropology." *General Anthropology Bulletin of the General Anthropology Division* 14, no. 2: 1–7. https://doi.org/10.1525/ga.2007.14.2.1a.

Durham, William H. 2008. "Fishing for Solutions: Ecotourism and Conservation in Galápagos National Park." In *Ecotourism and Conservation in the Americas*, edited by Amanda Stronza and William H Durham, 66–89. Wallingford, Oxfordshire, UK: CABI.

Durham, William H. 2012. "What Darwin Found Convincing in Galápagos." In *The Role of Science for Conservation*, edited by Matthias Wolff and Mark Gardener, 3–15. London: Routledge.

Dvorak, Michael, Birgit Fessl, Erwin Nemeth, Sonia Kleindorfer, and Sabine Tebbich. 2012. "Distribution and Abundance of Darwin's Finches and Other Land Birds on Santa Cruz Island, Galápagos: Evidence for Declining Populations." *Oryx* 46, no. 1: 78–86. https://doi.org/10.1017/S0030605311000597.

Dvorak, Michael, Erwin Nemeth, Beate Wendelin, Patricio Herrera, Denis Mosquera, David Anchundia, Christian Sevilla, Sabine Tebbich, and Birgit Fessl. 2017. "Conservation Status of Landbirds on Floreana: The Smallest Inhabited Galápagos Island." *Journal of Field Ornithology* 88, no. 2: 132–145.

Eddleman, W. R., R. E. Flores, and M. Legare. 2020. Black Rail (*Laterallus jamaicensis*), Version 1.0. In *Birds of the World*, edited by A. F. Poole and F. B. Gill. Ithaca, NY: Cornell Lab of Ornithology.

Eddy, Tyler D., Alan M. Friedlander, and Pelayo Salinas de León. 2019. "Ecosystem Effects of Fishing & El Niño at the Galápagos Marine Reserve." *PeerJ* 7 (May): e6878. https://doi.org/10.7717/peerj.6878.

Edgar, Graham J., Stuart A. Banks, Margarita Brandt, Rodrigo H. Bustamante, Angel Chiriboga, Sylvia A. Earle, Lauren E. Garske, et al. 2010. "El Niño, Grazers and Fisheries Interact to Greatly Elevate Extinction Risk for Galápagos Marine Species." *Global Change Biology* 16, no. 10: 2876–90. https://doi.org/10.1111/j.1365-2486.2009.02117.x.

Edgar, G. J., R. H. Bustamante, J-M. Fariña, M. Calvopiña, C. Martínez, and M. V. Toral-Granda. 2004. "Bias in Evaluating the Effects of Marine Protected Areas: The Importance of Baseline Data for the Galápagos Marine Reserve." *Environmental Conservation* 31, no. 3: 212–18. https://doi.org/ 10.1017/S0376892904001584.

Edwards, Danielle L., Edgar Benavides, Ryan C. Garrick, James P. Gibbs, Michael A. Russello, Kirstin B. Dion, Chaz Hyseni, Joseph P. Flanagan, Washington Tapia, and Adalgisa Caccone. 2013. "The Genetic Legacy of Lonesome George Survives: Giant Tortoises with Pinta Island Ancestry Identified in Galápagos." *Biological Conservation* 157 (January): 225–28. https://doi.org/10.1016/j.biocon.2012.10.014.

Ehrenberg, Rachel. 2018. "What Makes a Tree a Tree?" *Knowable Magazine*, March. https://doi.org/10.1146/knowable-033018-032602.

Eibl-Eibesfeldt, I. 1984. "The Large Iguanas of the Galápagos Islands." In *Key Environments—Galápagos*, edited by Roger Perry, 157–73. Oxford, England: Pergamon Press.

Eliasson, Uno. 1974. "Studies in Galápagos Plants. XIV. The Genus *Scalesia* Arnott." *Opera Botanica* 36: 1–117.

Eliasson, Uno. 1984. "Native Climax Forests." In *Key Environments—Galápagos*, edited by Roger Perry, 101–14. Oxford, England: Pergamon Press.

Ellis-Soto, Diego, Stephen Blake, Alaaeldin Soultan, Anne Guézou, Fredy Cabrera, and Stefan Lötters. 2017. "Plant Species Dispersed by Galápagos Tortoises Surf the Wave of Habitat Suitability under Anthropogenic Climate Change." *PLOS One* 12, no. 7: e0181333. https://doi.org/10. 1371/journal.pone.0181333.

Engie, Kim. 2015. "Adaptation and Shifting Livelihoods in the Small-Scale Fisheries of the Galápagos Marine Reserve, Ecuador." PhD diss., University of North Carolina at Chapel Hill.

Epler, Bruce. 2007. *Tourism, the Economy, Population Growth, and Conservation in Galápagos*. Puerto Ayora, Galápagos: Charles Darwin Foundation.

Epler, Bruce. 2013. *Galápagos: A Human History*. Campbell, CA: Fast Pencil.

Epstein, Graham, Jessica M. Vogt, Sarah K. Mincey, Michael Cox, and Burney Fischer. 2013. "Missing Ecology: Integrating Ecological Perspectives with the Social-Ecological System Framework." *International Journal of the Commons* 7, no. 2: 432–53. https://doi.org/10.18352/ijc.371.

Erazo, Gabriela, María Casafont, and Mariuxi Farías. 2017. "Analysis of Experiential Fishing as an Alternative Mode of Sustainable Tourism in Galápagos." In *Galápagos Report 2015–2016*, 98–102. Puerto Ayora, Galápagos: Galápagos National Park Directorate, Governing Council for the Special Regime of Galápagos, Charles Darwin Foundation, and Galápagos Conservancy.

Ernst, Carl H., and Jeffrey E. Lovich. 2009. *Turtles of the United States and Canada*. Baltimore: Johns Hopkins University Press.

Espin, Paola A., Carlos F. Mena, and Francesco Pizzitutti. 2019. "A Model-Based Approach to Study the Tourism Sustainability in an Island Environment: The Case of Galápagos Islands." In *Urban Galápagos: Transition to Sustainability in Complex Adaptive Systems*, edited by Thomaz Kvan and Justyna Karakiewicz, 97–114. Cham, Switzerland: Springer.

Estes, G., K. T. Grant, and P. R. Grant. 2000. "Darwin in the Galápagos: His Footsteps through the Archipelago." *Notes and Records of the Royal Society* 54, no. 3: 343–68. https://doi.org/10.1098/rsnr.2000.0117.

Evans, William R., and Kenneth V. Rosenberg. 2000. "Acoustic Monitoring of Night-Migrating Birds: A Progress Report." In *Strategies for Bird Conservation: The Partners in Flight Planning Process (Proceedings RMRS-P-16)*, edited by Rick Bonney, David N. Pashley, Robert J. Cooper, and Larry Niles, 151–59. Ogden, UT: Department of Agriculture, U.S. Forest Service, Rocky Mountain Research Station.

Fabinyi, Michael, Louisa Evans, and Simon J. Foale. 2014. "Social-Ecological Systems, Social Diversity, and Power: Insights from Anthropology and Political Ecology." *Ecology and Society* 19, no. 4: 28. https://doi.org/10.5751/ES-07029-190428.

Farrington, Heather L., Lucinda P. Lawson, Courtney M. Clark, and Kenneth Petren. 2014. "The Evolutionary History of Darwin's Finches: Speciation, Gene Flow, and Introgression in a Fragmented Landscape." *Evolution* 68, no. 10: 2932–44. https://doi.org/10.1111/evo.12484.

Feeny, David, Fikret Berkes, Bonnie J. McCay, and James M. Acheson. 1990. "The Tragedy of the Commons: Twenty-Two Years Later." *Human Ecology* 18, no. 1: 1–19. https://doi.org/10.1007/BF00889070.

Ferber, D. 2000. "Conservation Biology: Galápagos Station Survives Latest Attack by Fishers." *Science* 290, no. 5499: 2059–61. https://doi.org/10.1126/science.290.5499.2059.

Fernández-Mazuecos, Mario, Pablo Vargas, Ross A. McCauley, David Monjas, Ana Otero, Jaime A. Chaves, Juan Ernesto Guevara Andino, and Gonzalo Rivas-Torres. 2020. "The Radiation of Darwin's Giant Daisies in the Galápagos Islands." *Current Biology* 30. no. 24: 4989–4998. https://doi.org/10.1016/j.cub.2020.09.019.

Fessl, Birgit, David Anchundia, Jorge Carrión, Arno Cimadom, Javier Cotin, Francesca Cunninghame, Michael Dvorak, Denis Mosquera, Erwin Nemeth, Christian Sevilla, Sabine Tebbich, Beate Wendelin, and Charlotte Causton. 2017. "Galápagos Landbirds (Passerines, Cuckoos, and Doves): Status, Threats, and Knowledge Gaps." In *Galápagos Report 2015–2016*, 149–60. Puerto Ayora, Galápagos: Galápagos National Park Directorate, Governing Council for the Special Regime of Galápagos, Charles Darwin Foundation, and Galápagos Conservancy.

Fessl, Birgit, George E. Heimpel, and Charlotte E. Causton. 2018. "Invasion of an Avian Nest Parasite, *Philornis downsi*, to the Galápagos Islands: Colonization History, Adaptations to Novel Ecosystems, and Conservation Challenges." In *Disease Ecology*, edited by Patricia G. Parker. Cham, Switzerland: Springer. https://doi.org/10.1007/978-3-319-65909-1_9.

Fessl, Birgit, Glyn H. Young, Richard P. Young, Jorge Rodríguez-Matamoros, Michael Dvorak, Sabine Tebbich, and John E. Fa. 2010. "How to Save the Rarest Darwin's Finch from Extinction: The Mangrove Finch on Isabela Island." *Philosophical Transactions of the Royal Society of London. Series B, Biological Sciences* 365, no. 1543: 1019–30. https://doi.org/10.1098/rstb.2009.0288.

Finsen, Walter. n.d. "Debunking the Baroness." Accessed August 2, 2019. http://www.Galápagos.to/TEXTS/FINSEN.HTM.

Fitter, J. 2008. "Introduction to Albatrosses, Mollymawks and Gooneys." In *Albatross: Their World, Their Ways*, edited by T. De Roy, Mark Jones, and Julian Fitter, 186–89. Buffalo, New York: Firefly Books.

Fitter, Julian, Daniel Fitter, and David Hosking. 2016. *Wildlife of the Galápagos*. Princeton, NJ: Princeton University Press.

Fournier, Bertrand, Emily E. D. Coffey, W. O. van der Knaap, Leonardo D. Fernández, Anatoly Bobrov, and Edward A. D. Mitchell. 2016. "A Legacy of Human-Induced

Ecosystem Changes: Spatial Processes Drive the Taxonomic and Functional Diversities of Testate Amoebae in Sphagnum Peatlands of the Galápagos." *Journal of Biogeography* 43:533–43. https://doi.org/10.1111/jbi.12655.

Franklin, Alan B., Deborah A. Clark, and David B. Clark. 1979. "Ecology and Behavior of the Galápagos Rail." *The Wilson Bulletin* 91, no. 2: 202–21.

French, Susannah S., Dale F. DeNardo, Timothy J. Greives, Christine R. Strand, and Gregory E. Demas. 2010. "Human Disturbance Alters Endocrine and Immune Responses in the Galápagos Marine Iguana (*Amblyrhynchus cristatus*)." *Hormones and Behavior* 58, no. 5: 792–99. https://doi.org/10.1016/j.yhbeh.2010.08.001.

Frias-Soler, Roberto, Elizabeth Tindle, Georgina Espinosa Lopez, Simon Blomberg, Adelheid Studer-Thiersch, Michael Wink, and Robert Tindle. 2014. "Genetic and Phenotypic Evidence Supports Evolutionary Divergence of the American Flamingo (*Phoenicopterus ruber*) Population in the Galápagos Islands." *Waterbirds* 37, no. 4: 349–468.

Friesen, V. L., D. J. Anderson, T. E. Steeves, and E. A. Schreiber. 2002. "Molecular Support for Species Status of the Nazca Booby (*Sula granti*)." *The Auk* 119, no. 3: 820–26. https://doi.org/10.1642/0004-8038(2002)119[0820:MSFSSO]2.0.CO;2.

Frith, Clifford B. 2016. *Charles Darwin's Life with Birds: His Complete Ornithology.* New York: Oxford University Press.

Fritts, T. H. 1983. "Morphometrics of Galápagos Tortoises: Evolutionary Implications." In *Patterns of Evolution in Galápagos Organisms,* edited by R. I. Bowman, Margaret Berson, and Alan E. Leviton, 107–22. San Francisco: Pacific Division, American Association for the Advancement of Science.

Fritts, Thomas H. 1984. "Evolutionary Divergence of Giant Tortoises in Galápagos." *Biological Journal of the Linnean Society* 21, no. 1/2: 165–76. https://doi.org/10.1111/j.1095-8312.1984.tb02059.x.

Froyd, Cynthia A., Emily E. D. Coffey, Willem O. van der Knaap, Jacqueline F. N. van Leeuwen, Alan Tye, and Katherine J. Willis. 2014. "The Ecological Consequences of Megafaunal Loss: Giant Tortoises and Wetland Biodiversity." *Ecology Letters* 17, no. 2: 144–54. https://doi.org/10.1111/ele.12203.

Galápagos Conservancy. 2018. "Giant Tortoise Restoration Initiative." https://www.Galápagos.org/conservation/our-work/tortoise-restoration/.

Galápagos National Park Service. 2018. "Tourism Statistics between 1980 and 2015." Accessed September 3. http://www.Galápagos.gob.ec/estadistica-de-visitantes/.

García-Ramirez, Juan C., Gillian C. Gibb, and Steve A. Trewick. 2014. "Eocene Diversification of Crown Group Rails (Aves: Gruiformes: Rallidae)." *PLOS One* 9, no. 10: e109635. https://doi.org/10.1371/journal.pone.0109635.

García-Ramirez, Juan C., Emily Moriarty Lemmon, Alan R. Lemmon, and Nigel French. 2020. "Phylogenomic Reconstruction Sheds Light on New Relationships and Timescale of Rails (Aves: Rallidae) Evolution." *Diversity* 12, no. 70. https://doi.org/10.3390/d12020070.

Gardener, Mark R., Rachel Atkinson, and Jorge Luis Rentería. 2010. "Eradications and People: Lessons from the Plant Eradication Program in Galápagos." *Restoration Ecology* 18, no. 1: 20–29. https://doi.org/10.1111/j.1526-100X.2009.00614.x.

Gardener, Mark R., and Christophe Grenier. 2011. "Linking Livelihoods and Conservation: Challenges Facing the Galápagos Islands." In *Island Futures: Conservation and Development across the Asia-Pacific Region,* edited by

Godfrey Baldacchino and Daniel Niles, 73–85. Cham, Switzerland: Springer. https://doi.org/10.1007/978-4-431-53989-6.

Gavryushkina, Alexandra, Tracy A. Heath, Daniel T. Ksepka, Tanja Stadler, David Welch, and Alexei J. Drummond. 2017. "Bayesian Total-Evidence Dating Reveals the Recent Crown Radiation of Penguins." *Systematic Biology* 66, no. 1: 57–73. https://doi.org/10.1093/sysbio/syw060.

Geist, Dennis, Bridget A. Diefenbach, Daniel J. Fornari, Mark D. Kurz, Karen Harpp, and Jerzy Blusztajn. 2008. "Construction of the Galápagos Platform by Large Submarine Volcanic Terraces." *Geochemistry, Geophysics, Geosystems* 9, no. 3. https://doi.org/10.1029/2007GC001795.

Geist, Dennis J., Howard Snell, Heidi Snell, Charlotte Goddard, and Mark D. Kurz. 2014. "A Paleogeographical Model of the Galápagos Islands and Biogeographical and Evolutionary Implications." In *The Galápagos: A Natural Laboratory for the Earth Sciences*, edited by Karen S. Harpp, Eric Mittelstaedt, Noémi d'Ozouville, and David W. Graham, 145–66. Hoboken, NJ: John Wiley & Sons. https://doi.org/10.1002/9781118852538.ch8.

Gentile, Gabriele. 2016. "Land Iguanas: Emergence of a New Species." In *Galápagos: Preserving Darwin's Legacy*, edited by Tui De Roy, 114–21. London: Bloomsbury.

Gentile, Gabriele, Anna Fabiani, Cruz Marquez, Howard L. Snell, Heidi M. Snell, Washington Tapia, and Valerio Sbordoni. 2009. "An Overlooked Pink Species of Land Iguana in the Galápagos." *Proceedings of the National Academy of Sciences of the United States of America* 106, no. 2: 507–11. https://doi.org/10.1073/pnas.0806339106.

Gentile, Gabriele, Cruz Marquez, Howard L. Snell, Washington Tapia, and Arturo Izurieta. 2016. "Conservation of a New Flagship Species: The Galápagos Pink Land Iguana (*Conolophus marthae* Gentile and Snell, 2009)." In *Problematic Wildlife*, edited by Francesco M. Angelici, 315–36. Cham, Switzerland: Springer. https://doi.org/10.1007/978-3-319-22246-2_15.

Gentile, Gabriele, and Howard Snell. 2009. "*Conolophus marthae* sp. nov. (Squamata, Iguanidae), a New Species of Land Iguana from the Galápagos Archipelago." *Zootaxa* 2201, no. 1: 1–10. https://doi.org/10.5281/zenodo.189620.

Gergis, Joëlle L., and Anthony M. Fowler. 2009. "A History of ENSO Events since A.D. 1525: Implications for Future Climate Change." *Climatic Change* 92, no. 3/4: 343–87. https://doi.org/10.1007/s10584-008-9476-z.

Gibbons, John R. H. 1981. "The Biogeography of *Brachylophus* (Iguanidae) Including the Description of a New Species, *B. vitiensis*, from Fiji." *Journal of Herpetology* 15, no. 3: 255–73. https://doi.org/10.2307/1563429.

Gibbs, James. 2015. "Española: Creating Albatross 'Airstrips' and More." *Galápagos Conservancy*. May 21. https://www.Galápagos.org/blog/preparing-for-Espanola-2015/.

Gibbs, James P. 2017. "Galápagos Giant Tortoises Make a Comeback, Thanks to Innovative Conservation Strategies." *The Conversation*. February 15. http://theconversation.com/Galápagos-giant-tortoises-make-a-comeback-thanks-to-innovative-conservation-strategies-67591.

Gibbs, James P., Linda J. Cayot, and Washington Tapia A. 2020. *Galápagos Giant Tortoises*. London: Elsevier.

Gibbs, James P., Elizabeth A. Hunter, Kevin T. Shoemaker, Washington H. Tapia, and Linda J. Cayot. 2014. "Demographic Outcomes and Ecosystem Implications of Giant

Tortoise Reintroduction to Española Island, Galápagos." *PLOS One* 9, no. 10: e110742. https://doi.org/10.1371/journal.pone.0110742.

Gibbs, James P., Cruz Marquez, and Eleanor J. Sterling. 2008. "The Role of Endangered Species Reintroduction in Ecosystem Restoration: Tortoise–Cactus Interactions on Española Island, Galápagos." *Restoration Ecology* 16, no. 1: 88–93. https://doi.org/10.1111/j.1526-100X.2007.00265.x.

Gibbs, James P., W. Gregory Shriver, and Hernan Vargas. 2003. "An Assessment of a Galápagos Rail Population over Thirteen Years (1986 to 2000)." *Journal of Field Ornithology* 74, no. 2: 136–40. https://doi.org/10.1648/0273-8570-74.2.136

Gibbs, James P., Eleanor J. Sterling, and F. Javier Zabala. 2010. "Giant Tortoises as Ecological Engineers: A Long-Term Quasi-experiment in the Galápagos Islands." *Biotropica* 42, no. 2: 208–14. https://doi.org/10.1111/j.1744-7429.2009.00552.x.

Gibbs, James P., and Hara W. Woltz. 2010. "A Pilot Survey of the Central Colony of the Waved Albatross *Phoebastria irrorata* on Española Island." *Galápagos Research* 67: 18–20.

Gibson, Matthew J.S., Maria de Lourdes Torres, and Leonie C. Moyle. 2020. "Local extirpation is pervasive among historical populations of Galápagos endemic tomatoes." *Evolutionary Ecology* 34: 1–19.

Golden Gate Weather Services. 2017. "El Niño and La Niña Years and Intensities." September. http://ggweather.com/enso/oni.htm.

Goldfine, Dana, and Daniel Geller. 2013. *The Galápagos Affair: Satan Came to Eden.* New York: Zeitgeist Films.

González, José A., Carlos Montes, José Rodríguez, and Washington Tapia. 2008. "Rethinking the Galápagos Islands as a Complex Social-Ecological System: Implications for Conservation and Management." *Ecology and Society* 13, no. 2: 13. https://doi.org/10.5751/ES-02557-130213.

Google. 2019. "Google Earth." https://www.google.com/earth/.

Gorman, M. L. 1979. *Island Ecology.* Dordrecht: Springer Netherlands.

Gould, Stephen J. 1983. *Hen's Teeth and Horse's Toes: Further Reflections in Natural History.* New York: W. W. Norton.

Gould, Stephen J., and Elisabeth S. Vrba. 1982. "Exaptation—A Missing Term in the Science of Form." *Paleobiology* 8, no. 1: 4–15. https://doi.org/10.1017/S0094837300004310.

Grace, Jacquelyn K., and David J. Anderson. 2014. "Corticosterone Stress Response Shows Long-Term Repeatability and Links to Personality in Free-Living Nazca Boobies." *General and Comparative Endocrinology* 208 (November): 39–48. https://doi.org/10.1016/j.ygcen.2014.08.020.

Grace, Jacquelyn K., Karen Dean, Mary Ann Ottinger, and David J. Anderson. 2011. "Hormonal Effects of Maltreatment in Nazca Booby Nestlings: Implications for the 'Cycle of Violence'." *Hormones and Behavior* 60, no. 1: 78–85. https://doi.org/10.1016/j.yhbeh.2011.03.007.

Grant, Peter R. 1999. *Ecology and Evolution of Darwin's Finches.* Princeton, NJ: Princeton University Press.

Grant, Peter R., and Rosemary B. Grant. 2008. *How and Why Species Multiply: The Radiation of Darwin's Finches.* Princeton, NJ: Princeton University Press.

Grant, Peter R., and Rosemary B. Grant. 2014. *40 Years of Evolution: Darwin's Finches on Daphne Major Island.* Princeton, NJ: Princeton University Press.

Greenpeace. 2018. "Greenpeace USA." Accessed October 1. http://www.greenpeace.org/usa/.

Grehon, John. 2001. "Biogeography and Evolution of the Galápagos: Integration of the Biological and Geological Evidence." *Biological Journal of the Linnean Society* 74: 267–87. https://doi.org/10.1111/j.1095-8312.2001.tb01392.x.

Grémillet, David, Christophe Chauvin, Rory P. Wilson, Yvon Le Maho, and Sarah Wanless. 2005. "Unusual Feather Structure Allows Partial Plumage Wettability in Diving Great Cormorants *Phalacrocorax carbo*." *Journal of Avian Biology* 36, no. 1: 57–63. https://doi.org/10.1111/j.0908-8857.2005.03331.x.

Grenier, Christophe. 2000. *Conservation contre Nature: Les Iles Galápagos.* Vol. 1278. Paris, France: IRD Éditions.

Grenier, Christophe. 2012. "Nature and the World: A Geohistory of Galápagos." In *The Role of Science for Conservation*, edited by Matthias Wolff and Mark Gardener, 256–74. London: Routledge.

Grigg, Richard W. 2012. *In the Beginning: Archipelago, the Origin and Discovery of the Hawaiian Islands.* Waipahu, Hawaiʻi: Island Heritage.

Groc, Isabelle. 2014. "Should U.S. Government Kill Thousands of Birds to Save Salmon?" *National Geographic*, November. https://www.nationalgeographic.com/news/2014/11/141118-shooting-cormorants-columbia-river-salmon-endangered-species-environment/.

Grove, Jack Stern, and Robert J. Lavenberg. 1997. *The Fishes of the Galápagos Islands.* Stanford, CA: Stanford University Press.

Grubb, Peter. 1971. "The Growth, Ecology and Population Structure of Giant Tortoises on Aldabra." *Philosophical Transactions of the Royal Society of London. Series B, Biological Sciences.* 260: 327–72. https://doi.org/10.1098/rstb.1971.0018.

Guerrero, Ana Mireya, and Alan Tye. 2009. "Darwin's Finches as Seed Predators and Dispersers." *The Wilson Journal of Ornithology* 121, no. 4: 752–64. https://doi.org/10.1676/09-035.1.

Guerrero, Ana Mireya, and Alan Tye. 2011. "Native and Introduced Birds of Galápagos as Dispersers of Native and Introduced Plants." *Ornitología Neotropical* 22 (January): 207–227.

Gunther, A. 1902. "*Testudo galapagoensis*" ("Galapagos Tortoises"). *Novitates Zoologicae* 9: 184–92.

Guo, Jerry. 2006. "The Galápagos Islands Kiss Their Goat Problem Goodbye." *Science* 313, no. 5793: 1567. https://doi.org/10.1126/science.313.5793.1567.

Hamann, Ole. 1979. "Dynamics of a Stand of *Scalesia pedunculata* Hooker fil., Santa Cruz Island, Galápagos." *Botanical Journal of the Linnean Society* 78, no. 2: 67–84. https://doi.org/10.1111/j.1095-8339.1979.tb02186.x.

Hamann, Ole. 1984. "Plants Introduced into Galápagos—Not by Man, but by EI Niño?" *Noticias de Galápagos* 39: 15–19.

Hamann, Ole. 1993. "On Vegetation Recovery, Goats and Giant Tortoises on Pinta Island, Galápagos, Ecuador." *Biodiversity and Conservation* 2: 138–51. https://doi.org/10.1007/BF00056130.

Hamann, Ole. 2001. "Demographic Studies of Three Indigenous Stand-Forming Plant Taxa (*Scalesia, Opuntia*, and *Bursera*) in the Galápagos Islands, Ecuador." *Biodiversity and Conservation* 10, no. 2: 223–50. https://doi.org/10.1023/A:10089523.

Hamann, Ole. 2011. "Ecology, Demography, and Conservation in the Galápagos Islands Flora." In *The Biology of Island Floras*, edited by D. Bramwell and J. Caujapé-Castells,

385–424. Cambridge: Cambridge University Press. https://doi.org/10.1017/CBO9780511844270.017.

Hamann, Ole. 2016. "Sunflower Trees and Giant Cacti: Vegetation Changes over Time." In *Galápagos: Preserving Darwin's Legacy*, edited by Tui De Roy, 67–73. London: Bloomsbury.

Hardin, G. 1968. "The Tragedy of the Commons." *Science* 162, no. 3859: 1243–48. https://doi.org/10.1126/science.162.3859.1243.

Harpp, Karen S., Eric Mittelstaedt, Noémi d'Ozouville and David W. Graham, eds. 2014. *The Galápagos: A Natural Laboratory for the Earth Sciences.* Geophysical Monograph Series. Hoboken, NJ: John Wiley & Sons. https://doi.org/10.1002/9781118852538.ch15.

Harris, Donna B., and David W. Macdonald. 2007. "Interference Competition between Introduced Black Rats and Endemic Galápagos Rice Rats." *Ecology* 88, no. 9: 2330–44. https://doi.org/10.1890/06-1701.1.

Harris, Michael Philip. 1973. "The Biology of the Waved Albatross *Diomedea irrorata* of Hood Island, Galápagos." *Ibis* 115, no. 4: 483–510. https://doi.org/10.1111/j.1474-919X.1973.tb01988.x.

Harris, Michael Philip. 1974. "A Complete Census of the Flightless Cormorant (*Nannopterum harrisi*)." *Biological Conservation* 6, no. 3: 188–91. https://doi.org/10.1016/0006-3207(74)90066-4.

Harris, Michael Philip. 1979. "Population Dynamics of the Flightless Cormorant *Nannopterum harrisi*." *Ibis* 121, no. 2: 135–46. https://doi.org/10.1111/j.1474-919X.1979.tb04957.x.

Hazard, Lisa C. 2004. "Sodium and Potassium Secretion by Iguana Salt Glands." In *Iguanas: Biology and Conservation*, edited by Allison C. Alberts, Ronald L. Carter, William K. Hayes and Emilia P. Martins, 84–93. Berkeley, CA: University of California Press.

Hearn, Alex. 2008. "The Rocky Path to Sustainable Fisheries Management and Conservation in the Galápagos Marine Reserve." *Ocean & Coastal Management* 51, no. 8/9: 567–74.

Hedrick, Philip W. 2019. "Galápagos Islands Endemic Vertebrates: A Population Genetics Perspective." *The Journal of Heredity* 110, no. 2: 137–57. https://doi.org/10.1093/jhered/esy066.

Heleno, Ruben, Stephen Blake, Patricia Jaramillo, Anna Traveset, Pablo Vargas, and Manuel Nogales. 2011. "Frugivory and Seed Dispersal in the Galápagos: What Is the State of the Art?" *Integrative Zoology* 6, no. 2: 110–29. https://doi.org/10.1111/j.1749-4877.2011.00236.x.

Heller, Edmund. 1903. "Papers from the Hopkins Stanford Galápagos Expedition, 1898–1899. XIV. Reptiles." *Proceedings of the Washington Academy of Sciences* 5:39–98.

Helsen, Philippe, Robert A. Browne, David J. Anderson, Peter Verdyck, and Stefan Van Dongen. 2009. "Galápagos' *Opuntia* (Prickly Pear) Cacti: Extensive Morphological Diversity, Low Genetic Variability." *Biological Journal of the Linnean Society* 96, no. 2: 451–61. https://doi.org/10.1111/j.1095-8312.2008.01141.x.

Hendry, Andrew P., Peter R. Grant, B. Rosemary Grant, Hugh A. Ford, Mark J. Brewer, and Jeffrey Podos. 2006. "Possible Human Impacts on Adaptive Radiation: Beak Size Bimodality in Darwin's Finches." *Proceedings of the Royal Society. Series B, Biological Sciences* 273, no. 1596: 1887–94. https://doi.org/10.1098/rspb.2006.3534.

Hennessy, Elizabeth. 2015. "The Molecular Turn in Conservation: Genetics, Pristine Nature, and the Rediscovery of an Extinct Species of Galápagos Giant Tortoise." *Annals of the Association of American Geographers* 105, no. 1: 87–104. https://doi.org/10.1080/00045608.2014.960042.

Hennessy, Elizabeth. 2018. "The Politics of a Natural Laboratory: Claiming Territory and Governing Life in the Galápagos Islands." *Social Studies of Science* 48, no. 4: 483–506. https://doi.org/10.1177/0306312718788179.

Hennessy, Elizabeth. 2019. *On the Backs of Tortoises: Darwin, the Galápagos, and the Fate of an Evolutionary Eden*. New Haven, CT: Yale University Press.

Hennessy, E., and A. L. McCleary. 2011. "Nature's Eden? The Production and Effects of 'Pristine' Nature in the Galápagos Islands." *Island Studies Journal* 6, no. 2: 131–56.

Heylings, P., and M. Bravo. 2007. "Evaluating Governance: A Process for Understanding How Co-management Is Functioning, and Why, in the Galápagos Marine Reserve." *Ocean & Coastal Management* 50, no. 3/4: 174–208. https://doi.org/10.1016/j.ocecoaman.2006.09.003.

Heylings, P., and F. Cruz. 1998. "Common Property, Conflict and Participatory Management in the Galápagos Islands. I." In *Proceedings of the International Association for the Study of Common Property (IASCP) Conference 1998*. Vancouver, British Columbia, Canada. Bloomington, IN: Digital Library of the Commons, Indiana University. http://dlc.dlib.indiana.edu/dlc/handle/10535/1670.

Hickman, John. 1985. *The Enchanted Islands: The Galápagos Discovered*. Owestry, Shropshire, England: Anthony Nelson.

Hieronymus, Tobin L., and Lawrence M. Witmer. 2010. "Homology and Evolution of Avian Compound Rhamphothecae." *The Auk* 127, no. 3: 590–604. https://doi.org/10.1525/auk.2010.09122.

Hill, E. 2020. Galápagos Rail (*Laterallus spilonota*), Version 1.0. In *Birds of the World*, edited by T. S. Schulenberg. Ithaca, NY: Cornell Lab of Ornithology.

Hoeck, H. N. 1984. "Introduced Fauna." In *Key Environments: Galápagos*, edited by R. Perry, 233–45. Oxford, England: Pergamon Press.

Höglund, Jacob. 2009. *Evolutionary Conservation Genetics*. Oxford, England: Oxford University Press. https://doi.org/10.1093/acprof:oso/9780199214211.001.0001.

Holthuis, L.B. and H. Loesch. 1967. "The Lobsters of the Galápagos Islands (Decapoda, Palinuridea)." *Crustaceana* 12, no. 2: 214–22. https://doi.org/10.1163/156854067X00620.

Hong, Pei-Ying, Emily Wheeler, Isaac K. O. Cann, and Roderick I. Mackie. 2011. "Phylogenetic Analysis of the Fecal Microbial Community in Herbivorous Land and Marine Iguanas of the Galápagos Islands Using 16S rRNA-Based Pyrosequencing." *The ISME (International Society for Microbial Ecology) Journal* 5:1461–70. https://doi.org/10.1038/ismej.2011.33.

Hugi, Jasmina, and Marcelo R. Sánchez-Villagra. 2012. "Life History and Skeletal Adaptations in the Galápagos Marine Iguana (*Amblyrhynchus cristatus*) as Reconstructed with Bone Histological Data—A Comparative Study of Iguanines." *Journal of Herpetology* 46, no. 3: 312–24. https://doi.org/10.1670/11-071.

Humphries, Courtney A., V. Danilo Arevalo, Karen N. Fischer, and David J. Anderson. 2006. "Contributions of Marginal Offspring to Reproductive Success of Nazca Booby (*Sula granti*) Parents: Tests of Multiple Hypotheses." *Oecologia* 147, no. 2: 379–90. https://doi.org/10.1007/s00442-005-0264-4.

Hunter, Elizabeth A., James P. Gibbs, Linda J. Cayot, and Washington Tapia. 2013. "Equivalency of Galápagos Giant Tortoises Used as Ecological Replacement Species to Restore Ecosystem Functions." *Conservation Biology* 27, no. 4: 701–9. https://doi.org/10.1111/cobi.12038.

Hunter, Elizabeth A., and James P. Gibbs. 2014. "Densities of Ecological Replacement Herbivores Required to Restore Plant Communities: A Case Study of Giant Tortoises on Pinta Island, Galápagos." *Restoration Ecology* 22, no. 2: 248–56. https://doi.org/10.1111/rec.12055.

Hutchinson, G. Evelyn. 1965. *The Ecological Theater and the Evolutionary Play.* Yale University Press.

Huyvaert, Kathryn P., and David J. Anderson. 2004. "Limited Dispersal by Nazca Boobies *Sula granti.*" *Journal of Avian Biology* 35, no. 1: 46–53. https://doi.org/10.1111/j.0908-8857.2004.03131.x.

Idrovo, H. 2005. *Galápagos: Footsteps in Paradise.* Quito, Ecuador: Ediciones Libri Mundi.

International Union for the Conservation of Nature (IUCN). 2018a. "The IUCN Red List of Threatened Species: *Laterallus spilonota.*" https://www.iucnredlist.org/species/22692366/93350608.

International Union for the Conservation of Nature (IUCN). 2018b. "The IUCN Red List of Threatened Species: *Sula nebouxii.*" https://www.iucnredlist.org/species/22696683/132588719#population.

Itow, Syuzo. 1995. "Phytogeography and Ecology of *Scalesia* (Compositae) Endemic to the Galápagos Islands." *Pacific Science* 49, no. 1: 17–30.

Itow, Syuzo. 2003. "Zonation Pattern, Succession Process and Invasion by Aliens in Species-Poor Insular Vegetation of the Galápagos Islands." *Global Environmental Research* 7, no. 1: 39–58.

Izurieta, Arturo, Byron Delgado, Nicolas Moity, Monica Calvopiña, Iván Cedeño, Gonzalo Banda-Cruz, Eliecer Cruz, et al. 2018. "A Collaboratively Derived Environmental Research Agenda for Galápagos." *Pacific Conservation Biology* 24, no. 2: 168–77. https://doi.org/10.1071/PC17053_CO.

Izurieta, Juan Carlos. 2017. "Behavior and Trends in Tourism in Galápagos between 2007 and 2015." In *Galápagos Report 2015–2016:* 83–89. Puerto Ayora, Galápagos: Galápagos National Park Directorate, Governing Council for the Special Regime of Galápagos, Charles Darwin Foundation, and Galápagos Conservancy.

Jabr, Ferris. 2010. "Fact or Fiction: Can a Squid Fly Out of Water?" *Scientific American,* August.

Jackson, Michael Hume. 1993. *Galápagos, a Natural History.* Calgary, Canada: University of Calgary Press.

Jacob, Francois. 1977. "Evolution and Tinkering." *Science* 196, no. 4295: 1161–66. https://doi.org/10.1126/science.860134.

Jäger, Heinke. 2015. "Biology and Impacts of Pacific Island Invasive Species. 11. *Cinchona pubescens* (Red Quinine Tree) (Rubiaceae)." *Pacific Science* 69, no. 2: 133–53. https://doi.org/10.2984/69.2.1.

Jäger, Heinke. 2018. "Quinine Tree Invasion and Control in Galápagos: A Case Study." In *Understanding Invasive Species in the Galápagos Islands,* edited by María de Lourdes Torres and Carlos F. Mena, 69–76. Cham, Switzerland: Springer. https://doi.org/10.1007/978-3-319-67177-2_5.

Jäger, Heinke, Ingo Kowarik, and Alan Tye. 2009. "Destruction without Extinction: Long-Term Impacts of an Invasive Tree Species on Galápagos Highland Vegetation." *Journal of Ecology* 97, no. 6: 1252–63. https://doi.org/10.1111/j.1365-2745.2009.01578.x.

Jahncke, Jaime. 2007. *Draft Action Plan for Waved Albatross* Phoebastria irrorata. AC3 Doc 28. Agenda Item 6.2. Hobart, Tasmania, Australia: Secretariat of the Agreement on the Conservation of Albatrosses and Petrels.

Jahncke, Jaime, Elisa Goya, and Alex Guillen. 2001. "Seabird By-Catch in Small-Scale Longline Fisheries in Northern Peru." *Waterbirds: The International Journal of Waterbird Biology* 24, no. 1: 137–41. https://doi.org/10.2307/1522255.

James, Matthew J. 2017. *Collecting Evolution: The Galápagos Expedition That Vindicated Darwin*. New York: Oxford University Press.

Janakiram, Naveena B., Altaf Mohammed, and Chinthalapally V. Rao. 2015. "Sea Cucumbers Metabolites as Potent Anti-cancer Agents." *Marine Drugs* 13, no. 5: 2909–23. https://doi.org/10.3390/md13052909.

Janzen, Daniel H. 1974. "Tropical Blackwater Rivers, Animals, and Mast Fruiting by the Dipterocarpaceae." *Biotropica* 6, no. 2: 69–103. https://doi.org/10.2307/2989823.

Jaramillo, Maricruz, Michelle Donaghy-Cannon, F. Hernán Vargas, and Patricia G. Parker. 2016. "The Diet of the Galápagos Hawk (*Buteo galapagoensis*) before and after Goat Eradication." *Journal of Raptor Research* 50, no. 1: 33–44. https://doi.org/10.3356/rapt-50-01-33-44.1.

Jaramillo, Patricia, Swen Lorenz, Gabriela Ortiz, Pablo Cueva, Estalin Jiménez, Jaime Ortiz, Danny Rueda, Max Freire, James Gibbs, and Washington Tapia. 2015. "Galápagos Verde 2050: An Opportunity to Restore Degraded Ecosystems and Promote Sustainable Agriculture in the Archipelago." In *Galápagos Report 2013–2014*, 133–43. Puerto Ayora, Galápagos: Galápagos National Park Directorate, Governing Council for the Special Regime of Galápagos, Charles Darwin Foundation, and Galápagos Conservancy.

Jetz, W., G. H. Thomas, J. B. Joy, K. Hartmann, and A. O. Mooers. 2012. "The Global Diversity of Birds in Space and Time." *Nature* 491, no. 7424: 444–48. https://doi.org/10.1038/nature11631.

Jiménez-Uzcátegui, G., J. F. Freile, T. Santander G., L. Carrasco, D. F. Cisneros-Heredia, E. A. Guevara y B. A. Tinoco. 2019. *Lista Roja de las Aves de Galápagos* (Red List of the Birds of Galápagos). Quito, Ecuador: Ministerio del Ambiente, Aves y Conservación, Comité Ecuatoriano de Registros Ornitológicos, Fundación Charles Darwin, Universidad del Azuay, Red Aves Ecuador y Universidad San Francisco de Quito.

Jiménez-Uzcátegui, Gustavo, Bryan Milstead, Cruz Márquez, Javier Zabala, and Javi Zabala-Albizua. 2008. "Galápagos Vertebrates: Endangered Status and Conservation Actions." In *Galápagos Report 2006–2007,* 104–10. Puerto Ayora, Galápagos: Charles Darwin Foundation, Galápagos National Park, and National Galápagos Institute.

Jiménez-Uzcátegui, Gustavo, and Howard Snell. 2016. *CDF Checklist of Galápagos Mammals*. Puerto Ayora, Galápagos: Charles Darwin Foundation.

Jones, Clive G., John H. Lawton, and Moshe Shachak. 1996. "Organisms as Ecosystem Engineers." In *Ecosystem Management*, 130–47. New York: Springer. https://doi.org/10.1007/978-1-4612-4018-1_14.

Jordan, Chris. 2017. "Midway Journey". Presented at the TEDxRainier, January 7. https://www.youtube.com/watch?v=MjK0cvbm20M.

Jørgensen, C. B. 1998. "Role of Urinary and Cloacal Bladders in Chelonian Water Economy: Historical and Comparative Perspectives." *Biological Reviews of the*

Cambridge Philosophical Society 73, no. 4: 347–66. https://doi.org/10.1017/
S0006323198005210.

Juola, Frans A., Kevin McGraw, and Donald C. Dearborn. 2008. "Carotenoids and Throat
Pouch Coloration in the Great Frigatebird (*Fregata minor*)." *Comparative Biochemistry
and Physiology. Part B, Biochemistry & Molecular Biology* 149, no. 2: 370–77. https://
doi.org/10.1016/j.cbpb.2007.10.010.

Karnauskas, K. B., S. Jenouvrier, C. W. Brown, and R. Murtugudde. 2015. "Strong Sea
Surface Cooling in the Eastern Equatorial Pacific and Implications for Galápagos
Penguin Conservation." *Geophysical Research Letters* 42, no. 15: 6432–37. https://doi.
org/10.1002/2015GL064456.

Karnauskas, Kristopher B., Eric Mittelstaedt, and Raghu Murtugudde. 2017.
"Paleoceanography of the Eastern Equatorial Pacific over the Past 4 Million Years and
the Geologic Origins of Modern Galápagos Upwelling." *Earth and Planetary Science
Letters* 460 (February): 22–28. https://doi.org/10.1016/j.epsl.2016.12.005.

Kastdalen, Alf. 1982. "Changes in the Biology of Santa Cruz Island between 1935 and
1965." *Noticias de Galápagos* 45: 7–12.

Keller, Evelyn Fox. 1984. *A Feeling for the Organism, 10th Anniversary Edition: The Life
and Work of Barbara McClintock*. New York: Macmillan.

Kelley, Daniel, Kevin Page, Diego Quiroga, Raul Salazar, and Raul Salazar Herrera. 2019.
*In the Footsteps of Darwin: Geoheritage, Geotourism and Conservation in the Galápagos
Islands*. Cham, Switzerland: Springer.

Kennedy, Martyn, Carlos A. Valle, and Hamish G. Spencer. 2009. "The Phylogenetic
Position of the Galápagos Cormorant." *Molecular Phylogenetics and Evolution* 53, no.
1: 94–98. https://doi.org/10.1016/j.ympev.2009.06.002.

Keynes, Richard D. 1988. *Charles Darwin's Beagle Diary*. Cambridge: Cambridge
University Press.

King, Richard J. 2013. *The Devil's Cormorant: A Natural History*. Lebanon, NH: University
of New Hampshire Press.

Kirchman, Jeremy J. 2012. "Speciation of Flightless Rails on Islands: A DNA-Based
Phylogeny of the Typical Rails of the Pacific." *The Auk* 129, no. 1: 56–69. https://doi.org/
10.1525/auk.2011.11096.

Kleindorfer, Sonia, Georgina Custance, Katharina J. Peters, and Frank J. Sulloway. 2019.
"Introduced Parasite Changes Host Phenotype, Mating Signal and Hybridization
Risk: *Philornis downsi* Effects on Darwin's Finch Song." *Proceedings of the Royal
Society. Series B, Biological Sciences* 286, no. 1904: 20190461. https://doi.org/10.1098/
rspb.2019.0461.

Konecny, Michael J. 1987. "Food Habits and Energetics of Feral House Cats in the
Galápagos Islands." *Oikos* 50, no. 1: 24. https://doi.org/10.2307/3565398.

Kricher, J. 2006. *Galápagos: A Natural History*. Princeton, NJ: Princeton University Press.

Kruuk, Hans, and Howard Snell. 1981. "Prey Selection by Feral Dogs from a Population
of Marine Iguanas (*Amblyrhynchus cristatus*)." *The Journal of Applied Ecology* 18, no.
1: 197–204. https://doi.org/10.2307/2402489.

Lack, David. 1947. *Darwin's Finches*. Cambridge: Cambridge University Press.

Lamichhaney, Sangeet, Jonas Berglund, Markus Sällman Almén, Khurram Maqbool,
Manfred Grabherr, Alvaro Martinez-Barrio, Marta Promerová, et al. 2015. "Evolution
of Darwin's Finches and Their Beaks Revealed by Genome Sequencing." *Nature* 518,
no. 7539: 371–75. https://doi.org/10.1038/nature14181.

Lamichhaney, Sangeet, Fan Han, Matthew T. Webster, Leif Andersson, B. Rosemary Grant, Peter R. Grant. 2018. "Rapid Hybrid Speciation in Darwin's Finches." *Science* 359:224–28. https://doi.org/10.1126/science.aao4593.

Lanner, Ronald M. 1996. *Made for Each Other: A Symbiosis of Birds and Pines.* New York: Oxford University Press.

Larson, Edward J. 2001. *Evolution's Workshop: God and Science on the Galápagos Islands.* London: Allen Lane.

Laurie, W. A., and D. Brown. 1990. "Population Biology of Marine Iguanas (*Amblyrhynchus cristatus*). II. Changes in Annual Survival Rates and the Effects of Size, Sex, Age and Fecundity in a Population Crash." *Journal of Animal Ecology* 59, no. 2: 529–44. https://doi.org/10.2307/4879.

Lavoie, Christian, Felipe Cruz, G. Victor Carrion, Karl Campbell, C. Josh Donlan, Sylvia Harcourt, and Marycarmen Moya. 2007. *The Thematic Atlas of Project Isabela: An Illustrative Document Describing, Step-by-step, the Biggest Successful Goat Eradication Project on the Galápagos Islands, 1998–2006.* Puerto Ayora, Galápagos: Charles Darwin Foundation.

Lawesson, Jonas Erik. 1988. "Stand-Level Dieback and Regeneration of Forests in the Galápagos Islands." *Vegetatio* 77, no. 1/3: 87–93. https://doi.org/10.1007/BF00045754.

Lawson, Lucinda P., Birgit Fessl, F. Hernán Vargas, Heather L. Farrington, H. Francesca Cunninghame, Jakob C. Mueller, Erwin Nemeth, P. Christian Sevilla, and Kenneth Petren. 2017. "Slow Motion Extinction: Inbreeding, Introgression, and Loss in the Critically Endangered Mangrove Finch (*Camarhynchus heliobates*)." *Conservation Genetics* 18, no. 1: 159–70. https://doi.org/ 10.1007/s10592-016-0890-x.

Lawson, Lucinda P., John Niedzwiecki, and Kenneth Petren. 2019. "Darwin's finches: a model of landscape effects on metacommunity dynamics in the Galápagos Archipelago." *Ecography* 42, no. 10: 1636–47.

Le, Minh, Christopher J. Raxworthy, William P. McCord, and Lisa Mertz. 2006. "A Molecular Phylogeny of Tortoises (Testudines: Testudinidae) Based on Mitochondrial and Nuclear Genes." *Molecular Phylogenetics and Evolution* 40, no. 2: 517–31. https:// doi.org/10.1016/j.ympev.2006.03.003.

Leck, Charles F. 1980. "Establishment of New Population Centers with Changes in Migration Patterns." *Journal of Field Ornithology* 51, no. 2: 168–73.

Lethier, Hervé, and Paula Bueno. 2018. *Report on the Reactive Monitoring Mission to Galápagos Islands World Heritage Site (Ecuador) 21–25 August 2017.* IUCN.

Litt, Josef. 2018. *Galápagos.* Twyford, Berkshire, UK: Mostly Underwater Books.

Liu, Yu, Kim M. Cobb, Huiming Song, Qiang Li, Ching-Yao Li, Takeshi Nakatsuka, Zhisheng An, et al. 2017. "Recent Enhancement of Central Pacific El Niño Variability Relative to Last Eight Centuries." *Nature Communications* 8 (May): 15386. https://doi. org/10.1038/ncomms15386.

Liu, Jianguo, Thomas Dietz, Stephen R. Carpenter, Carl Folke, Marina Alberti, Charles L. Redman, Stephen H. Schneider, et al. 2007. "Coupled Human and Natural Systems." *AMBIO: A Journal of the Human Environment* 36, no. 8: 639–49. https://doi.org/ 10.1579/0044-7447(2007)36[639:CHANS]2.0.CO;2.

Livezey, Bradley C. 1992. "Flightlessness in the Galápagos Cormorant (*Compsohalieus [Nannopterum] harrisi*): Heterochrony, Giantism and Specialization." *Zoological Journal of the Linnean Society* 105, no. 2: 155–224. https://doi.org/10.1111/j.1096-3642.1992.tb01229.x.

Livezey, Bradley C. 2003. "Evolution of Flightlessness in Rails (Gruiformes, Rallidae)." *Ornithological Monographs* 53: 1–654. https://doi.org/10.2307/40168337.

Lu, Flora, Gabriela Valdivia, and Wendy Wolford. 2013. "Social Dimensions of 'Nature at Risk' in the Galápagos Islands, Ecuador." *Conservation and Society* 11, no. 1: 83. https://doi.org/10.4103/0972-4923.110945.

Lundh, Jacob P. 1999. *The Galápagos: A Brief History.* http://www.lundh.no/jacob/Galápagos/pg05.htm.

Lynch, Wayne. 2018. *Galápagos: A Traveler's Introduction.* Buffalo, New York: Firefly Books.

MacArthur, Robert H., and Edmund O. Wilson. 1967. *The Theory of Island Biogeography.* Princeton, NJ: Princeton University Press.

Macdonald, T. 1997. *Conflict in the Galápagos Islands: Analysis and Recommendations for Management.* Cambridge, MA: Harvard University, Weatherhead Center for International Affairs.

MacFarland, Craig G., and W. G. Reeder. 1974. "Cleaning Symbiosis Involving Galápagos Tortoises and Two Species of Darwin's Finches." *Zeitschrift Für Tierpsychologie* 34, no. 5: 464–83. https://doi.org/10.1111/j.1439-0310.1974.tb01816.x.

Macia, S., Michael P. Robinson, Paul Craze, Robert Dalton, and James D. Thomas. 2004. "New Observations on Airborne Jet Propulsion (Flight) in Squid, with a Review of Previous Reports." *Journal Molluscan Studies* 70, no. 3: 297–99. https://doi.org/10.1093/mollus/70.3.297.

Mackie, Roderick I., Mathew Rycyk, Rebecca L. Ruemmler, Rustam I. Aminov, and Martin Wikelski. 2004. "Biochemical and Microbiological Evidence for Fermentative Digestion in Free-Living Land Iguanas (*Conolophus pallidus*) and Marine Iguanas (*Amblyrhynchus cristatus*) on the Galápagos Archipelago." *Physiological and Biochemical Zoology* 77, no. 1: 127–38. https://doi.org/10.1086/383498.

MacLeod, Amy, Ariel Rodríguez, Miguel Vences, Pablo Orozco-terWengel, Carolina García, Fritz Trillmich, Gabriele Gentile, Adalgisa Caccone, Galo Quezada, and Sebastian Steinfartz. 2015. "Hybridization Masks Speciation in the Evolutionary History of the Galápagos Marine Iguana." *Proceedings of the Royal Society. Series B, Biological Sciences* 282, no. 1809: 20150425. https://doi.org/10.1098/rspb.2015.0425.

Macleod, Amy, and Sebastian Steinfartz. 2016. "The Conservation Status of the Galápagos Marine Iguanas, *Amblyrhynchus cristatus*: A Molecular Perspective." *Amphibia-Reptilia* 37, no. 1: 91–109. https://doi.org/10.1163/15685381-00003035.

Maldonado, Roberto, and Elvis Llerena. 2018. *La Colonización de Galápagos: Historias Humanas.* Puerto Ayora, Galápagos: Dirección del Parque Nacional Galápagos.

Malone, Catherine L., Víctor Hugo Reynoso, and Larry Buckley. 2017. "Never Judge an Iguana by Its Spines: Systematics of the Yucatan Spiny Tailed Iguana, *Ctenosaura defensor* (Cope, 1866)." *Molecular Phylogenetics and Evolution* 115:27–39.

Malthus, T. R. 1798. *An Essay on the Principle of Population as It Affects the Future Improvement of Society, with Remarks on the Speculations of Mr. Godwin, M. Condorcet, and Other Writers.* London: J. Johnson.

Manzi, Maya, and Oliver T. Coomes. 2002. "Cormorant Fishing in Southwestern China: A Traditional Fishery under Siege." *Geographical Review* 92, no. 4: 597–603. https://doi.org/10.2307/4140937.

Marlow, Ronald William, and James L. Patton. 1981. "Biochemical Relationships of the Galápagos Giant Tortoises (*Geochelone elephantopus*)." *Journal of Zoology* 195, no. 3: 413–22. https://doi.org/10.1111/j.1469-7998.1981.tb03474.x.

Márquez, Cruz, David A. Wiedenfeld, Sixto Naranjo, and Washington Llerena. 2008. "1997–8 El Niño and the Galápagos Tortoises *Geochelone vandenburghi* on Alcedo Volcano, Galápagos." *Aquatic Commons*, December.

Martin, John H., Michael Gordon, and Steve E. Fitzwater. 1991. "The Case for Iron." *Limnology and Oceanography Letters* 36, no. 8: 1793–1802. https://doi.org/10.4319/lo.1991.36.8.1793.

Mauchamp, André, and Rachel Atkinson. 2010. "Rapid, Recent and Irreversible Habitat Loss: *Scalesia* Forest on the Galápagos Islands." In *Galápagos Report 2009–2010,* 108–12. Puerto Ayora, Galápagos: Charles Darwin Foundation, Galápagos National Park, and Governing Council of Galápagos.

Mayr, Ernst. 1982. *The Growth of Biological Thought: Diversity, Evolution, and Inheritance.* Cambridge, MA: Harvard University Press.

McPhaden, Michael J. 1999. "Genesis and Evolution of the 1997–98 El Niño." *Science* 283:950–54.

McMullen, Conley K. 1999. *Flowering Plants of the Galápagos.* Ithaca, NY: Cornell University Press.

McNew, Sabrina M., Daniel Beck, Ingrid Sadler-Riggleman, Sarah A. Knutie, Jennifer AH Koop, Dale H. Clayton, and Michael K. Skinner. 2017. "Epigenetic Variation between Urban and Rural Populations of Darwin's finches." *BMC Evolutionary Biology* 17, no. 1: 1–14.

McNew, Sabrina M., and Dale H. Clayton. 2018. "Alien Invasion: Biology of *Philornis* Flies Highlighting *Philornis downsi*, an Introduced Parasite of Galápagos Birds." *Annual Review of Entomology* 63 (January): 369–87. https://doi.org/10.1146/annurev-ento-020117-043103.

Meile, Robert J., Robert C. Lacy, F. Hernán Vargas, and Patricia G. Parker. 2013. "Modeling *Plasmodium* Parasite Arrival in the Galápagos Penguin (*Spheniscus mendiculus*)." *The Auk* 130, no. 3: 440–48. https://doi.org/10.1525/auk.2013.12147.

Meltzoff, Sarah K. 2013. *Listening to Sea Lions: Currents of Change from Galápagos to Patagonia.* Lanham, Maryland: Altamira Press.

Melville, Herman, and Lynn Michelsohn. 2011. *In the Galápagos Islands with Herman Melville: The Encantadas or Enchanted Isles.* Santa Fe, NM: Cleanan Press.

Mercier, Annie, Roberto Ycaza Hidalgo, and Jean-François Hamel. 2004. "Aquaculture of the Galápagos Sea Cucumber, *Isostichopus fuscus*." In *Advances in Sea Cucumber Aquaculture and Management*, edited by Allessandro Lovatelli, Chantal Conand, Steve Purcell, Sven Uthicke, Jean-François Hamel, and Annie Mercier, 347–58. FAO Fisheries Technical Paper, No. 463. Rome, Italy: FAO.

Mercier, A., R. H. Ycaza, and J. F. Hamel. 2007. "Long-Term Study of Gamete Release in a Broadcast-Spawning Holothurian: Predictable Lunar and Diel Periodicities." *Marine Ecology Progress Series* 329 (January): 179–89. https://doi.org/10.3354/meps329179.

Merlen, Godfrey. 1993. "Pepino War, 1992" ("Sea Cucumber War, 1992"). *Noticias de Galápagos* no. 52 (May): 3.

Merlen, Godfrey. 1998. "Scavenging Behavior of the Waved Albatross in Galápagos: A Potential Problem with Increasing Longlining." *Noticias de Galápagos* no. 59 (April): 20–23.

Merlen, Godfrey. 2013. *Gone, Gone . . . Going: The Fate of the Vermilion Flycatcher on Darwin's Islands.* In *Galápagos Report 2011–2012*, 180–88. Puerto Ayora, Galápagos: Galápagos National Park Service, Governing Council for the Special Regime of Galápagos, Charles Darwin Foundation, and Galápagos Conservancy.

Milinkovitch, Michel C., Daniel Monteyne, James P. Gibbs, Thomas H. Fritts, Washington Tapia, Howard L. Snell, Ralph Tiedemann, Adalgisa Caccone, and Jeffrey R. Powell. 2004. "Genetic Analysis of a Successful Repatriation Programme: Giant Galápagos Tortoises." *Proceedings of the Royal Society. Series B, Biological Sciences* 271, no. 1537: 341–45. https://doi.org/10.1098/rspb.2003.2607.

Miller, Joshua M., Maud C. Quinzin, Nikos Poulakakis, James P. Gibbs, Luciano B. Beheregaray, Ryan C. Garrick, Michael A. Russello, et al. 2017. "Identification of Genetically Important Individuals of the Rediscovered Floreana Galápagos Giant Tortoise (*Chelonoidis elephantopus*) Provide Founders for Species Restoration Program." *Scientific Reports* 7, no. 1: 11471. https://doi.org/10.1038/s41598-017-11516-2.

Mills, Kyra L., and Hernan Vargas. 1997. "Current Status, Analysis of Census Methodology, and Conservation of the Galápagos Penguin, *Spheniscus mendiculus.*" *Noticias de Galápagos*, no. 58: 8–15.

Miralles, Aurélien, Amy Macleod, Ariel Rodríguez, Alejandro Ibáñez, Gustavo Jiménez-Uzcategui, Galo Quezada, Miguel Vences, and Sebastian Steinfartz. 2017. "Shedding Light on the Imps of Darkness: An Integrative Taxonomic Revision of the Galápagos Marine Iguanas (Genus *Amblyrhynchus*)." *Zoological Journal of the Linnean Society* 181, no. 3: 678–710. https://doi.org/10.1093/zoolinnean/zlx007.

Moity, Nicolas. 2018. "Evaluation of No-Take Zones in the Galápagos Marine Reserve, Zoning Plan 2000." *Frontiers in Marine Science* 5 (July). https://doi.org/10.3389/fmars.2018.00244.

Moolna, Randy. 2007. Preliminary Observations Indicate That Giant Tortoise Ingestion Improves Seed Germination for an Endemic Ebony Species in Mauritius. *African Journal of Ecology* 46:217–19.

Moore, Randy, and Roslyn Cameron. 2019. *Galápagos Revealed: Finding the Places That Most People Miss*. Fairfax, Virginia: Galápagos Conservancy.

Moore, Randy, and Sehoya Cotner. 2014. *Understanding Galápagos: What You'll See and What It Means*. New York: McGraw-Hill.

Mueller-Dombois, Dieter, and Francis R. Fosberg. 1998. *Vegetation of the Tropical Pacific Islands*. Edited by Francis R. Fosberg. Cham, Switzerland: Springer Science & Business Media. https://doi.org/10.1007/978-1-4419-8686-3.

Mukherjee, S. 2016. *The Gene: An Intimate History*. New York: Scribner.

Müller, Martina S., Julius F. Brennecke, Elaine T. Porter, Mary Ann Ottinger, and David J. Anderson. 2008. "Perinatal Androgens and Adult Behavior Vary with Nestling Social System in Siblicidal Boobies." *PLOS One* 3, no. 6: e2460. https://doi.org/10.1371/journal.pone.0002460.

Müller, Martina S., Elaine T. Porter, Jacquelyn K. Grace, Jill A. Awkerman, Kevin T. Birchler, Alex R. Gunderson, Eric G. Schneider, Mark A. Westbrock, and David J. Anderson. 2011. "Maltreated Nestlings Exhibit Correlated Maltreatment as Adults: Evidence of a 'Cycle of Violence' in Nazca Boobies (*Sula granti*)." *The Auk* 128, no. 4: 615–19. https://doi.org/10.1525/auk.2011.11008.

Naranjo, Luis Germán, Luis Fernando Castillo, Richard Johnston-González, Carlos Hernández, Carlos Ruiz, and Felipe Estela. 2006. "Waterbird Monitoring and Conservation in Protected Areas of the Colombian Pacific." In *Waterbirds around the World*, edited by G.C. Boere, C.A. Galbraith and D.A. Stroud, 177–80. Inverness: Scottish Natural Heritage.

NASA. 2017. "What Is La Niña." Accessed September 5, 2017. http://www.pmel.noaa.gov/pubs/outstand/mcph2029/images/fig03.jpg.

NASA. 2018. "Sea Surface Temperature, Height, Chlorophyll Visualizer." *Ocean Motion and Surface Currents.* Accessed September 18, 2018. http://oceanmotion.org/html/resources/ssedv.htm.

NASA Scientific Visualization Studio. 1998. "SeaWIFS Phytoplankton around the Galápagos Islands in May 1998." NASA/Goddard Space Flight Center, The SeaWiFS Project and GeoEye. https://svs.gsfc.nasa.gov/205.

Nash, Steve. 2009. "Ecotourism and Other Invasions." *BioScience* 59:106–10. https://doi.org/ 10.1525/bio.2009.59.2.3.

Nathan, Stephen A., and R. Mark Leckie. 2009. "Early History of the Western Pacific Warm Pool during the Middle to Late Miocene (~13.2–5.8 Ma): Role of Sea-Level Change and Implications for Equatorial Circulation." *Palaeogeography, Palaeoclimatology, Palaeoecology* 274, no. 3/4: 140–59. https://doi.org/10.1016/j.palaeo.2009.01.007.

National Science Teachers Association. 2017. "Galápagos Education." https://www.nsta.org/publications/interactive/Galápagos/activities/pdf/letters.pdf.

Neira, C. E. 2016. "Case Study on the Galápagos Islands: Balance for Biodiversity & Migration." *Earth Jurisprudence and Environmental Justice Journal.*6, no. 1: 131–52.

Nelson, Bryan. 1968. *Galápagos: Islands of Birds.* New York: William Morrow.

Nelson, Bryan. 1978. *The Sulidae: Gannets and Boobies.* Oxford, England: Oxford University Press.

Neuman-Lee, Lorin A., and Susannah S. French. 2017. "Endocrine-Reproductive-Immune Interactions in Female and Male Galápagos Marine Iguanas." *Hormones and Behavior* 88 (February): 60–69. https://doi.org/10.1016/j.yhbeh.2016.10.017.

Neves, Felipe Machado, Patrícia Luciano Mancini, Fernanda Pinto Marques, Guilherme Tavares Nunes, and Leandro Bugoni. 2015. "Cannibalism by Brown Booby (*Sula leucogaster*) at a Small Tropical Archipelago." *Brazilian Journal of Ornithology* 23, no. 3: 299–304.

Nicholls, Henry. 2006. *Lonesome George: The Life and Loves of a Conservation Icon.* London: Macmillan.

Nicholls, Henry. 2014. *The Galápagos: A Natural History.* New York: Basic Books.

Nielsen, Lene Rostgaard, Marianne Philipp, and Hans R. Siegismund. 2002. "Selective Advantage of Ray Florets in *Scalesia affinis* and *S. pedunculata* (Asteraceae), Two Endemic Species from the Galápagos." *Evolutionary Ecology* 16, no. 2: 139–53. https://doi.org/10.1023/A:1016301027929.

Nogué, Sandra, Lea de Nascimento, Cynthia A. Froyd, Janet M. Wilmshurst, Erik J. de Boer, Emily E. D. Coffey, Robert J. Whittaker, José María Fernández-Palacios, and Kathy J. Willis. 2017. "Island Biodiversity Conservation Needs Palaeoecology." *Nature Ecology & Evolution* 1, no. 7: 181. https://doi.org/10.1038/s41559-017-0181.

Nolan, John M., Stephen Beatty, Katie A. Meagher, Alan N. Howard, David Kelly, and David I. Thurnham. 2014. "Verification of Meso-zeaxanthin in Fish." *Journal of Food Processing & Technology* 5, no. 6: 335. https://doi.org/10.4172/2157-7110.1000335.

Noltie, H. J. 2012. "The Generic Name *Scalesia* (Compositae)—An Etymological Blunder." *Archives of Natural History* 39, no. 1: 167–69. https://doi.org/10.3366/anh.2012.0071.

Nowak, J. B. 1986. "Isla de La Plata and the Galápagos." *Noticias de Galápagos* 44:17.

O'Connor, John M., Peter Stoffers, Jan R. Wijbrans, and Tim J. Worthington. 2007. "Migration of Widespread Long-Lived Volcanism across the Galápagos Volcanic

Province: Evidence for a Broad Hotspot Melting Anomaly?" *Earth and Planetary Science Letters* 263, no. 3/4: 339–54. https://doi.org/10.1016/j.epsl.2007.09.007.

O'Connor, J. A., F. J. Sulloway, J. Robertson, and S. Kleindorfer. 2010. "*Philornis downsi* Parasitism Is the Primary Cause of Nestling Mortality in the Critically Endangered Darwin's Medium Tree Finch (*Camarhynchus pauper*)." *Biodiversity and Conservation* 19, no. 3: 853–66. https://doi.org/10.1007/s10531-009-9740-1.

Ochi, Hiroki, Paul Baldock, and Shu Takeda. 2019. "Central Neuronal Control of Bone Remodeling." In *Primer on the Metabolic Bone Diseases and Disorders of Mineral Metabolism, Ninth Edition*, edited by John P. Bilezikian, 1020–27. Hoboken, NJ: John Wiley & Sons.

Odling-Smee, John F., Kevin N. Laland, and Marcus W. Feldman. 2003. *Niche Construction: The Neglected Process in Evolution No. 37*. Princeton, NJ: Princeton University Press.

Okajima, Yasuhisa, and Yoshinori Kumazawa. 2009. "Mitogenomic Perspectives into Iguanid Phylogeny and Biogeography: Gondwanan Vicariance for the Origin of Madagascan Oplurines." *Gene* 441, no. 1/2: 28–35. https://doi.org/10.1016/j.gene.2008.06.011.

Oleas, Reyna. 2008. "The Galápagos National Park Entrance Fee: A Global Perspective and Options for the Future." In *Galápagos Report 2007–2008*, 75–80. Puerto Ayora, Galápagos: Charles Darwin Foundation, Galápagos National Park, and National Galápagos Institute.

Olesen, Jens M., Christian F. Damgaard, Francisco Fuster, Ruben H. Heleno, Manuel Nogales, Beatriz Rumeu, Kristian Trøjelsgaard, Pablo Vargas, and Anna Traveset. 2018. "Disclosing the Double Mutualist Role of Birds on Galápagos." *Scientific Reports* 8, no. 1: 57. https://doi.org/10.1038/s41598-017-17592-8.

Olson, Storrs L. 1980. "A New Genus of Penguin-like Pelecaniform Bird from the Oligocene of Washington (Pelecaniformes: Plotopteridae)." In *Papers in Avian Paleontology Honoring Hildegarde Howard*, edited by Kenneth E. Campbell, Jr., 51–57. *Contributions in Science*, 330. Los Angeles: Natural History Museum of Los Angeles County.

Olson, Storrs L. 2014. "The Early Scientific History of Galápagos Iguanas." *Archives of Natural History* 41, no. 1: 141–53. https://doi.org/10.3366/anh.2014.0217.

Olson, S. L. 2017. "The Early Scientific History of Galápagos Tortoises." *Archives of Natural History* 44, no. 2: 241–58. https://doi.org/10.3366/anh.2017.0447.

Ortega Pacheco, Daniel, and Walter Bustos. 2016. *Conservation Status Report of the Galápagos Archipelago*. Puerto Ayora, Galápagos: Galápagos National Park Directorate.

Ospina, Pablo. 2006. *Galápagos, Naturaleza y Sociedad: Actores Sociales y Conflictos Ambientales En Las Islas Galápagos* (Galapagos, Nature, and Society: Social Actors and Environmental Conflicts in the Galápagos Islands). Quito, Ecuador: Corporación Editora Nacional.

Ostrom, Elinor. 2005. *Understanding Institutional Diversity*. Princeton, NJ: Princeton University Press.

Ostrom, Elinor. 2009. "A General Framework for Analyzing Sustainability of Social-Ecological Systems." *Science* 325, no. 5939: 419–22. https://doi.org/10.1126/science.1172133.

Ostrom, Elinor, Christina Chang, Mark Pennington, and Vlad Tarko. 2012. *The Future of the Commons—Beyond Market Failure and Government Regulation*. London: Institute of Economic Affairs. https://doi.org/10.2139/ssrn.2267381.

Palmer, Jamie L., Thomas F. McCutchan, F. Hernan Vargas, Sharon L. Deem, Marilyn Cruz, Daniel A. Hartman, and Patricia G. Parker. 2013. "Seroprevalence of Malarial Antibodies in Galápagos Penguins (*Spheniscus mendiculus*)." *The Journal of Parasitology* 99, no. 5: 770–76. https://doi.org/10.1645/12-57.1.

Pangestuti, Ratih, and Zainal Arifin. 2018. "Medicinal and Health Benefit Effects of Functional Sea Cucumbers." *Journal of Traditional and Complementary Medicine*, July. https://doi.org/10.1016/j.jtcme.2017.06.007.

Paredes, Rosana, Carlos B. Zavalaga, and Daryl J. Boness. 2002. "Patterns of Egg Laying and Breeding Success in Humboldt Penguins (*Spheniscus humboldti*) at Punta San Juan, Peru." *The Auk* 119, no. 1: 244–50. https://doi.org/10.2307/4090031.

Parent, Christine E., Adalgisa Caccone, and Kenneth Petren. 2008. "Colonization and Diversification of Galápagos Terrestrial Fauna: A Phylogenetic and Biogeographical Synthesis." *Philosophical Transactions of the Royal Society. Series B, Biological Sciences* 363:3347–61. https://doi.org/10.1098/rstb.2008.0118

Parent, Christine, and Guy Coppois. 2016. "On the Snails' Trail: Evolution and Speciation among a Vanishing Tribe." In *Galápagos: Preserving Darwin's Legacy*, edited by Tui De Roy, 74–81. London: Bloomsbury.

Parker, Patricia G., ed. 2018. *Disease Ecology: Galápagos Birds and Their Parasites*. Cham, Switzerland: Springer.

Parker, Patricia. 2016. "Parasites and Pathogens: Threats to Native Birds." In *Galápagos: Preserving Darwin's Legacy*, edited by Tui De Roy, 177–83. London: Bloomsbury.

Parker, Patricia G., Elizabeth L. Buckles, Heather Farrington, Kenneth Petren, Noah K. Whiteman, Robert E. Ricklefs, Jennifer L. Bollmer, and Gustavo Jiménez-Uzcátegui. 2011. "110 Years of Avipoxvirus in the Galápagos Islands." *PLOS One* 6, no. 1: e15989. https://doi.org/10.1371/journal.pone.0015989.

Parker, Steve. 2015. *Evolution: The Whole Story*. Buffalo, New York: Firefly Books.

Patterson, S. A., J. A. Morris-Pocock, and V. L. Friesen. 2011. "A Multilocus Phylogeny of the Sulidae (Aves: Pelecaniformes)." *Molecular Phylogenetics and Evolution* 58, no. 2: 181–91. https://doi.org/10.1016/j.ympev.2010.11.021.

PBS. 2000. "Tracking El Niño." *NOVA Online*.

Peck, Stewart B. 2001. *Smaller Orders of Insects of the Galápagos Islands, Ecuador: Evolution, Ecology, and Diversity*. Ottawa, Ontario, Canada: NRC Research Press.

Pennycuick, C. J. 1982. "The Flight of Petrels and Albatrosses (Procellariiformes), Observed in South Georgia and Its Vicinity." *Philosophical Transactions of the Royal Society. Series B, Biological Sciences* 300, no. 1098: 75–106. https://doi.org/10.1098/rstb.1982.0158.

Peterson, Roger Tory. 1967. "The Galápagos: Eerie Cradle of New Species." *National Geographic* 131, no. 4: 541–85.

Philipp, Marianne, and Lene Rostgaard Nielsen. 2010. "Reproductive Ecology of *Scalesia cordata* (Asteraceae), an Endangered Species from the Galápagos Islands." *Botanical Journal of the Linnean Society* 162, no. 3: 496–503. https://doi.org/10.1111/j.1095-8339.2010.01034.x.

Phillips, R. A., R. Gales, G. B. Baker, M. C. Double, M. Favero, F. Quintana, M. L. Tasker, H. Weimerskirch, M. Uhart, and A. Wolfaardt. 2016. "The Conservation Status and Priorities for Albatrosses and Large Petrels." *Biological Conservation* 201 (September): 169–83. https://doi.org/10.1016/j.biocon.2016.06.017.

Phillips, R. Brand, David A. Wiedenfeld, and Howard L. Snell. 2012. "Current Status of Alien Vertebrates in the Galápagos Islands: Invasion History, Distribution, and

Potential Impacts." *Biological Invasions* 14, no. 2: 461–80. https://doi.org/10.1007/s10530-011-0090-z.

Pink Floyd. 1986. "Learning to Fly." *A Momentary Lapse of Reason.*

Pitman, Robert L., and Lisa T. Ballance. 2002. "The Changing Status of Marine Birds Breeding at San Benedicto Island, Mexico." *The Wilson Journal of Ornithology* 114, no. 1: 11–19.

Pitman, Robert L., and Joseph R. Jehl. 1998. "Geographic Variation and Reassessment of Species Limits in the 'Masked' Boobies of the Eastern Pacific Ocean." *The Wilson Bulletin* 110, no. 2: 155–70.

Porter, D. 1815. *Journal of a Cruise Made to the Pacific Ocean in the United States Frigate Essex: In the Years 1812, 1813, and 1814.* Philadelphia: Bradsford and Inskeep. https://doi.org/10.5962/bhl.title.153613.

Porter, David. 1823. *A Voyage in the South Seas in the Years 1812, 1813, and 1814: With Particular Details of the Gallipagos [sic] and Washington Islands.* London: Sir Richard Phillips. https://doi.org/10.5962/bhl.title.96900.

Poulakakis, Nikos, Danielle L. Edwards, Ylenia Chiari, Ryan C. Garrick, Michael A. Russello, Edgar Benavides, Gregory J. Watkins-Colwell, et al. 2015. "Description of a New Galápagos Giant Tortoise Species (*Chelonoidis;* Testudines: Testudinidae) from Cerro Fatal on Santa Cruz Island." *PLOS One* 10, no. 10: e0138779. https://doi.org/10.1371/journal.pone.0138779.

Poulakakis, Nikos, Scott Glaberman, Michael Russello, Luciano B. Beheregaray, Claudio Ciofi, Jeffrey R. Powell, and Adalgisa Caccone. 2008. "Historical DNA Analysis Reveals Living Descendants of an Extinct Species of Galápagos Tortoise." *Proceedings of the National Academy of Sciences of the United States of America* 105, no. 40: 15464–469. https://doi.org/10.1073/pnas.0805340105.

Poulakakis, Nikos, Joshua M. Miller, Evelyn L. Jensen, Luciano B. Beheregaray, Michael A. Russello, Scott Glaberman, Jeffrey Boore, and Adalgisa Caccone. 2020. "Colonization History of Galápagos Giant Tortoises: Insights from Mitogenomes Support the Progression Rule." *Journal of Zoological Systematics and Evolutionary Research* 58, no. 4:1262–75. https://doi.org/10.1111/jzs.12387.

Poulakakis, Nikos, Michael Russello, Dennis Geist, and Adalgisa Caccone. 2012. "Unravelling the Peculiarities of Island Life: Vicariance, Dispersal and the Diversification of the Extinct and Extant Giant Galápagos Tortoises." *Molecular Ecology* 21, no. 1: 160–73. https://doi.org/10.1111/j.1365-294X.2011.05370.x.

Powell, Robert B., and S. H. Ham. 2008. "Can Ecotourism Interpretation Really Lead to Pro-conservation Knowledge, Attitudes and Behaviour? Evidence from the Galápagos Islands." *Journal of Sustainable Tourism* 16, no. 4: 467–89. https://doi.org/10.1080/09669580802154223.

Pregill, Gregory K., and Tom Dye. 1989. "Prehistoric Extinction of Giant Iguanas in Tonga." *Copeia* 1989, no. 2: 505–508. https://doi.org/10.2307/1445455.

Pregill, Gregory K., and David W. Steadman. 2004. "South Pacific Iguanas: Human Impacts and a New Species." *Journal of Herpetology* 38, no. 1: 15–21. https://doi.org/10.1670/73-03A.

Pritchard, P. C. H. 1996. "The Galápagos Tortoises: Nomenclatural and Survival Status." *Chelonian Research Monographs* 1: 1–85.

Prum, Richard O. 2017. *The Evolution of Beauty: How Darwin's Forgotten Theory of Mate Choice Shapes the Animal World—And Us.* New York: Knopf Doubleday.

Purcell, Steven W., Annie Mercier, Chantal Conand, Jean-François Hamel, M. Verónica Toral-Granda, Alessandro Lovatelli, and Sven Uthicke. 2013. "Sea

Cucumber Fisheries: Global Analysis of Stocks, Management Measures and Drivers of Overfishing." *Fish and Fisheries* 14, no. 1: 34–59. https://doi.org/10.1111/j.1467-2979.2011.00443.x.

Purcell, Steven W., Beth A. Polidoro, Jean-François Hamel, Ruth U. Gamboa, and Annie Mercier. 2014. "The Cost of Being Valuable: Predictors of Extinction Risk in Marine Invertebrates Exploited as Luxury Seafood." *Proceedings of the Royal Society. Series B, Biological Sciences* 281, no. 1781: 20133296. https://doi.org/10.1098/rspb.2013.3296.

Purvis, Andrew, John L. Gittleman, and Thomas Brooks. 2005. *Phylogeny and Conservation.* Cambridge: Cambridge University Press.

Putnam, Brian Seth. 2014. "Modeling Flightless Galápagos Seabirds as Impacted by El Nino and Climate Change." Master's thesis, Nova Southeastern University.

Quammen, David. 2018. *The Tangled Tree: A Radical New History of Life.* New York: Simon and Schuster.

Quiroga, Diego. 2009. "Crafting Nature: The Galápagos and the Making and Unmaking of a Natural Laboratory." *Journal of Phase Equilibria* 16, no. 1: 123–40. https://doi.org/10.2458/v16i1.21695.

Quiroga, Diego. 2014. "Ecotourism in the Galápagos: Management of a Dynamic Emergent System." *The George Wright Forum* 31, no. 3: 280–89.

Quiroga, Diego, and Ana Sevilla. 2017. "Darwin's Galápagos Myth." In *Darwin, Darwinism and Conservation in the Galápagos Islands*, edited by Diego Quiroga and Ana Sevilla, 1–7. Cham, Switzerland: Springer. https://doi.org/10.1007/978-3-319-34052-4_1.

Ramírez-González, Jorge, Gonzalo Banda-Cruz, Jerson Moreno, Dan Ovando, Harry Reyes, Patricia Rosero, and Isabel Timpe. 2018. "Implementation of a Multiple Indicator System for Fisheries with Limited Information in a Context of Co-management, Case Study: Spiny Lobster Fishery in the Galápagos Marine Reserve." *Ocean & Coastal Management* 154 (March): 20–25. https://doi.org/10.1016/j.ocecoaman.2017.12.027.

Rassmann, K., D. Tautz, F. Trillmich, and C. Gliddon. 1997. "The Microevolution of the Galápagos Marine Iguana *Amblyrhynchus cristatus* Assessed by Nuclear and Mitochondrial Genetic Analyses." *Molecular Ecology* 6, no. 5: 437–52. https://doi.org/10.1046/j.1365-294X.1997.00209.x.

Raup, D. M. 1994. "The Role of Extinction in Evolution." *Proceedings of the National Academy of Sciences of the United States of America* 91, no. 15: 6758–63. https://doi.org/10.1073/pnas.91.15.6758.

Rauzon, Mark J. 2016. *Isles of Amnesia: The History, Geography, and Restoration of America's Forgotten Pacific Islands.* Honolulu: University of Hawai'i Press. https://doi.org/10.21313/hawaii/9780824846794.001.0001.

Reck, Günther. 2014. "Development of the Galápagos Marine Resources Reserve." In *The Galápagos Marine Reserve*, edited by Judith Denkinger and Luis Vinueza, 139–58. Cham, Switzerland: Springer. https://doi.org/10.1007/978-3-319-02769-2.

Reid, Ian R., Paul A. Baldock, and Jillian Cornish. 2018. "Effects of Leptin on the Skeleton." *Endocrine Reviews* 39, no. 6: 938–59.

Reilly, Pauline. 1994. *Penguins of the World.* New York: Oxford University Press.

Rentería, Jorge Luis, Mark R. Gardener, F. Dane Panetta, Rachel Atkinson, and Mick J. Crawley. 2012a. "Possible Impacts of the Invasive Plant *Rubus niveus* on the Native Vegetation of the *Scalesia* Forest in the Galápagos Islands." *PLOS One* 7, no. 10: e48106. https://doi.org/10.1371/journal.pone.0048106.

Rentería, Jorge Luis, Mark R. Gardener, F. Dane Panetta, and Mick J. Crawley. 2012b. "Management of the Invasive Hill Raspberry (*Rubus niveus*) on Santiago Island,

Galápagos: Eradication or Indefinite Control?" *Invasive Plant Science and Management* 5 (1): 37–46. https://doi.org/10.1614/IPSM-D-11-00043.1.

Reyes, Harry, Jorge Ramírez, and Anna Schuhbauer. 2012. "Evaluation of the Spiny Lobster Fishery in the Galápagos Marine Reserve." In *Galápagos Report 2011–2012,* 149–55. Puerto Ayora, Galápagos: Galápagos National Park Service, Governing Council for the Special Regime of Galápagos, Charles Darwin Foundation, and Galápagos Conservancy.

Richardson, Philip L. 2011. "How Do Albatrosses Fly around the World without Flapping Their Wings?" *Progress in Oceanography* 88, no. 1/4: 46–58. https://doi.org/10.1016/j.pocean.2010.08.001.

Rick, Charles M., and Robert I. Bowman. 1961. "Galápagos Tomatoes and Tortoises." *Evolution* 15, no. 4: 407–17. https://doi.org/10.1111/j.1558-5646.1961.tb03171.x.

Ripley, S. Dillon. 1976. "Rails of the World." *American Scientist* 64:628–35.

Rivera-Parra, Jose L., Kenneth M. Levenstein, James C. Bednarz, F. Hernan Vargas, Victor Carrión, and Patricia G. Parker. 2012. "Implications of Goat Eradication on the Survivorship of the Galápagos Hawk." *The Journal of Wildlife Management* 76, no. 6: 1197–1204. https://doi.org/10.1002/jwmg.372.

Rodhouse, P., R. W. A. Barling, W. I. C. Clark, Ann-Louise Kinmonth, Elizabeth M. Mark, D. Roberts, Lesley E. Armitage, et al. 1975. "The Feeding and Ranging Behaviour of Galápagos Giant Tortoises (*Geochelone elephantopus*)." *Journal of Zoology* 176, no. 3: 297–310. https://doi.org/10.1111/j.1469-7998.1975.tb03203.x.

Rodo, Xavier, Mercedes Pascual, George Fuchs, and A. S. G. Faruque. 2002. "ENSO and Cholera: A Nonstationary Link Related to Climate Change?" *Proceedings of the National Academy of Sciences of the United States of America* 99, no. 20: 12901–6. https://doi.org/10.1073/pnas.182203999.

Romero, L. Michael, and Martin Wikelski. 2010. "Stress Physiology as a Predictor of Survival in Galápagos Marine Iguanas." *Proceedings of the Royal Society. Series B, Biological Sciences* 277, no. 1697: 3157–62. https://doi.org/10.1098/rspb.2010.0678.

Romero, Michael, and Martin Wikelski. 2016. "Marine Iguanas: Life on the Edge." In *Galápagos: Preserving Darwin's Legacy*, edited by Tui De Roy, 106–13. London: Bloomsbury.

Roots, Clive. 2006. *Flightless Birds*. Westport, CT: Greenwood Publishing Group.

Rose, Ruth. 1924. "Man and the Galápagos." In *Galápagos: World's End,* by William Beebe, 332–417. New York: Putnam's Sons.

Rosenberg, Daniel K. 1990. "The Impact of Introduced Herbivores on the Galápagos Rail (*Laterallus spilonotus*)." *Monographs in Systematic Botany from Missouri Botanical Gardens* 32, no. 1990: 169–78.

Rosindell, J., L. Harmon, Y. Wong, H. Morlon, K. Zhong, and J. Sutton. 2017. "OneZoom Tree of Life Explorer." http://www.onezoom.org/.

Rueda, D., V. Carrion, P. A. Castaño, F. Cunninghame, P. Fisher, E. Hagen, J. B. Ponder, et al. 2019 "Preventing Extinctions: Planning and Undertaking Invasive Rodent Eradication from Pinzón Island, Galápagos." *Island Invasives: Scaling Up to Meet the Challenge* 62: 51–56.

Ruiz, Diego J., and Matthias Wolff. 2011. "The Bolivar Channel Ecosystem of the Galápagos Marine Reserve: Energy Flow Structure and Role of Keystone Groups." *Journal of Sea Research* 66, no. 2: 123–34. https://doi.org/10.1016/j.seares.2011.05.006.

Ruttenberg, Benjamin I. 2001. "Effects of Artisanal Fishing on Marine Communities in the Galápagos Islands." *Conservation Biology* 15, no. 6: 1691–99. https://doi.org/10.1046/j.1523-1739.2001.99556.x.

Ryan, Sadie J., Catherine A. Lippi, Ryan Nightingale, Gabriela Hamerlinck, Mercy J. Borbor-Cordova, Marilyn Cruz B, Fernando Ortega, Renato Leon, Egan Waggoner, and Anna M. Stewart-Ibarra. 2019. "Socio-Ecological Factors Associated with Dengue Risk and *Aedes aegypti* Presence in the Galápagos Islands, Ecuador." *International Journal of Environmental Research and Public Health* 16, no. 5: 682. https://doi.org/10.3390/ijerph16050682.

Safina, Carl. 2002. *Eye of the Albatross: Visions of Hope and Survival.* New York: Henry Holt.

Santiago-Alarcon, Diego, and Jane Merkel. 2018. "New Host–Parasite Relationships by Host-Switching." In *Disease Ecology*, edited by Patricia G. Parker, 157–77. Cham, Switzerland: Springer. https://doi.org/10.1007/978-3-319-65909-1_7.

Scally, Aylwyn. 2016. "The Mutation Rate in Human Evolution and Demographic Inference." *Current Opinion in Genetics & Development* 41 (August): 36–43. https://doi.org/10.1016/j.gde.2016.07.008.

Schep, Stijn, Mirte Ruesen, Viviana Gallegos, Pieter van Beukering, and Wouter Botzen. 2014. *Does Tourism Growth on the Galápagos Islands Contribute to Sustainable Economic Development? An Ecosystem Valuation from a Tourist Perspective and a Cost–Benefit Analysis of Tourism Growth Scenarios.* Amsterdam: IVM Institute for Environmental Studies, Vrije Universiteit Amsterdam.

Schiller, Laurenne, Juan José Alava, Jack Grove, Günther Reck, and Daniel Pauly. 2013. *A Reconstruction of Fisheries Catches for the Galápagos Islands, 1950–2010.* Working Paper #2013-11. Vancouver, B.C.: Fisheries Centre, the University of British Columbia.

Schiller, Laurenne, Juan José Alava, Jack Grove, Günther Reck, and Daniel Pauly. 2015. "The Demise of Darwin's Fishes: Evidence of Fishing Down and Illegal Shark Finning in the Galápagos Islands." *Aquatic Conservation: Marine and Freshwater Ecosystems* 25, no. 3: 431–46. https://doi.org/10.1002/aqc.2458.

Schilling, Edward E., and Jose L. Panero. 2011. "A Revised Classification of Subtribe Helianthinae (Asteraceae: Heliantheae) II. Derived Lineages." *Botanical Journal of the Linnean Society* 167, no. 3: 311–31.

Schilling, Edward E., José L. Panero, and Uno H. Eliasson. 1994. "Evidence from Chloroplast DNA Restriction Site Analysis on the Relationships of *Scalesia* (Asteraceae: Heliantheae)." *American Journal of Botany* 81, no. 2: 248–54.

Schofield, Eileen K. 1989. "Effects of Introduced Plants and Animals on Island Vegetation: Examples from Galápagos Archipelago." *Conservation Biology* 3, no. 3: 227–39. https://doi.org/10.1111/j.1523-1739.1989.tb00081.x.

Shepherd, S. A., P. Martinez, M. V. Toral-Granda, and G. J. Edgar. 2004. "The Galápagos Sea Cucumber Fishery: Management Improves as Stocks Decline." *Environmental Conservation* 31, no. 2: 102–10.

Shimizu, Y. 1997. "Competitive Relationship between Tree Species of *Scalesia* and Introduced Plants with Reference to Regeneration Mechanisms of *Scalesia* Forests in the Galápagos Islands." *Regional Review* 11: 1–131.

Shriver, W. Gregory, James P. Gibbs, Hara W. Woltz, Nicole P. Schwarz, and Margaret A. Pepper. 2011. "Galápagos Rail, *Laterallus spilonotus,* Population Change Associated with Habitat Invasion by the Red-barked Quinine Tree, *Cinchona pubescens.*" *Bird Conservation International* 21, no. 2: 221–27. https://doi.org/10.1017/S0959270910000481.

Sierra Club. 2018. "Sierra Club." https://www.sierraclub.org/home.

Smith, Mike U. 2017. "How Does Evolution Explain Blindness in Cavefish?" *The American Biology Teacher* 79, no. 2: 95–101.

Snell, Howard L., Heidi M. Snell, and C. Richard Tracy. 1984. "Variation among Populations of Galápagos Land Iguanas (*Conolophus*): Contrasts of Phylogeny and Ecology." *Biological Journal of the Linnean Society* 21, no. 1/2: 185–207. https://doi.org/10.1111/j.1095-8312.1984.tb02061.x.

Snow, Barbara K. 1966. "Observations on the Behaviour and Ecology of the Flightless Cormorant *Nannopterum harrisi*." *Ibis* 108, no. 2: 265–80. https://doi.org/10.1111/j.1474-919X.1966.tb07270.x.

Sonnenholzner, J. I., L. B. Ladah, and K. D. Lafferty. 2009. "Cascading Effects of Fishing on Galápagos Rocky Reef Communities: Reanalysis Using Corrected Data." *Marine Ecology Progress Series* 375 (January): 209–18. https://doi.org/10.3354/meps07890.

Spear, Larry B., David G. Ainley, and Sophie W. Webb. 2003. "Distribution, Abundance and Behaviour of Buller's, Chatham Island and Salvin's Albatrosses off Chile and Peru." *Ibis* 145, no. 2: 253–69. https://doi.org/10.1046/j.1474-919X.2003.00151.x.

Stacey, Lucia, and Vlada Fuks. 2007. *Struggling for the Golden Egg: Conservation Politics in the Galápagos*. Master's thesis, Roskilde Universitetscenter. https://rucforsk.ruc.dk/ws/files/57776152/light_version_without_pictures.pdf

Steadman, David W. 2006. *Extinction and Biogeography of Tropical Pacific Birds*. Chicago: University of Chicago Press.

Steadman, David W., Thomas W. Stafford, Douglas J. Donahue, and A. J. Jull. 1991. "Chronology of Holocene Vertebrate Extinction in the Galápagos Islands." *Quaternary Research* 36, no. 1: 126–33. https://doi.org/10.1016/0033-5894(91)90021-V.

Steinfurth, Antje, F. Hernan Vargas, Rory P. Wilson, Michael Spindler, and David W. Macdonald. 2008. "Space Use by Foraging Galápagos Penguins During Chick Rearing." *Endangered Species Research* 4, no. 1–2: 105–12. https://doi.org/10.3354/esr00046.

Stephens, P. A., W. J. Sutherland, and R. P. Freckleton. 1999. "What Is the Allee Effect?" *Oikos* 87, no. 1: 185–90. https://doi.org/10.2307/3547011.

Stervander, Martin, Peter G. Ryan, Martim Melo, and Bengt Hansson. 2019. "The Origin of the World's Smallest Flightless Bird, the Inaccessible Island Rail, *Atlantisia rogersi* (Aves: Rallidae)." *Molecular Phylogenetics and Evolution* 130:92–98. https://doi.org/10.1016/j.ympev.2018.10.007.

Stewart, Paul D. 2006. *Galápagos: The Islands That Changed the World*. New Haven, CT: Yale University Press.

Strahm, Wendy, and Marc Patry. 2010. *Reactive Monitoring Mission Report, Galápagos Islands, 27 April—6 May 2010*. Paris: UNESCO.

Street, Phillip A. 2013. "Abundance, Survival, and Breeding Probabilities of the Critically Endangered Waved Albatross." PhD diss., Colorado State University.

Stronza, Amanda L., Carter A. Hunt, and Lee A. Fitzgerald. 2019. "Ecotourism for Conservation?" *Annual Review of Environment and Resources* 44:229–53. https://doi.org/10.1146/annurev-environ-101718-033046

Sulloway, Frank J. 1982a. "Darwin's Conversion: The Beagle Voyage and Its Aftermath." *Journal of the History of Biology* 15, no. 3: 325–96. https://doi.org/10.1007/BF00133143.

Sulloway, Frank J. 1982b. "Darwin and His Finches: The Evolution of a Legend." *Journal of the History of Biology* 15, no. 1: 1–53. https://doi.org/10.1007/BF00132004.

Sulloway, Frank J. 2005. "The Evolution of Charles Darwin." *Smithsonian*, December.

354	REFERENCES

Sulloway, Frank J. 2006. "Surviving the Galápagos: A Letter from the Field." *Brick* 77:72–82.

Sulloway, Frank J. 2009. "Tantalizing Tortoises and the Darwin-Galápagos Legend." *Journal of the History of Biology* 42, no. 1: 3–31. https://doi.org/10.1007/s10739-008-9173-9.

Sulloway, Frank J. 2015. "The Mystery of the Disappearing *Opuntia*." *Galápagos Matters*, Autumn/Winter: 8–9.

Sundevall, C. J. 1871. "On Birds from the Galápagos Islands." *Proceedings of the Zoological Society of London* 1871: 124–129.

Suryan, Robert M., David J. Anderson, Scott A. Shaffer, Daniel D. Roby, Yann Tremblay, Daniel P. Costa, Paul R. Sievert, et al. 2008. "Wind, Waves, and Wing Loading: Morphological Specialization May Limit Range Expansion of Endangered Albatrosses." *PLOS One* 3, no. 12: e4016. https://doi.org/10.1371/journal.pone.0004016.

Swinton, Tilda. 2007. *Galápagos: Islands That Changed the World*. BBC Videos.

Tapia, Washington, Pablo Ospina, Diego Quiroga, Günther Reck, José Antonio González Novoa, Carlos Montes del Olmo, Eliécer Cruz et al. 2008. "Hacia una visión compartida de Galápagos: el archipiélago como un sistema socioecológico." ("Toward a Shared Vision of Galapagos: The Archipelago as a Social-Ecological System".) In *Informe Galápagos 2007–2008*, 11–16. Puerto Ayora, Galápagos, Ecuador: Fundación Charles Darwin, Parque Nacional Galápagos, y Instituto Nacional Galápagos.

Tapia, Washington, James P. Gibbs, Danny Rueda, Jorge Carrión, Fredy Villalba, Jeffreys Málaga, Galo Quezada, Daniel Lara, Adalgisa Caccone, and Linda J. Cayot. 2017. "Giant Tortoise Restoration Initiative: Beyond Rescue to Full Recovery." In *Galápagos Report 2015–2016*, 173–80. Puerto Ayora, Galápagos: Galápagos National Park Directorate, Governing Council for the Special Regime of Galápagos, Charles Darwin Foundation, and Galápagos Conservancy.

Tarazona, J., and W. Arntz. 2001. "The Peruvian Coastal Upwelling System." In *Coastal Marine Ecosystems of Latin America*, edited by Ulrich Seeliger and Björn Kjerfve, 229–44. Berlin: Springer-Verlag. https://doi.org/10.1007/978-3-662-04482-7_17.

Taylor, B., and C. J. Sharpe (2020). "Inaccessible Island Rail (*Atlantisia rogersi*), Version 1.0." In *Birds of the World*, edited by J. del Hoyo, A. Elliott, J. Sargatal, D. A. Christie, and E. de Juana. Ithaca, NY: Cornell Lab of Ornithology. Online at birdsoftheworld.org.

Taylor, J. Edward, George A. Dyer, Micki Stewart, Antonio Yunez-Naude, and Sergio Ardila. 2003. "The Economics of Ecotourism: A Galápagos Islands Economy-wide Perspective." *Economic Development and Cultural Change* 51, no. 4: 977–97. https://doi.org/10.1086/377065.

Taylor, J. Edward, Jared Hardner, and Micki Stewart. 2009. "Ecotourism and Economic Growth in the Galápagos: An Island Economy-wide Analysis." *Environment and Development Economics* 14, no. 2: 139–62. https://doi.org/10.1017/S1355770X08004646.

Taylor, Scott A., David J. Anderson, and Vicki L. Friesen. 2013a. "Evidence for Asymmetrical Divergence-Gene Flow of Nuclear Loci, But Not Mitochondrial Loci, Between Seabird Sister Species: Blue-Footed (*Sula nebouxii*) and Peruvian (*S. variegata*) Boobies." *PLOS One* 8, no. 4: e62256. https://doi.org/10.1371/journal.pone.0062256.

Taylor, Scott A., James A. Morris-Pocock, Bernie R. Tershy, J. A. Castillo-Guerrero, and V. L. Friesen. 2013b. "Hybridization from Possible Sexual Mis-imprinting: Molecular Characterization of Hybridization between Brown *Sula leucogaster* and Blue-footed Boobies *S. nebouxii*." *Marine Ornithology* 41: 113–19.

Taylor, Scott A., Laura Maclagan, David J. Anderson, and Vicki L. Friesen. 2011. "Could Specialization to Cold-Water Upwelling Systems Influence Gene Flow and Population Differentiation in Marine Organisms? A Case Study Using the Blue-Footed Booby, *Sula nebouxii.*" *Journal of Biogeography* 38, no. 5: 883–93.

Taylor, Scott A., Carlos B. Zavalaga, and Vicki L. Friesen. 2010. "Hybridization Between Blue-Footed (*Sula nebouxii*) and Peruvian (*Sula variegata*) Boobies in Northern Peru." *Waterbirds* 33, no. 2: 251–57. https://doi.org/10.1675/063.033.0215.

Tebbich, Sabine, Michael Taborsky, Birgit Fessl, Michael Dvorak, and Hans Winkler. 2004. "Feeding Behavior of Four Arboreal Darwin's Finches: Adaptations to Spatial and Seasonal Variability." *The Condor* 106, no. 1: 95–105. https://doi.org/10.1093/condor/106.1.95

Tershy, Bernie R., Dawn Breese, and Donald Croll. 2000. "Insurance Eggs Versus Additional Eggs: Do Brown Boobies Practice Obligate Siblicide?" *The Auk* 117, no. 3: 817–20. https://doi.org/10.1642/0004-8038(2000)117[0817:IEVAED]2.0.CO;2.

Thisleton-Dyer, William T. 1905. *Hooker's Icones Plantarum*, Vol. xxviii, Plate 2717. London: Dulau & Co.

Thompson, Diane M., Jessica L. Conroy, Aaron Collins, Stephan R. Hlohowskyj, Jonathan T. Overpeck, Melanie Riedinger-Whitmore, Julia E. Cole, et al. 2017. "Tropical Pacific Climate Variability over the Last 6000 Years as Recorded in Bainbridge Crater Lake, Galápagos." *Paleoceanography* 32, no. 8: 903–22. https://doi.org/10.1002/2017PA003089.

Tindle, R. 1984. "The Evolution of Breeding Strategies in the Flightless Cormorant (*Nannopterum harrisi*) of the Galápagos." *Biological Journal of the Linnean Society* 21, no. 1/2: 157–64. https://doi.org/10.1111/j.1095-8312.1984.tb02058.x.

Tindle, Robert W., Elizabeth Tindle, Dimitrios Vagenas, and Michael P. Harris. 2013. "Population Dynamics of the Galápagos Flightless Cormorant *Phalacrocorax harrisi* in Relation to Sea Temperature." *Marine Ornithology* 41, no. 2: 121–33.

Tindle, Robert William, Arnaldo Tupiza, Simon Blomberg, and Elizabeth Tindle. 2016. "The Biology of an Isolated Population of American Flamingo *Phoenicopterus ruber* in the Galápagos Islands." *Galápagos Research* 68:15–27.

Tompkins, Emily M., Howard M. Townsend, and David J. Anderson. 2017. "Decadal-Scale Variation in Diet Forecasts Persistently Poor Breeding under Ocean Warming in a Tropical Seabird." *PLOS One* 12, no. 8: e0182545. https://doi.org/10.1371/journal.pone.0182545.

Tonnis, Brandon, Peter R. Grant, B. Rosemary Grant, and Kenneth Petren. 2005. "Habitat Selection and Ecological Speciation in Galápagos Warbler Finches (*Certhidea olivacea* and *Certhidea fusca*)." *Proceedings of the Royal Society B: Biological Sciences* 272, no. 1565: 819–26. https://doi.org/10.1098/rspb.2004.3030.

Toral-Granda, V. 2008. "Galápagos Islands: A Hotspot of Sea Cucumber Fisheries in Latin America and the Caribbean." In *Sea Cucumbers: A Global Review of Fisheries and Trade*, edited by V. Toral-Granda, A. Lovatelli, and M. Vasconcellos, 231–53. FAO Fisheries and Aquaculture Technical Paper No. 516. Rome, Italy: FAO.

Toral-Granda, M. Verónica, Charlotte E. Causton, Heinke Jäger, Mandy Trueman, Juan Carlos Izurieta, Eddy Araujo, Marilyn Cruz, Kerstin K Zander, Arturo Izurieta, and Stephen T. Garnett. 2017. "Alien Species Pathways to the Galápagos Islands, Ecuador." *PLOS One* 12, no. 9: e0184379. https://doi.org/10.1371/journal.pone.0184379.

Toral-Granda, M. Verónica, and Priscilla C. Martinez. 2004. *Population Density and Fishery Impacts on the Sea Cucumber (*Isostichopus fuscus*) in the Galápagos Marine Reserve*. Santa Cruz, Galápagos: Charles Darwin Research Station.

Toral-Granda, M. Verónica, and Priscilla C. Martínez. 2007. "Reproductive Biology and Population Structure of the Sea Cucumber *Isostichopus fuscus* (Ludwig, 1875) (Holothuroidea) in Caamaño, Galápagos Islands, Ecuador." *Marine Biology* 151, no. 6: 2091–98. https://doi.org/10.1007/s00227-007-0640-1.

Torres, Roxana, and Alberto Velando. 2003. "A Dynamic Trait Affects Continuous Pair Assessment in the Blue-Footed Booby, *Sula nebouxii*." *Behavioral Ecology and Sociobiology* 55, no. 1: 65–72. https://doi.org/10.1007/s00265-003-0669-1.

Torres, Roxana, and Alberto Velando. 2005. "Male Preference for Female Foot Colour in the Socially Monogamous Blue-Footed Booby, *Sula nebouxii*." *Animal Behaviour* 69, no. 1: 59–65. https://doi.org/10.1016/j.anbehav.2004.03.008.

Torres, Roxana, and Alberto Velando. 2010. "Color in a Long-Lived Tropical Seabird." In *Behavioral Ecology of Tropical Animals*, 155–88. Advances in the Study of Behavior, Vol. 42. Burlington, VT: Academic Press. https://doi.org/10.1016/S0065-3454(10)42005-7.

Toulkeridis, T. 2011. *Volcanic Galápagos*. Quito, Ecuador: Ediecuatorial.

Townsend, Charles H. 1925. "The Galápagos Tortoises in Their Relation to the Whaling Industry: A Study of Old Logbooks." *Zoologica* 4, no. 3: 55–135.

Townsend, Howard, Kathryn Huyvaert, Peter Hodum, and David Anderson. 2002. "Nesting Distributions of Galápagos Boobies (Aves: Sulidae): An Apparent Case of Amensalism." *Oecologia* 132, no. 3: 419–27. https://doi.org/10.1007/s00442-002-0992-7.

Tracy, C. R., and K. A. Christian. 1985. "Are Marine Iguana Tails Flattened?" *British Journal of Herpetology* 6:434–35.

Traveset, Anna, Manuel Nogales, Pablo Vargas, Beatriz Rumeu, Jens M. Olesen, Patricia Jaramillo, and Ruben Heleno. 2016. "Galápagos Land Iguana (*Conolophus subcristatus*) as a Seed Disperser." *Integrative Zoology* 11, no. 3: 207–13. https://doi.org/10.1111/1749-4877.12187.

Traveset, Anna, Jens M. Olesen, Manuel Nogales, Pablo Vargas, Patricia Jaramillo, Elena Antolín, María Mar Trigo, and Ruben Heleno. 2015. "Bird–Flower Visitation Networks in the Galápagos Unveil a Widespread Interaction Release." *Nature Communications* 6 (March): 6376. https://doi.org/10.1038/ncomms7376.

Treherne, John. 1983. *The Galápagos Affair*. New York: Random House.

Trillmich, Fritz. 2016. "Sea Lions and Fur Seals: Cold Water Species on the Equator." In *Galápagos: Preserving Darwin's Legacy*, edited by Tui De Roy, 170–76. London: Bloomsbury.

Trillmich, Fritz, and Martin Wikelski. 1994. "Foraging Strategies of the Galápagos Marine Iguana (*Amblyrhynchus cristatus*): Adapting Behavioral Rules to Ontogenetic Size Change." *Behaviour* 128, no. 3: 255–79. https://doi.org/10.1163/156853994X00280.

Trillmich, Krisztina G.K. 1983. "The Mating System of the Marine Iguana (*Amblyrhynchus cristatus*)." *Zeitschrift für Tierpsychologie* 63, no. 2–3: 141–72.

Troyer, K. 1982. "Transfer of Fermentative Microbes between Generations in a Herbivorous Lizard." *Science* 216, no. 4545: 540–42. https://doi.org/10.1126/science.216.4545.540.

Tukey, John W. 1960. "Conclusions vs Decisions." *Technometrics* 2, no. 1: 423–33. https://doi.org/10.1080/00401706.1960.10489909.

Tullis, Paul. 2016. "Galápagos Stampede." *Scientific American* 314, no. 4: 52–57. https://doi.org/10.1038/scientificamerican0416-52.

Tye, Alan. 2006. "Can We Infer Island Introduction and Naturalization Rates from Inventory Data? Evidence from Introduced Plants in Galápagos." *Biological Invasions* 8, no. 2: 201–15.

Tye, Alan, and Javier Francisco-Ortega. 2011. "Origins and Evolution of Galápagos Endemic Vascular Plants." In *The Biology of Island Floras*, edited by David Bramwell and Juli Caujapé-Castells, 89–153. Cambridge: Cambridge University Press. https://doi.org/10.1017/CBO9780511844270.006.

Tzika, Athanasia C., Sabrina F. P. Rosa, Anna Fabiani, Howard L. Snell, Heidi M. Snell, Cruz Marquez, Washington Tapia, Kornelia Rassmann, Gabriele Gentile, and Michel C Milinkovitch. 2008. "Population Genetics of Galápagos Land Iguana (Genus *Conolophus*) Remnant Populations." *Molecular Ecology* 17, no. 23: 4943–52. https://doi.org/10.1111/j.1365-294X.2008.03967.x.

Union of Concerned Scientists. Accessed August 27, 2018. https://www.ucsusa.org/.

U.S. Department of Agriculture. 2016a. "USDA Announces Conservation Reserve Program Results." https://www.usda.gov/media/press-releases/2016/05/05/usda-announces-conservation-reserve-program-results.

U.S. Department of Agriculture. 2016b. "USDA Announces Additional Financial Incentives for Conservation Reserve Program Participants to Improve Forest Health and Enhance Wildlife Habitat." https://www.fsa.usda.gov/news-room/news-releases/2016/nr_20161209_rel_208.

Valente, Luis M., Albert B. Phillimore, and Rampal S. Etienne. 2015. "Equilibrium and Non-equilibrium Dynamics Simultaneously Operate in the Galápagos Islands." *Ecology Letters* 18, no. 8: 844–52. https://doi.org/10.1111/ele.12461.

Valle, C. A. 1994a. "The Ecology and Evolution of Sequential Polyandry in Galápagos Cormorants (*Compsohalieus nannopterum harrisi*)." PhD diss., Princeton University.

Valle, Carlos A. 1994b. "'Pepino War, 1992'—Is Conservation just a Matter for the Elite? A Galapagueños Viewpoint." *Noticias de Galápagos* no. 53 (April): 2.

Valle, Carlos A. 1995. "Effective Population Size and Demography of the Rare Flightless Galápagos Cormorant." *Ecological Applications* 5, no. 3: 601–17. https://doi.org/10.2307/1941970.

Valle, Carlos A. 2013. "Ecological Selection and the Evolution of Body Size and Sexual Size Dimorphism in the Galápagos Flightless Cormorant." In *Evolution from the Galápagos*, edited by Gabriel Trueba and Carlos Montúfar, 143–58. Cham, Switzerland: Springer. https://doi.org/10.1007/978-1-4614-6732-8_12.

Valle, Carlos A. 2016. "The Flightless Cormorant: The Evolution of Female Rule." In *Galápagos: Preserving Darwin's Legacy*, edited by Tui De Roy, 162–69. London: Bloomsbury.

Valle, Carlos A., and Patricia G. Parker. 2012. "Research on Evolutionary Principles in Galápagos: An Overview of the Past 50 Years." In *The Role of Science for Conservation*, edited by Matthias Wolff and Mark Gardener, 16–34. London: Routledge.

Van Denburgh, John. 1914. "*The Gigantic Land Tortoises of the Galápagos Archipelago.*" *Proceedings of the California Academy of Sciences, Fourth Series* 2, no. 1: 203–374, Plates 12–124.

Vargas, Hernán. 1996. "What Is Happening with the Avifauna of San Cristóbal?" *Noticias de Galápagos* 57:23–24.

Vargas, F. Hernán, Scott Harrison, Solanda Rea, and David W. Macdonald. 2006. "Biological Effects of El Niño on the Galápagos Penguin." *Biological Conservation* 127, no. 1: 107–14. https://doi.org/10.1016/j.biocon.2005.08.001.

Vargas, F. Hernán, Robert C. Lacy, Paul J. Johnson, Antje Steinfurth, Robert J. M. Crawford, P. Dee Boersma, and David W. Macdonald. 2007. "Modelling the Effect of El Niño on the Persistence of Small Populations: The Galápagos Penguin as a Case Study." *Biological Conservation* 137, no. 1: 138–48. https://doi.org/10.1016/j.biocon.2007.02.005.

Vargas, F. Hernán, S. Barlow, Tom Hart, G. Jiménez-Uzcátegui, J. Chávez, S. Naranjo, and D. W. Macdonald. 2008. "Effects of Climate Variation on the Abundance and Distribution of Flamingos in the Galápagos Islands." *Journal of Zoology* 276, no. 3: 252–65.

Velando, Alberto, René Beamonte-Barrientos, and Roxana Torres. 2006. "Pigment-Based Skin Colour in the Blue-Footed Booby: An Honest Signal of Current Condition Used by Females to Adjust Reproductive Investment." *Oecologia* 149, no. 3: 535–42. https://doi.org/10.1007/s00442-006-0457-5.

Velando, Alberto, Hugh Drummond, and Roxana Torres. 2009. "Senescing Sexual Ornaments Recover after a Sabbatical." *Biology Letters* 6, no. 2: 194–96. https://doi.org/10.1098/rsbl.2009.0759.

Villacis, Byron, and Daniela Carrillo. 2013. "The Socioeconomic Paradox of Galápagos." In *Science and Conservation in the Galápagos Islands*, edited by Stephen J. Walsh and Carlos F. Mena, 69–85. Cham, Switzerland: Springer. https://doi.org/10.1007/978-1-4614-5794-7_4.

Vitousek, Maren N., James S. Adelman, Nathan C. Gregory, and James J. H. St. Clair. 2007. "Heterospecific Alarm Call Recognition in a Non-vocal Reptile." *Biology Letters* 3, no. 6: 632–34. https://doi.org/10.1098/rsbl.2007.0443.

Vora, Shivani. 2018. "Is Land Tourism Threatening the Galápagos?" *The New York Times*, June 1.

Wagner, Eric L., and P. Dee Boersma. 2011. "Effects of Fisheries on Seabird Community Ecology." *Reviews in Fisheries Science* 19, no. 3: 157–67. https://doi.org/10.1080/10641262.2011.562568.

Walsh, Stephen J., Kim Engie, Philip H. Page, and Brian G. Frizzelle. 2019. "Demographics of Change: Modeling the Transition of Fishers to Tourism in the Galápagos Islands." In *Urban Galápagos: Transition to Sustainability in Complex Adaptive Systems*, edited by Thomas Kvan and Justyna Karakiewicz, 61–83. Cham, Switzerland: Springer. https://doi.org/10.1007/978-3-319-99534-2_5.

Walsh, Stephen J., and Carlos F. Mena. 2013. *Science and Conservation in the Galápagos Islands*. Vol. 1. Cham, Switzerland: Springer. https://doi.org/10.1007/978-1-4614-5794-7.

Watanabe, Yuuki Y., Akinori Takahashi, Katsufumi Sato, Morgane Viviant, and Charles-André Bost. 2011. "Poor Flight Performance in Deep-Diving Cormorants." *The Journal of Experimental Biology* 214 (Pt 3): 412–21. https://doi.org/10.1242/jeb.050161.

Watkins, Graham, and Felipe Cruz. 2007. *Galápagos at Risk: A Socioeconomic Analysis of the Situation in the Archipelago*. Puerto Ayora, Galápagos: Charles Darwin Foundation.

Watkins, Graham, and Pete Oxford. 2009. *Galápagos: Both Sides of the Coin*. Quito, Ecuador: Enfoque Ediciones.

Watson, James D., Andrew Berry, and Kevin Davies. 2017. *DNA: The Story of the Genetic Revolution*. New York: Knopf.

Wei, Jianwen, and Patricia Ducy. 2010. "Co-dependence of Bone and Energy Metabolisms." *Archives of Biochemistry and Biophysics* 503, no. 1: 35–40. https://doi.org/10.1016/j.abb.2010.05.021.

Weiner, J. 1995. *The Beak of the Finch: A Story of Evolution in Our Time.* New York: Vintage.

Weiss, Madeline C., Filipa L. Sousa, Natalia Mrnjavac, Sinje Neukirchen, Mayo Roettger, Shijulal Nelson-Sathi, and William F. Martin. 2016. "The Physiology and Habitat of the Last Universal Common Ancestor." *Nature Microbiology*, no. 9: 1–8. https://doi.org/10.1038/nmicrobiol.2016.116.

Werner, R., and K. Hoernle. 2003. "New Volcanological and Volatile Data Provide Strong Support for the Continuous Existence of Galápagos Islands over the Past 17 Million Years." *International Journal of Earth Sciences* 92, no. 6: 904–11. https://doi.org/10.1007/s00531-003-0362-7.

White, Fred N. 1973. "Temperature and the Galápagos Marine Iguana—Insights into Reptilian Thermoregulation." *Comparative Biochemistry and Physiology Part A: Physiology* 45, no. 2: 503–13. https://doi.org/10.1016/0300-9629(73)90459-3.

White, William M., Alexander R. McBirney, and Robert A. Duncan. 1993. "Petrology and Geochemistry of the Galápagos Islands: Portrait of a Pathological Mantle Plume." *Journal of Geophysical Research* 98, no. B11: 19533–63. https://doi.org/10.1029/93JB02018.

Wiggins, Ira L., and Duncan M. Porter. 1971. *Flora of the Galápagos Islands.* Stanford, CA: Stanford University Press.

Wikelski, M. 1999. "Influences of Parasites and Thermoregulation on Grouping Tendencies in Marine Iguanas." *Behavioral Ecology* 10, no. 1: 22–29. https://doi.org/10.1093/beheco/10.1.22.

Wikelski, Martin. 2005. "Evolution of Body Size in Galápagos Marine Iguanas." *Proceedings of the Royal Society. Series B, Biological Sciences* 272, no. 1576: 1985–93. https://doi.org/10.1098/rspb.2005.3205.

Wikelski, Martin, and Karin Nelson. 2004. "Conservation of Galápagos Marine Iguanas (*Amblyrhynchus cristatus*)." *Journal of the International Iguana Society* 11, no. 4: 191–97.

Wikelski, Martin, and L. Michael Romero. 2003. "Body Size, Performance and Fitness in Galápagos Marine Iguanas." *Integrative and Comparative Biology* 43, no. 3: 376–86. https://doi.org/10.1093/icb/43.3.376.

Wikelski, M., and C. Thom. 2000. "Marine Iguanas Shrink to Survive El Niño." *Nature* 403, no. 6765: 37–38. https://doi.org/10.1038/47396.

Wikelski, Martin, and Fritz Trillmich. 1997. "Body Size and Sexual Size Dimorphism in Marine Iguanas Fluctuate as a Result of Opposing Natural and Sexual Selection: An Island Comparison." *Evolution* 51, no. 3: 922–36. https://doi.org/10.2307/2411166.

Wikelski, Martin, Vanessa Wong, Brett Chevalier, Niels Rattenborg, and Howard L. Snell. 2002. "Marine Iguanas Die from Trace Oil Pollution." *Nature* 417, no. 6889: 607–8. https://doi.org/10.1038/417607a.

Wikelski, M., and Peter H. Wrege. 2000. "Niche Expansion, Body Size, and Survival in Galápagos Marine Iguanas." *Oecologia* 124, no. 1: 107–15. https://doi.org/10.1007/s004420050030.

Wildscreen Arkive. 2017. "Waved Albatross (*Phoebastria irrorata*)." http://www.arkive.org/waved-albatross/phoebastria-irrorata/.

Williams, George C. 1966. *Adaptation and Natural Selection: A Critique of Some Current Evolutionary Thought.* Princeton, NJ: Princeton University Press. https://doi.org/10.1515/9781400820108.

Wilson, Rory P., Peter G. Ryan, Andrew James, and Marie-Pierre T. Wilson. 1987. "Conspicuous Coloration May Enhance Prey Capture in Some Piscivores." *Animal Behaviour* 35, no. 5: 1558–60. https://doi.org/10.1016/S0003-3472(87)80028-3.

Wilson, Rory P., F. Hernán Vargas, Antje Steinfurth, Philip Riordan, Yan Ropert-Coudert, and David W. Macdonald. 2008. "What Grounds Some Birds for Life? Movement and Diving in the Sexually Dimorphic Galápagos Cormorant." *Ecological Monographs* 78, no. 4: 633–52. https://doi.org/10.1890/07-0677.1.

Wingfield, John C., Robert E. Hegner, Alfred M. Dufty, Jr., and Gregory F. Ball. 1990. "The 'Challenge Hypothesis': Theoretical Implications for Patterns of Testosterone Secretion, Mating Systems, and Breeding Strategies." *The American Naturalist* 136, no. 6: 829–46. https://doi.org/10.1086/285134.

Wittmer, Margret. 2013. *Floreana: A Woman's Pilgrimage to the Galápagos*. Mount Kisco, NY: Moyer Bell.

Wolf, Jochen B. W., Diethard Tautz, and Fritz Trillmich. 2007. "Galápagos and Californian Sea Lions Are Separate Species: Genetic Analysis of the Genus *Zalophus* and Its Implications for Conservation Management." *Frontiers in Zoology* 4 (September): 20. https://doi.org/10.1186/1742-9994-4-20.

Wolfe, Kennedy, and Maria Byrne. 2016. "Population Biology and Recruitment of a Vulnerable Sea Cucumber, *Stichopus herrmanni*, on a Protected Reef." *Marine Ecology* 38, no. 1: e12397. https://doi.org/10.1111/maec.12397.

Wolff, Matthias, and Mark Gardener. 2012. *The Role of Science for Conservation*. London: Routledge.

Wolff, M., D. J. Ruiz, and M. Taylor. 2012. "El Niño Induced Changes to the Bolívar Channel Ecosystem (Galápagos): Comparing Model Simulations with Historical Biomass Time Series." *Marine Ecology Progress Series* 448 (February): 7–22. https://doi.org/10.3354/meps09542.

Woram, J. 2005. *Charles Darwin Slept Here: Tales of Human History at World's End*. Rockville Centre, NY: Rockville Press.

Yaemsiri, S., N. Hou, M. M. Slining, and K. He. 2010. "Growth Rate of Human Fingernails and Toenails in Healthy American Young Adults." *Journal of the European Academy of Dermatology and Venereology* 24, no. 4: 420–23. https://doi.org/10.1111/j.1468-3083.2009.03426.x.

Takahiro Yonezawa, Naoki Kohno, and Masami Hasegawa. 2009. The Monophyletic Origin of Sea Lions and Furseals (Carnivora; Otariidae) in the Southern Hemisphere. *Gene* 441, no. 1/2: 89–99. https://doi.org/10.1016/j.gene.2009.01.022.

Young, Lindsay C., Eric A. Vanderwerf, David G. Smith, John Polhemus, Naomi Swenson, Chris Swenson, Brent R. Liesemeyer, Betsy H. Gagne, and Sheila Conant. 2009. "Demography and Natural History of Laysan Albatross on Oahu, Hawaii." *The Wilson Journal of Ornithology* 121, no. 4: 722–29. https://doi.org/10.1676/08-150.1.

Zaher, Hussam, Mario H. Yánez-Muñoz, Miguel T. Rodrigues, Roberta Graboski, Fabio A. Machado, Marco Altamirano-Benavides, Sandro L. Bonatto, and Felipe G. Grazziotin. 2018. "Origin and Hidden Diversity within the Poorly Known Galápagos Snake Radiation (Serpentes: Dipsadidae)." *Systematics and Biodiversity*, 16:7. https://doi.org/10.1080/14772000.2018.1478910.

Zapata, Carlos. 2007. "Evaluation of the Quarantine and Inspection System for Galápagos (SICGAL) after Seven Years." In *Galápagos Report 2006–2007*, 60–66. Puerto Ayora, Galápagos: Charles Darwin Foundation, Galápagos National Park, and National Galápagos Institute.

Zavalaga, Carlos B., Steven D. Emslie, Felipe A. Estela, Martina S. Müller, Giacomo Dell'Omo, and David J. Anderson. 2012. "Overnight Foraging Trips by Chick-Rearing Nazca Boobies *Sula granti* and the Risk of Attack by Predatory Fish." *Ibis* 154, no. 1: 61–73. https://doi.org/10.1111/j.1474-919X.2011.01198.x.

Zhang, Zhaohui, Guillaume Leduc, and Julian P. Sachs. 2014. "El Niño Evolution during the Holocene Revealed by a Biomarker Rain Gauge in the Galápagos Islands." *Earth and Planetary Science Letters* 404 (October): 420–34. https://doi.org/10.1016/j.epsl.2014.07.013.

Zimmer, Carl, and Douglas J. Emlen. 2019. *Evolution: Making Sense of Life.* 3rd ed. New York: Macmillan Learning.

Index

For the benefit of digital users, indexed terms that span two pages (e.g., 52–53) may, on occasion, appear on only one of those pages.

Appendices are not included in this index. Figures are indicated by *f* following the page number